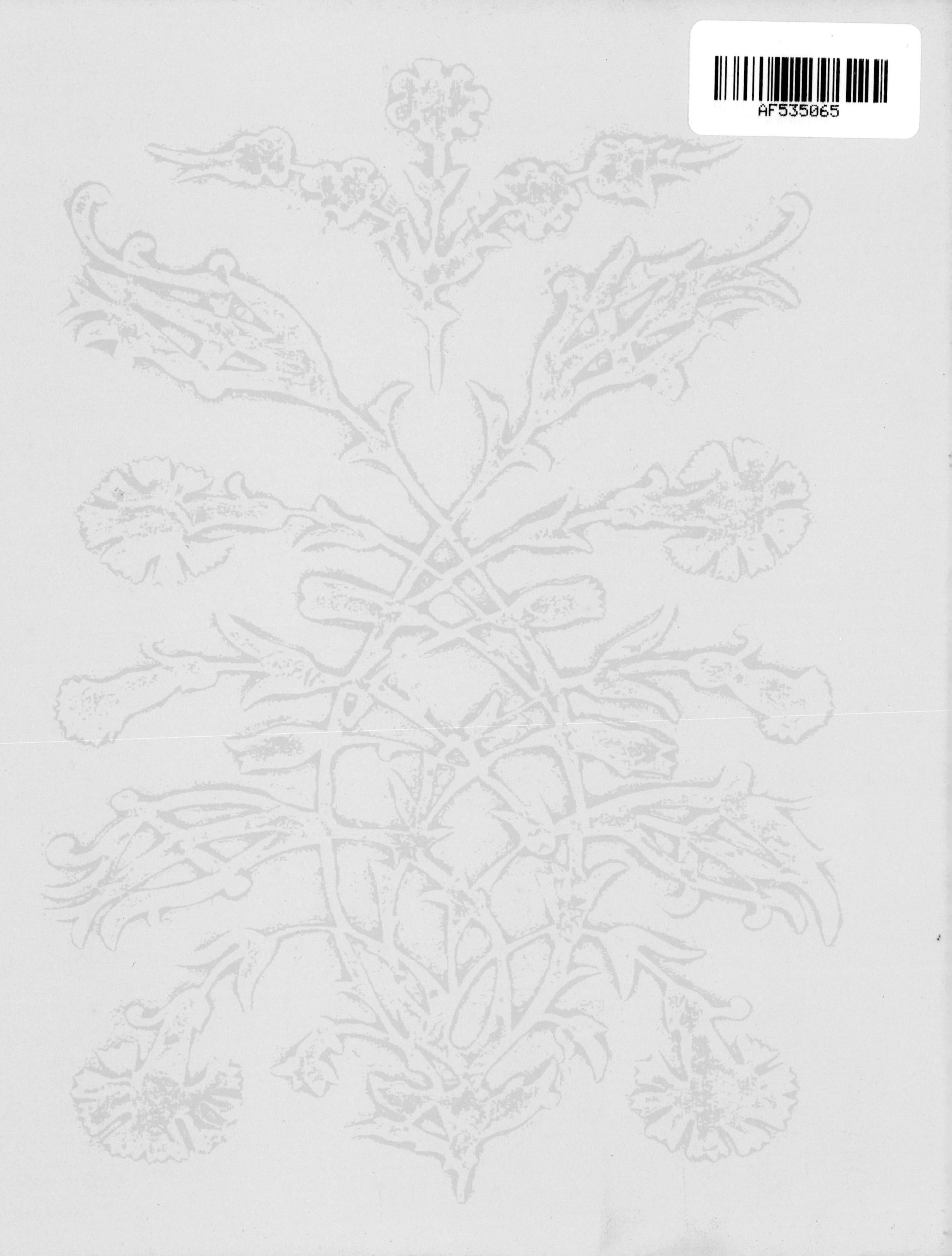
AF535065

Wendell August Forge

Seventy Five Years of Artistry in Metal

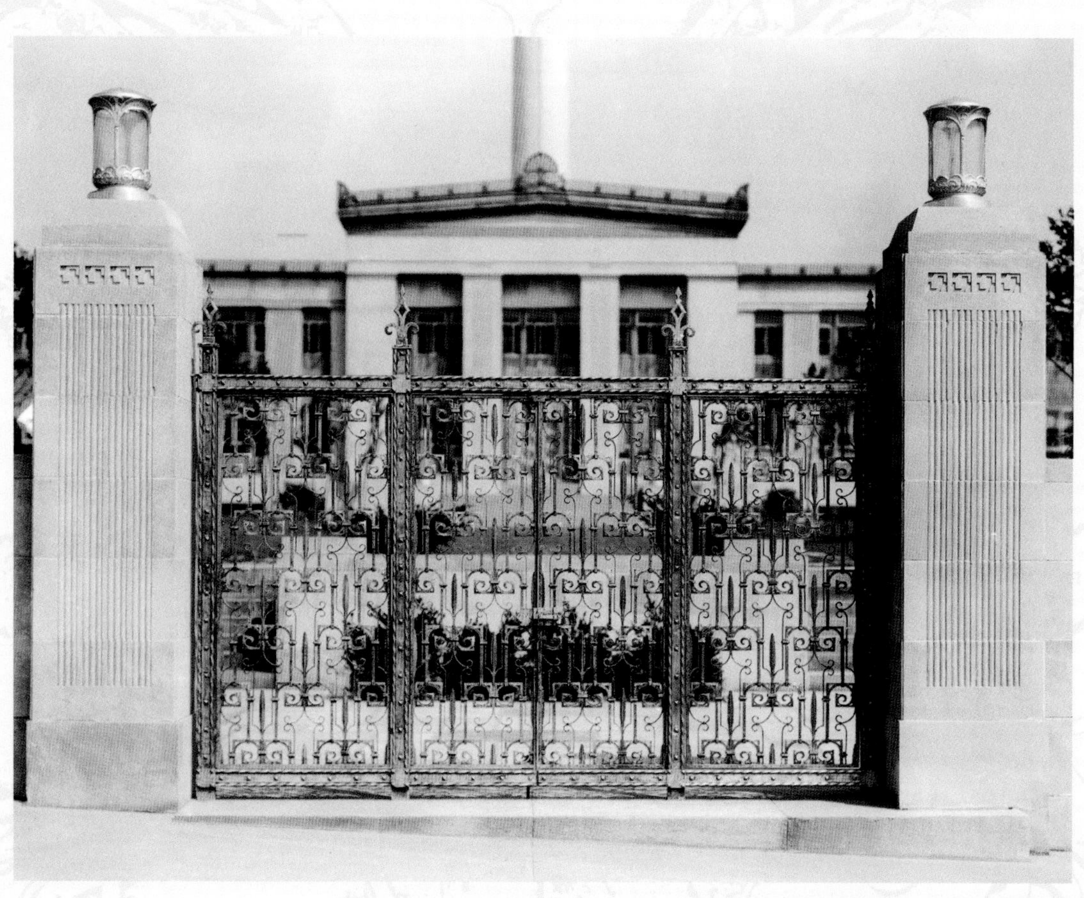

Wendell August Forge

Seventy Five Years of Artistry in Metal

Bonita J. Campbell

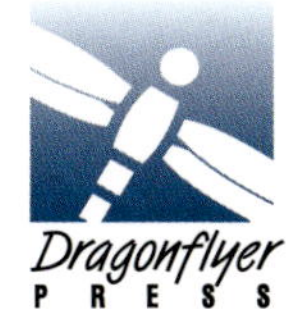

Jacket front: Wendell August Forge Craftsman Paul Staff at the forge in Grove City, Pennsylvania. Bottom, left to right, covered bon bon of aluminum with *Bittersweet* motif, *circa* 1940; center medallion of commissioned SALT II bronze plate, 1979; hinged cigarette box of aluminum with *Hobnail* motif, *circa* 1937.
Photography by Richard Smaltz.

Jacket back: Top, portion of aluminum architectural "calling card," *circa* 1931; left, aluminum and bronze architectural sample, *circa* 1935; lower right, bowl and pitcher of aluminum with *Coffee Bean* motif, *circa* 1954.
Photography by Richard Smaltz.

Copyright © 1999 by Bonita J. Campbell

All rights reserved. No part of this book may be used or reproduced in any manner whatsoever without written permission except for brief quotations embodied in attributed reviews and articles.

Published by
Dragonflyer Press
521 North Mountain Avenue, Suite A
Upland, California 91786

Publisher's Cataloging-in-Publication
(Provided by Quality Books, Inc.)

Campbell, Bonita J.
Wendell August Forge : seventy five years of artistry in metal / Bonita J. Campbell – 1st ed.
p. cm.
Includes index.
Preassigned LCCN: 98-86660
ISBN: 0-944933-07-6

1. Wendell August Forge–Collectibles. 2. Aluminum forgings. 3. Aluminum giftware–Collectibles–Pennsylvania–Grove City. I: Title.

NK7700.C36 1998 739.5'7'075
QBI98-921

First Edition

10 9 8 7 6 5 4 3 2 1

Printed in Hong Kong

Contents

Acknowledgments

I am indebted to the many individuals who assisted in making this book possible by generously providing primary source materials, such as documents, interviews, product documentation, and photographs, and to the several institutions that provided access to primary sources.

Individual contributors include Bonnie J. (Deniker) Adams, Steven and Heidi Adams, Thomas and Lois Armour, Susan (Sutherland) Babb, Evelyn Beebe, Aldour Bilodeau, Gust Boleen, Susan (Mrs. Gordon) Brown, David A. Bruck, John Carlson, John S. "Jack" Carruthers, Jessie Cole, Clare Cornell, Sylvia Cornell, Laurie Dean, Arthur DePonceau, James V. DePonceau, Marian Dick, Dot Donato, Everl Donato, Maxine Evans, Mary Ann Felegy, Teresa Forese, William Freeman, Mrs. Bill Fustine, Lynne Goldin, Lisa Grossenbacher, Alan Harris, Charlene Hasenplug, Alice (Marshall) Hoffman, Linda Hoover, Carol Hughes, Raymond E. and Pat Iadonato, Violet Iadonato, Winnie Inscoe, Helen King, F.W. "Bill" Knecht III, F.W. "Will" Knecht IV, Elsa Kramer, Diane W. Lacey, Mary Lou Allen Lambert, Joseph and Eleanor Leone, Mike and Jody Mancuso, Joseph A. McCandless, Sarah Jane McCandless, Mrs. Wayne McGinnis, Eleanor (Mrs. Milford) McLaughlin, Mary Lou (McConnell) McNaughton, James McWilliams, Charles Oberlin, Gwen Oberlin, Mr. and Mrs. Everett Palmer, Fred Parquette, John Patrick, Lucy (Mrs. Joseph) Pisoni, Father Otto B. Pisoni, Anthony "Tony" and Elaine Pompa, Nicholas "Nick" and Kathy Pompa, Mrs. John Potts, Adaline (Rossi) Reese, Christopher Rossi, Frank and Lenora Rossi, Irene Rossi, Adam Sabatose, Michael Sanderson, Barbara Schaefer, James W. Serio, James Sewall, David Shively, Monsignor Robert J. Shuda, James Sterrett, Leo M. Stevenson, Marilyn Sullivan, Doug and Ann Sutherland, Edna (Mrs. Lester) Sutherland, Henry Swyers, Mary Charlotte (August) Taylor, Cecelia Terry, Eugene and Josephine Town, Patty Trasin, Ruth Trimm, Carolyn Hayes Uber, Theodore Uber, Bruce H. Vinton, James M. "Jimmy" Vinton, Dennis and Rita Wildnauer, Mrs. Leroy Williams, Reverend Thomas R. Wilson, Ronald E. Winder, Nancy and Jack Withrow, Catherine Youngo, Leonard Youngo, and Joseph and Joan Ziccardi.

Institutional contributors include Brockway Area Historical Society; California State University Northridge Library; Grove City Community Library; Kemerer Museum of Decorative Arts, Bethlehem, Pennsylvania; Library of Congress; Los Angeles Public Library; Mercer County Genealogical Society; Mercer County Historical Society; Our Lady, Queen of the Most Holy Rosary Cathedral, Toledo, Ohio; and the National Museum of American History of the Smithsonian Institution.

Among the individuals identified above are a few whose contributions were particularly notable, and thus special acknowledgments are extended to: Bill Knecht and Cathy Youngo of the Wendell August Forge, who made it possible to access some Forge source materials, document portions of the Forge collection, and bring the research effort to fruition; former Wendell August Forge personnel James V. Deponceau, Joseph Leone, Mary Lou McNaughton, Frank Rossi, and Joseph Ziccardi, who gave endlessly of their time, expertise, and rich memories for numerous interviews, letters, and manuscript reviews; collectors Joseph A. McCandless, Sarah Jane McCandless, Dennis and Rita Wildnauer, and Nancy Withrow, who provided access to their extensive collections for photographs and product documentation of Wendell August Forge items; Henry Swyers and Mary Charlotte (August) Taylor, who shared priceless photographs and other documentation from family archives; the irreplaceable Adam Sabatose, whose congenial and professional assistance with Brockway-related research was without bound; William Freeman, who, in addition to tracking down and documenting several hundred Wendell August Forge artifacts, shared with me the investment of hundreds of hours perched on hard library chairs seeking out primary source documentation in print and on microfilm; and Gwen Oberlin, who proved to the world that good primary records research can be accomplished in the Grove City Community Library by a dedicated and professional 12-year-old girl.

To all, and to any whom I may have inadvertently omitted, my sincere gratitude, for both your assistance and for enriching my life.

Bonita J. Campbell
May, 1998

Preface

Summers in the 1940s meant visiting my grandparents in Illinois and treasure hunting in their basement. Secreted away in boxes and hidden under protective covers were the remnants of various activities of my father and grandparents, and other family and friends of generations past. As a young child I whiled away hours, and even days, creating from my treasure hunts my own fantasies of the future.

About 50 years later, no longer a child, my treasure hunting expeditions now took place in the basement of my parents' home in Colorado. My mother, her youth scarred by the Great Depression, had matured into an accumulator of some note, particularly if the artifacts involved had sentimental value. I was now old enough to be able to connect many of the objects with my own past. Thus the fantasies of the future that my treasure hunts had once engendered were replaced by memories of the past.

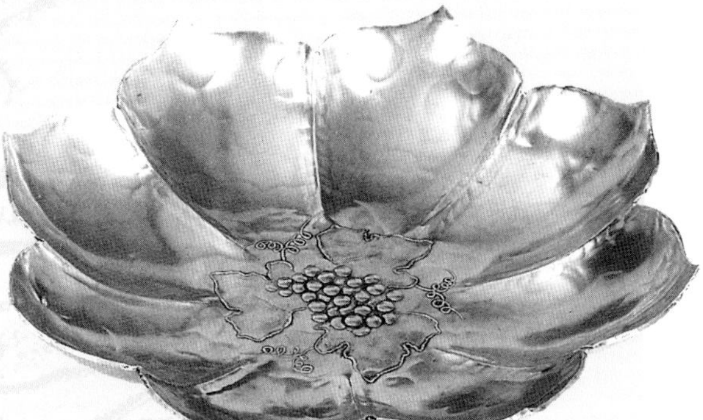

On one such expedition, intending to look at glassware, I encountered a metal serving tray wrapped carefully in tissue paper. With hazy recollections of its use on special occasions, I asked my mother about it. The tray, made of aluminum, had been a wedding gift, but she had no knowledge of its genesis. I was sufficiently taken by the hand craftsmanship evidenced by the piece that I thought it would be interesting, and probably not too difficult, to find out more about it.

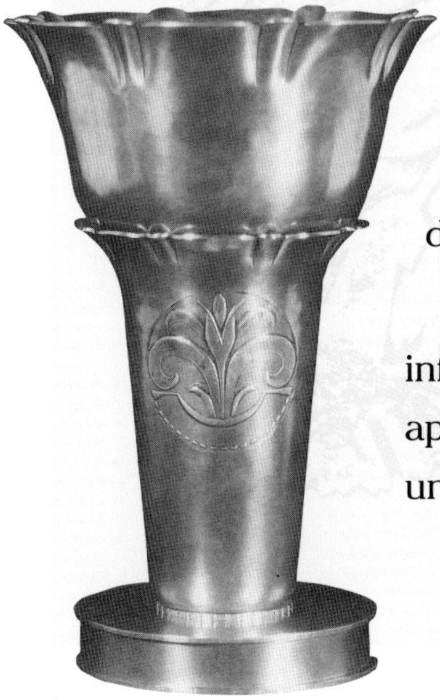

Thus began a very different type of hunt, one in which information was the treasure to be had. It soon became apparent that there was little information to be found. Quite unlike my experiences in seeking information about certain types of glassware, with its richly documented history, research and published work pertaining to decorative art aluminum was sparse. An entire era of fine hand craftsmanship and consumer infatuation seemed to have vanished with little historical trace.

Curiousity yielded to obsession. Artifacts were more accessible than information, and so the house began to fill with decorative art aluminum wares, perhaps with the fanciful notion that if I surrounded myself with them and looked at them long enough, the pieces would divulge their life stories. The search for data continued, but much more slowly and painfully, with many library trips to search blindly through magazines, newspapers, microfilm, trademark registrations, and old city directories. Blind telephone calls were made, and letters written, to people who might know something or perhaps know someone else who might know something. People who had been connected with the industry, however remote the association, were interviewed.

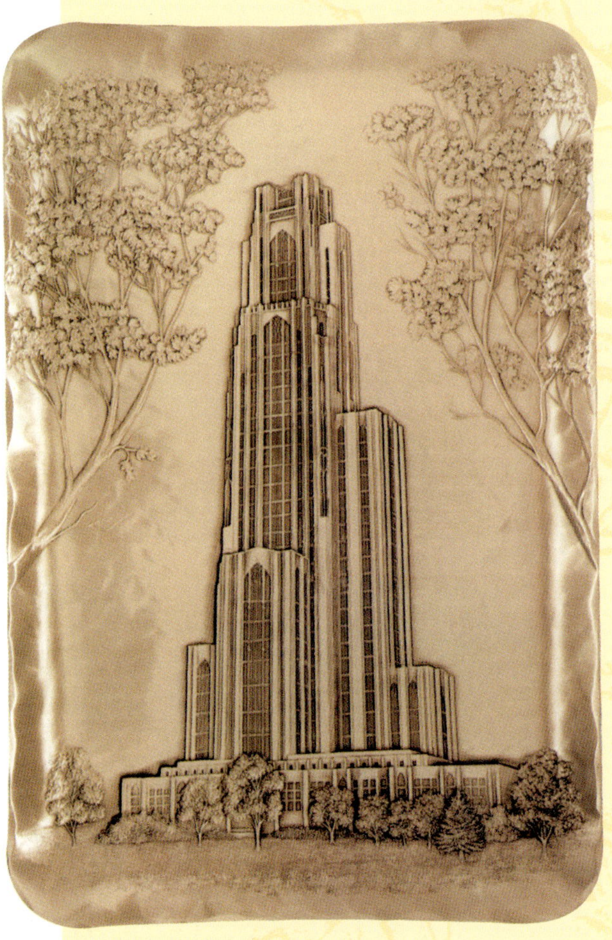

When the opportunity serendipitously arose to research the history and metal wares of the Wendell August Forge, I jumped at the chance to embark on a significant treasure hunt. The Wendell August Forge had founded the mid-20th Century hand forged aluminum industry in the early 1930s, but its heritage remained murky. What emerged from this treasure hunt was a rich story, during an historically important era, of the development of a uniquely American decorative art form, encompassing such modest items as small coasters and such magnificent ones as a statue of St. John the Baptist.

Objects, particularly hand crafted ones of unique design, tend to gain value from information about their origins. Who made an object and how? Who designed it and why? When was it produced? How many were produced? The answers to these questions allow us to more fully appreciate an object and its place in our lives and our history. Thus it is rewarding to know that the Wendell August Forge aluminum creamer that you hold in your hands was designed by James McCausland (who was trained as an architect) and was embellished using a die engraved by Louis Donato (who studied painting and sculpture). It is rewarding to know that the motif on your creamer was named *Grecian* and that a repoussé process was used to render it in relief. It is rewarding to know that your creamer was made *circa* 1935-1937, and that not many of them were created. It is rewarding to know that your creamer, picked up for less than a dollar at a yard sale a few years ago, now brings many multiples of that amount. And, it is rewarding to know that your small creamer is part of a larger tradition of families of craftsmen and artisans who survived the Great Depression while ornamenting grand cathedrals. Yes, your creamer is now a real treasure.

The Wendell August Forge was a primary influence in creating different uses for the first modern metal, aluminum, and for establishing a new industry based on the hand forging methods that it pioneered. While the rest of the industry has long since faded away, the Wendell August Forge has adapted to the changing times, surviving the lean years and returning to healthy ones. Through it all, it has remained devoted to its tradition of families and hand craftsmanship. The story of the Forge is a story of its people, and its role in American culture. As such, this history should prove of interest to collectors, museums, decorative art historians, students of American Culture, and the people of the Wendell August Forge.

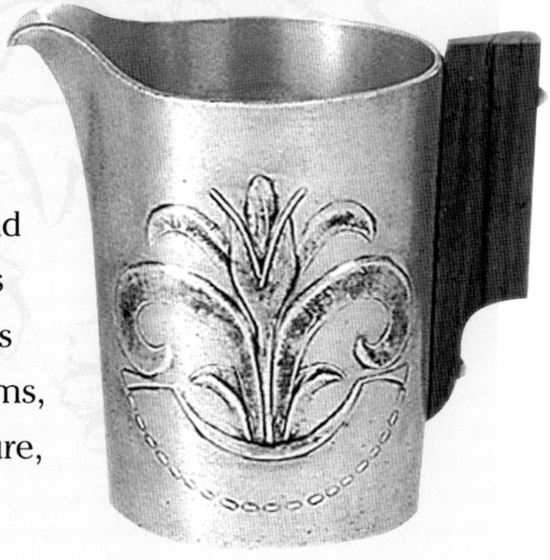

An effort has been made to make this book as useful and readable as possible to a wide range of people having different levels of interest in the topic. The text has been arranged as a time-sequenced narrative. Some pertinent topics not in the mainstream of the narrative have been set apart as separate entities; for example, profiles of former Wendell August Forge craftsmen who initiated their own decorative art aluminum lines. Chapter end notes have been used for both source citations and content, the latter convention allowing for the provision of additional details without cluttering the narrative. Because the research has been principally dependent on primary sources, references that would normally be considered for bibliographic citation are so few that a formal bibliography has been omitted. Of the six appendixes, the first two pertain to Wendell August Forge personnel and the Wendell August Forge utility patent. The remaining four appendixes are devoted to the presentation of information pertinent to dating Wendell August Forge metal wares of the past.

Although every effort has been made to be as complete and accurate as possible, no publication is ever perfect, and there is no reason to believe that this one will be the exception. For any errors there may be, I remain responsible. Corrections and additional information will always be welcome.

Bonita J. Campbell
May, 1998

Chapter One

Founding an Industry

About 65 miles east of the Ohio-Pennsylvania border, heavy-laden trucks crowd the eastbound lanes of Interstate 80 as the off-ramp for State Highway 68 and the town of Clarion, Pennsylvania appears. It was to the north of Clarion and along the Clarion River that Chief Cornplanter and the Seneca peoples once flourished. Settlers first began entering this region after the land that now comprises northwestern Pennsylvania was acquired from the Six Nations in the late 1700s for the paltry sum of about $5,000. The Seneca were among those banished to the Allegheny Indian Reservation, which was flooded in the 1960s to create the Allegheny Reservoir, bringing about the end of the continuing presence of the Seneca in Pennsylvania.

Heading southeasterly out of Clarion toward Brookville, the remnants of the clay, coal and oil works that dominated the region in the early part of this century remain evident. Abandoned oil rigs and the scars of bituminous coal mines can still be seen as the main ranges of the Allegheny Mountains, cutting diagonally across the state, draw near. Just past Brookville, State Highway 28 heads northwest, parallel with the Allegheny range and the tracks of the now-defunct Pittsburgh and Shawmut Railroad. The road follows the Seneca Trail and, in earlier times, finding arrowheads, skinning knives, and primitive mortars along the way was not uncommon.[1]

Passing through Hazen and Sugar Hill, you soon begin the descent into Brockway. Intersected by Little Toby Creek and railroad tracks, the borough of Brockway is surrounded by wooded hillsides, with houses climbing slopes on both sides of its quintessential small town Main Street. Settled in 1822 by Alonzo and Chauncey Brockway, the community then known as Brockwayville was initially established as an early lumbering center. Taking advantage of the course of the Little Toby to the Clarion River, and thence to the Allegheny and the Ohio, enterprising settlers and entrepreneurs farmed the Allegheny watershed for its rich stands of white pine then so highly coveted in the New England states. As the timber was depleted, the arrival of the railroads provided access to more remote stands and, not the least, the rich coal fields of the region.

It was the imaginative promotion of one John Keys of Brockwayville that brought the coal mining industry to the region. In 1876, Keys extracted an enormous one-ton cube of high-quality bituminous coal from the Sugar Hill area to be exhibited at the Centennial Exhibition in Philadelphia. Although the coal cube went on to the American capital city for further exhibition at the Smithsonian Institution, the business investors and speculators headed toward Brockwayville. Within a decade, Brockwayville and the surrounding region had been inundated with mine workers from both the United States and Europe and was enslaved to the erratic cycles of the coal market. Alleviated somewhat by the steadily growing glass manufacturing and clay industries, the coal industry remained the mainstay of the region until World War I when it, like the timber industry before it, began to decline in opportunity and influence.[2]

In the post-Civil War American economy, Pennsylvania was the chief producer of the basic ingredients of industrialism, with much of the raw materials wealth emanating from the northwestern part of the state. It was near the end of this period that Wendell McMinn "Gus" August arrived in Brockwayville in 1885 at the age of three weeks. August had been born in Rew, to the north near the Pennsylvania-New York border, but his mother, Charlotte (McMinn) August, daughter of John and Margaret (McGhee) McMinn, had died in childbirth. The motherless boy was taken in by his uncle and aunt, Daniel D. and Ellen (McMinn) Groves, who were well-established in the area.[3] His elder sister, Myrtle, was adopted by another uncle and aunt.

Thus embraced by an extended and relatively well-to-do family, Wendell August made his way through the local Brockwayville schools and one year of attendance at a military school in New York. Returning to Brockwayville, he clerked in the post office for two years, and then continued his formal education at Bucknell University, where he studied science and engineering, completing his Bachelor of Science degree in 1907. The next two years found him quietly teaching mathematics, and then, in marked contrast, he spent two years traversing the relative wilds of Colorado, Idaho, Montana, Oregon, and Utah, working at a wide range of jobs in mining, construction, and the railroads. Returning home again to Brockwayville, he embarked on a multitude of business ventures and, in 1912, married Jessie McVean Palmer, the daughter of Dr. William and Mary (Howell) Palmer of Johnsonburg.[4] By the time of World War I, three sons (Wendell McMinn, Jr., Robert Edward, and Donald Dick) had made their appearances, and Wendell August was a coal broker who also held interests in the L. M. Groves Mercantile Company, the Fox Ranch, and the Black Diamond Mine of the Toby Coal Mining Company.

In the early 1920s, August undertook the building of a fine new home at the corner of Main Street and Twelfth Avenue.[5] Appropriate to the colonial architecture of the residence was the lavish use of ornamental hand wrought iron. According to oral history, August found it desirable to make certain changes as construction progressed, one of which entailed the addition of two more doors. Natale Rossi, who joined the Wendell August Forge as a metalworker about four years after its establishment, would later relate what he had been told about the event, writing "So in this home, all the doors had hand forged door latches . . . Well, two doors were added and Mr. August couldn't find two door latches that would match the ones he had. So he took one of them to a blacksmith at his mine [Ottone Pisoni], and asked him if he could make two like that."[6]

Ottone "Tony" Pisoni had been apprenticed to an ornamental blacksmith in Italy, and since his arrival in the Brockwayville area some 36 years before, had applied his skills primarily to the mundane tasks of shoeing horses and mules, and sharpening and repairing tools, for the local farmers and miners. Pisoni took on the task of producing the latches, and August soon had two finely-made door latches to match the others — and at a fraction of the cost of the originals. Natale Rossi related his understanding as follows: "He [Pisoni] hammered two latches, which took him about two and a half hours. Mr. August said, 'You know, I could buy any of the ones we bought before, which cost $35 each. Mr. Pisoni's hourly rate at that time was 65 cents. For about $4, which included the materials, we could produce something which would sell for $70.' Mr. August asked Mr. Pisoni if he would consider working for him, if he started as a blacksmith, making hand-wrought iron items."[7]

A different and potentially more romantic

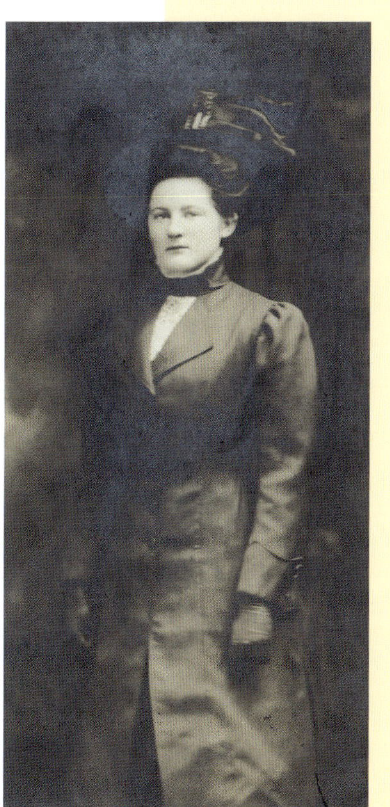

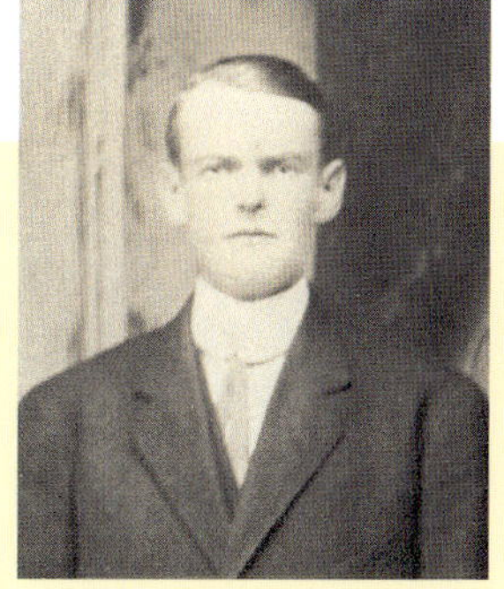

Far left, Jessie Palmer (note 4)
Left, Wendell August (note 4)
Below, August residence in Brockway (note 5)

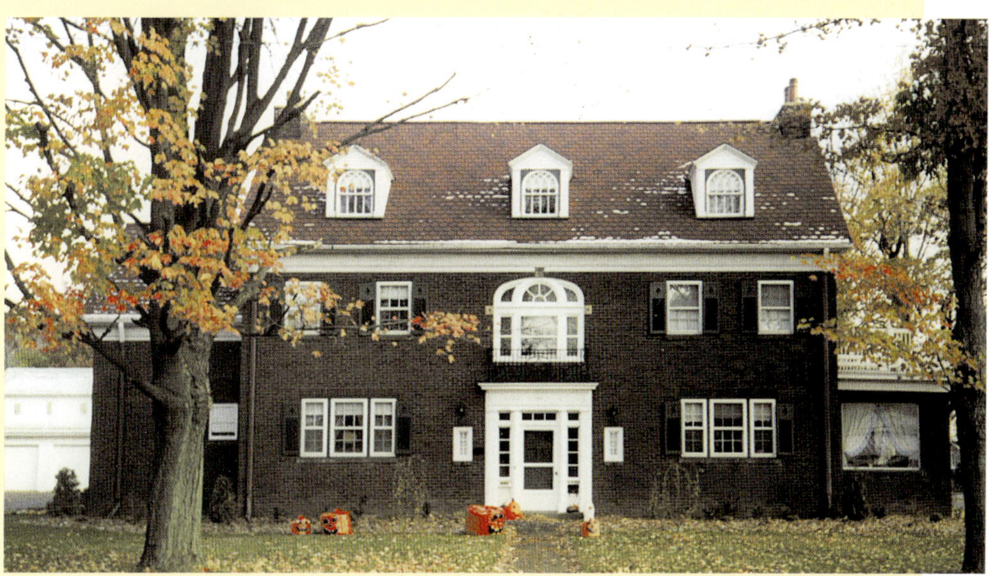

version of Wendell August's initial encounter with the production of ornamental wrought iron seems to have been prepared for a wider range of public consumption. As related in a 1936 issue of *The Jewelers' Circular-Keystone*, August "had a hobby — that of making forged ornamental iron work by hand in his spare time in the blacksmith shop at his mine near Brockway, Pa.," and made the lighting fixtures for his home there. On one occasion, his wife returned from a shopping trip to Buffalo with a bridge lamp having a forged base, which she had purchased for $30. Wendell thought the price high, and made a bet with her that "he could make a dozen stands like the one she had purchased for the same amount of money. So he had 'Tony,' an employee, make up several stands and keep a record of his time and the cost of materials. The result was that 'Tony' turned out several stands at a cost of $2 each."[8]

The United States in the early 1920s was awash with wealth. On the average, Americans were better housed and clothed than any other people in history. Real income was increasing for all economic classes, and the rate of income growth for the upper economic classes was phenomenal. Being well acquainted with the upper economic classes and their need for fine residences, businesses, and retreats, August sensed yet another opportunity to expand his business holdings. Thus, in October of 1923,[9] he established the Wendell August Forge, with four blacksmiths to make ornamental wrought iron, including Ottone Pisoni as the lead blacksmith.

The Forge was initially built at the corner of Main and McCain,[10] behind what was then the Johnson Garage, where Pontiac, Oakland, and Maxwell automobiles were sold. Retaining McCreery & Co. of Pittsburgh and Wm. Hungerer Co. of Buffalo as district representatives, and working with the architectural firm of Howard and Hatcher of

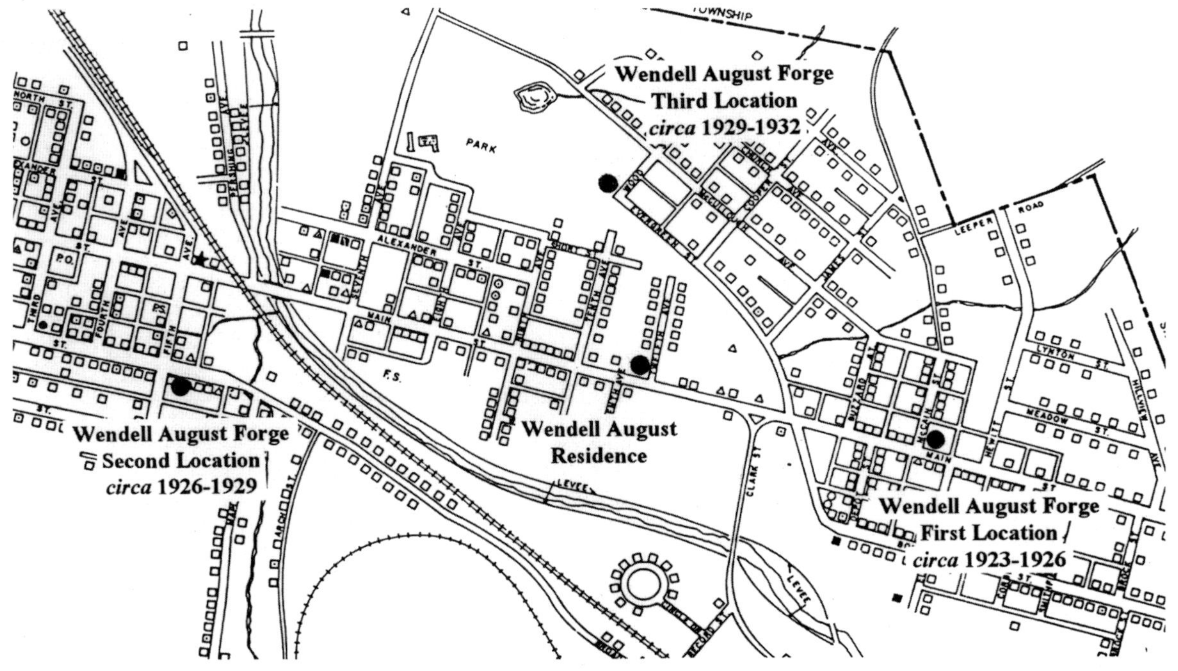

Wendell August residence and Forge sites in Brockway (note 10)

DuBois and Pittsburgh, the Wendell August Forge proceeded to produce a regular line of ornamental hand wrought iron and to acquire contracts for the production of commissioned architectural embellishments. Among the items touted in a Forge brochure were fireplace andirons, candlesticks, lighting standards, door knockers, latches, railings, and window and door grills.[11]

In late July of 1926, the fledgling business in the recently renamed town of Brockway was dealt a brutal blow when the Johnson Garage burned to the ground and, along with it, the Wendell August Forge and its recently completed orders of wrought iron.[12] August moved his business into temporary quarters in Sam Keller's unused blacksmith shop by the intersection of Broad Street and Fifth Avenue. Three years after the fire, he purchased property near the corner of Wood and Evergreen Streets and constructed a new facility, at which time the *Brockway Record* noted that the company "has attained an enviable reputation for the manufacture of ornamental iron work and are enjoying a good business."[13]

Little information about the Forge's ornamental wrought iron work has been found. Architectural installations for which the Wendell August Forge is currently known to have produced wrought iron ornamentation include the Consistory in Buffalo, New York[14] and, in Pennsylvania, the Latonia Theatre in Oil City,[15] the Ambassador Apartments in Pittsburgh,[16] and banks in Brockway, Clarion, New Bethlehem, Oil City, Patton, and St. Marys.[17] Some of these installations were apparently designed by architect James A. McCausland, after he joined the Wendell August Forge in 1928. The involvement with bank installations probably stemmed from Howard J. Chapin, a long-time friend of Wendell August, who had joined the Forge as a salesman and whose father was affiliated with the First National Bank of Brockway.

The wonderfully prosperous world of the 1920s came to an abrupt end within just a few months of the completion of the Forge's new facility. On September 7, 1929, a reaction to the excessive stock market increases set in and a downward spiral in prices began, culminating on October 24, which came to be known as Black Thursday, when more than 13 million shares of stock were sold. The crash of the stock market ultimately wiped out two thirds of the value of all listed securities and led to immediate demands for repayment of loans, setting off an enormous contraction of the financial markets, thus bringing about the Great Depression.

Wendell August apparently lost everything but the Forge. The price of coal dropped so low that it was worth less than the shipping charges, and so the coal that August had been shipping was simply dumped out by the side of the tracks.[18] Six workers struggled on at the Forge, laboring at one point for six months without wages to complete an architectural contract so that the Forge could remain in business and the employees could be paid.[19]

While August was developing his small ornamental wrought iron business in Brockway, events were taking place to the southwest in Pittsburgh that would bring the Wendell August Forge and the Aluminum Company of America (Alcoa) together, a

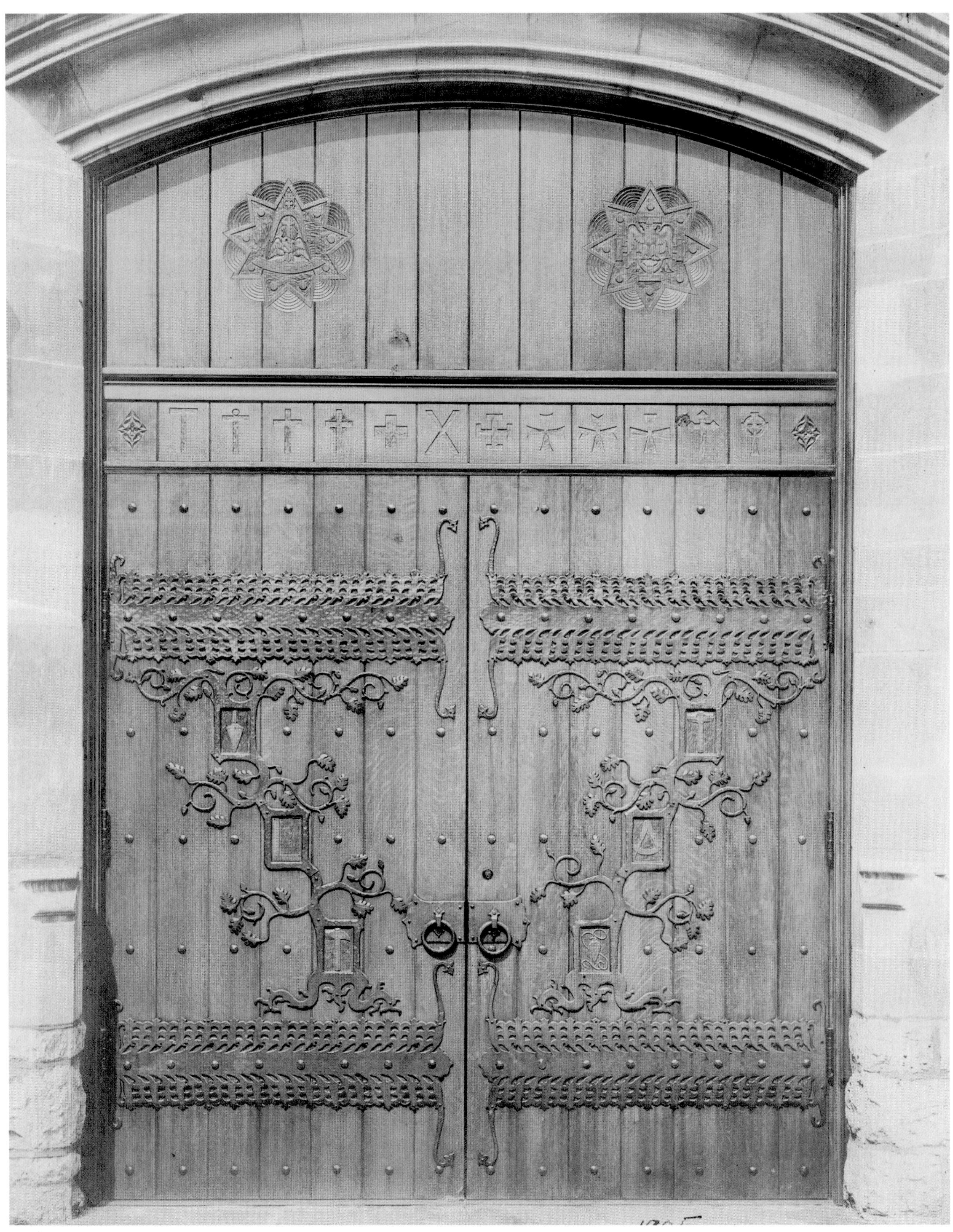

Ornamental wrought iron on doors at the Consistory *(note 14)*

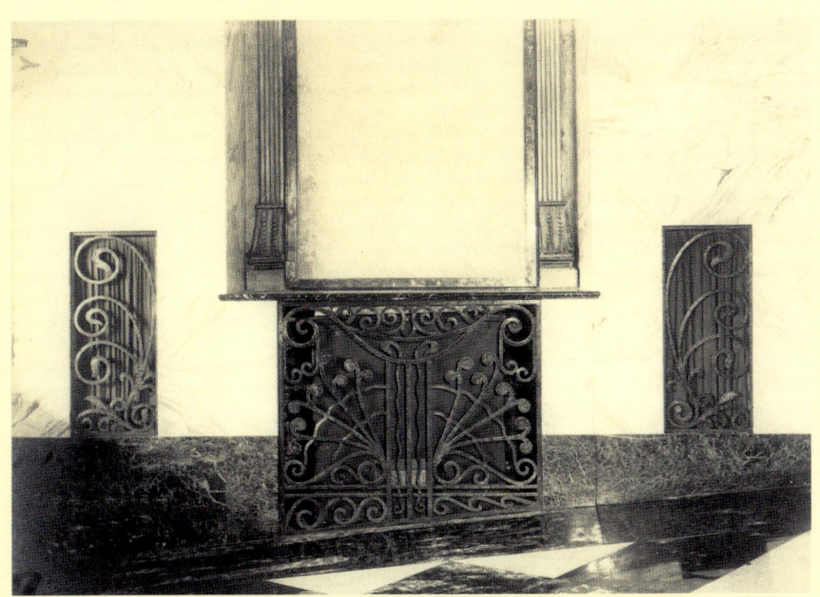

Above, Ornamental wrought iron grills at the Latonia Theater (note 15)

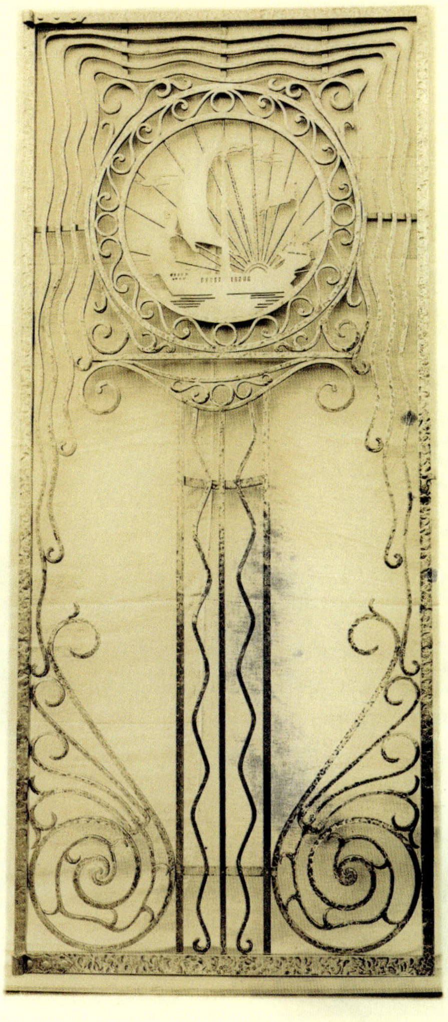

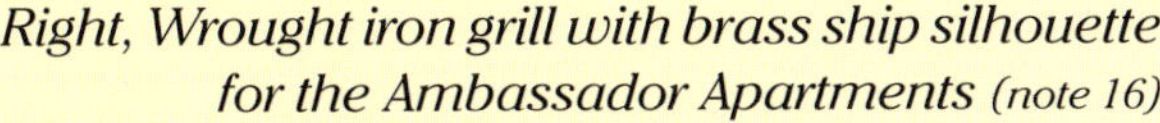

Right, Wrought iron grill with brass ship silhouette for the Ambassador Apartments (note 16)

meeting that would ultimately lead to a new consumer products industry.

Following World War I, Alcoa had begun to invest heavily in the research that was essential to the creation and sustenance of new markets for the relatively new wonder metal.[20] Driven by business requirements, research became a key priority at Alcoa, with various operations scattered around at major operating facilities. In about 1928, Arthur Vining Davis, Alcoa's Chairman, made a personal tour of the research facilities. Appalled by the circumstances, he ordered the erection of a new facility, to be named the Aluminum Research Laboratories, on a hill overlooking the New Kensington production works east of Pittsburgh.[21]

Pittsburgh architect Henry Hornbostel was retained to design the building. As befitting the future home of Alcoa's eminent materials science research organization, the structure was to be lavishly ornamented with aluminum. Hornbostel had already successfully undertaken several architectural designs incorporating aluminum ornamentation, such as the Warren G. Harding Memorial in Marion, Ohio and, in Pittsburgh, the Grant Building and the German Evangelical Protestant Church, and thus was a logical choice as the architect for Alcoa's new facility.[22]

The Wonder Metal

No other metal in history has been subjected to as many rags-to-riches cycles as has aluminum. As we approach the year 2000, it is difficult to conceive of a time when aluminum was considered to be the most precious of metals. Although aluminum is the third most abundant element on earth, it does not occur in usable metallic form in nature. As a result, it was not until the early 1800s that aluminum was even proved to exist.

It was in 1809 that Sir Humphry Davy of England showed that an aluminum-iron alloy could be produced and aluminum oxide obtained by dissolving the alloy. Chemical reactions were used in 1825 by Hans Christian Oersted of Denmark and in 1827 by Friedrich Wöhler of Germany to produce minute amounts of the metal. In 1854, H. Sainte-Claire Deville of France and Robert Von Bunsen of Germany improved Wöhler's method, allowing for the development of a successful but extremely expensive commercial process for the production of the metal.

When aluminum in its metallic form was first introduced to the public at the 1855 Paris Exposition, its silvery sheen, light weight, and rarity caused a sensation. More costy than gold, metallic aluminum was accessible to only the wealthy, its use restricted to such exotic items as a jewel-encrusted aluminum and gold rattle for the son of Napoleon III, an aluminum and gold helmet worn by King Frederick VII of Denmark, and aluminum and gold jewelry suites for Queen Victoria of England.

Because aluminum does not rust, thereby staining marble-faced structures, it was the metal of choice in 1884 for the 100-ounce cap on the Washington Monument. Being at that time the largest quantity of aluminum to have been cast for a single item, the small pyramid was exhibited to admiring crowds in the window of Tiffany's in New York City.

It was the development of technology sufficient for the generation of high-amperage electric current that allowed the invention of practical electrolytic processes for producing metallic aluminum. In 1886, working independently of one another, Charles Martin Hall of the United States and Paul T. Héroult of France both discovered the same electrolytic process for the cost-effective production of metallic aluminum. In the United States, Hall's invention led to the formation of the Pittsburgh Reduction Company which, in 1907, was renamed the Aluminum Company of America.

In less than a generation, the public perception of aluminum changed from that of a precious metal to that of a decidedly utilitarian one. During the early part of the 1900s, consumer-related applications were largely limited to cast or machine-formed cookware. As the 1920s drew to a close, aluminum gained the attention of the modernist industrial designers and others with interests in the decorative arts, and the handworking of the metal was explored. Increasingly-attractive items emerged, and aluminum decorative arts wares became extremely popular. The fascination with the metal was heightened due to its importance for military use, particularly for aircraft, during World War II. The saturation of the consumer goods markets with decorative aluminum products in the late 1940s through the mid-1950s led to a disdain for the familiar, and aluminum wares fell from favor. It was not until the 1980s that interest in vintage aluminum items emerged in the antiques and collectibles communities and the use of aluminum for decorative purposes began to revive.[20]

By 1929, aluminum was highly regarded for the purpose of architectural ornamentation. The relatively light weight made the aluminum ornamentation easier and less expensive to install than, for example, wrought iron. The fact that aluminum did not rust meant that imposing marble edifices would no longer be stained with reddish-brown streaks emanating from the building's ornamentation, and that paint would no longer be required to prevent deterioration. In Alcoa's words, "It is the permanency of aluminum, its defiance of the elements, that makes it the ideal architectural metal. It is light, strong; will not corrode; is easy to form and work; and will not stain."[23]

With fortuitous timing for the Wendell August Forge, a contract for the elaborate aluminum gates for the stone wall surrounding the Aluminum Research Laboratories became available. The bulk of the aluminum architectural ornamentation of the period, ranging from elaborate statuary and lighting standards to window grills and spandrels, was produced by casting, and it was quite reasonably assumed that this process would be used for the creation of the gates.

Individual recollections of the precise circumstances have varied, but regardless of who was responsible, there is agreement that some Alcoa aluminum bar stock arrived at the Wendell August Forge for the purpose of experimenting with the process of hand working the metal. According to Natale Rossi, writing in 1988, it was Howard Chapin who heard about the plans for the aluminum gates and acquired the sample of aluminum bar stock to work with in order to determine if the gates could be hand wrought.[24] In 1940, a decade after the event, it was reported that August learned of the plans to cast ornamental gates for the Aluminum Research Laboratories while visiting an architect's office and, returning to Brockway, had an aluminum sample forged. August recalled that "we made up a small section of a grill and discovered, among other things, that the coal tar from the fire changed the color of the metal so that it resembled highly-polished Swedish steel." August continued "I wish I would have preserved the look on the face of one of the aluminum company's high officials when he lifted that sample, thinking that it was made of steel. He would hardly believe it was aluminum."[25]

Recollections of the first experimentation of the Wendell August Forge with the hand forging of a sample of aluminum bar stock vary somewhat, but are consistently dramatic.[26] Applying an old art to a new metal, and accustomed to working with iron, blacksmith Ottone "Tony" Pisoni overheated the end of the aluminum bar — and it simply vaporized! Heating it again, but to a lower temperature, he then struck the heated end of the bar with his hammer and the metal splattered like drops of rain. Additional experimentation followed, and soon Pisoni could easily bend and twist the bar into extravagantly ornate forms almost at will. A hand wrought sample of one of the elements for the gates was then created for presentation to Hornbostel and the Alcoa executives. Later on, the Forge would prepare an architectural "calling card," illustrating some of the ornamental techniques that could be applied to aluminum bar stock.[27]

August bid for the contract to make the gates. With new product development as one of the research organization's primary charges, and the demonstrated ability of the Wendell August Forge to hand work the aluminum, it was more than fitting that the new approach be used to create the gates for Alcoa's Aluminum Research Laboratories. The contract was awarded to the Wendell August Forge.

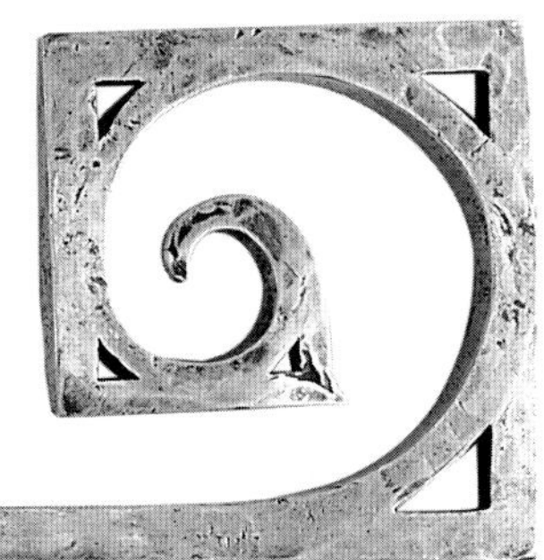

Aluminum element for Aluminum Research Laboratories gates (note 27)

The making of the gates involved its own sets of difficulties. In addition to further experimentation with the hand working of the metal into exquisite forms, the welding problems loomed large. August's original plan had been to use the rivets and straps customary with wrought iron work to join the segments of the gates together. However, his customer contended that aluminum might be too ductile for this method, and, insisting that the segments be welded, proceeded to hire two welders. According to August, when the welders worked on the gates, "the hundreds of bars . . . kept popping out of position, regardless of how tightly they were fastened," as a result of the expansion of the aluminum as it was heated. After several days of failures, the welders gave up, and the problems of joining were resolved as August had originally intended — by using rivets and aluminum straps.[28]

The resulting gates[29] provided exquisite evidence that aluminum could be hand wrought into the elaborate architectural ornamentation so popular at the time, and that the Wendell August Forge was the one place where such work could be accomplished. The gates were installed in 1930 and, although the facility now stands vacant, still remain in place and in excellent condition. The completion of the gates for the Aluminum Research Laboratories building would lead to desperately needed contracts for Wendell August Forge hand wrought aluminum architectural installations at several sites, including the Grove City National Bank and the St. Clair Savings and Trust Company in Pittsburgh.

Less clear in their genesis are the ornately decorated panels, "executed in repoussé by hand-hammering of aluminum sheets,"[30] for the elevator doors inside of the facility.[31] As the story goes, the Wendell August Forge acquired the contract to decorate the doors in accord with Hornbostel's design at the same time that it was awarded the contract for the gates. Natale Rossi wrote, "The way we made the design was to cut a piece of 1/8 inch iron and rivet it to a 1/4 inch piece."[32] According to James DePonceau, the Forge craftsman responsible for carbon-coloring

Aluminum architectural "calling card" (note 27)

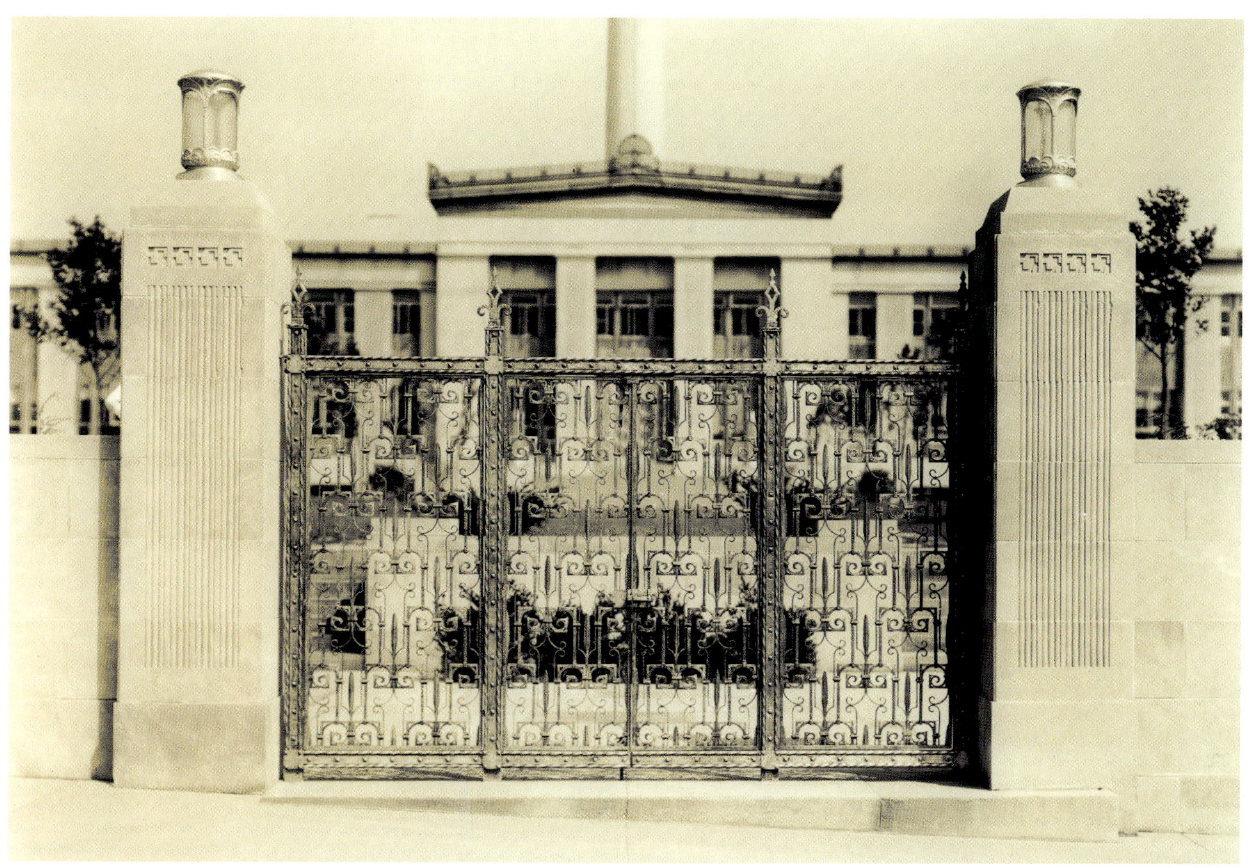

Aluminum Research Laboratories gates (note 29)

the doors, "There's quite a story of how the original designs were made. They depended on scrolls. They were cut into 1/8" sheet iron using a coping saw, to cut around these various scrolls. Then they would rivet them to a 1/4" thick, a heavier piece of iron, then they would hammer through that open area. They got beautiful patterns that way."[33] Later on, DePonceau would write that "discarded scrap aluminum sheet used *under* the . . . iron pattern served the purpose of keeping the embossed area unmarked . . . The embossing was simplified by using the exact architect's layout on top of the aluminum to be embossed; marking the openings with a pencil so that tools, punches, or wedged hammers could be used over the exact openings. If an area needed more embossing one simply rebolted this setup after marking the place or places needing additional embossing. The bolting of each section to be embossed would guarantee being able to retouch any area needing more embossing."[34]

When the doors were completed, Hornbostel apparently requested that two trays, using some elements of the motif from the elevator doors, be made for presentation as mementos to Alcoa's Chairman, Arthur V. Davis, and President, Roy A. Hunt.[35] According to Arthur Armour, "when the building was done, Henry Hornbostel wanted to have some mementos, to give to some of his friends, and for himself. So he asked to have parts of the elevator doors, in the research laboratory, made into trays."[36] The Alcoa executives were apparently quite taken

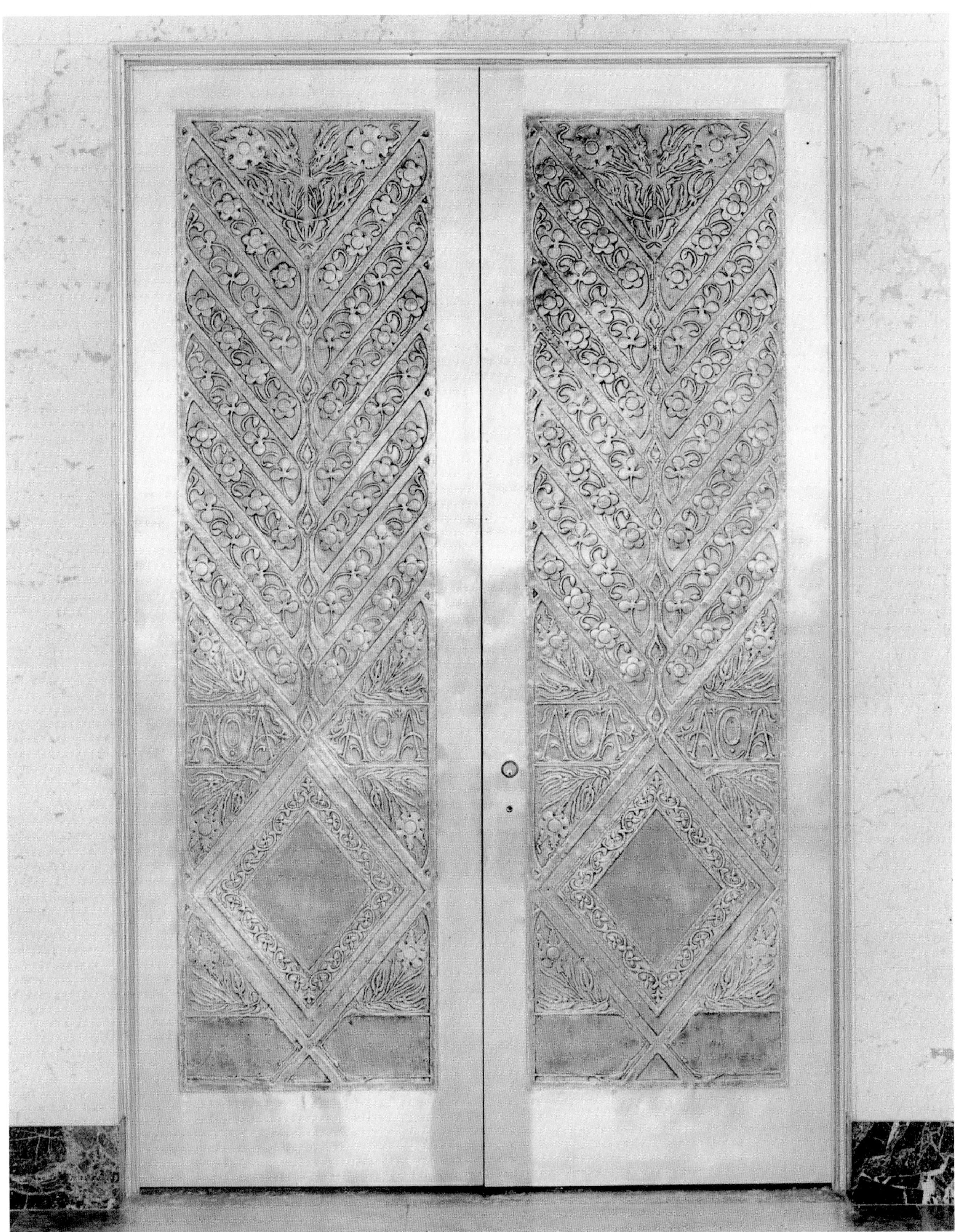

Aluminum Research Laboratories elevator doors (note 31)

with the beauty and uniqueness of the pieces, and additional trays were requested. Alcoa executives, ever mindful of new product opportunities, ensured that the items were brought to the attention of Edgar Kaufmann, the Pittsburgh department store magnate and patron of the arts, for whom Frank Lloyd Wright would design and build the renowned "Falling Water" retreat at Bear Run in 1936.

Kaufmann seems to have commissioned the Forge to make a tray having a motif of his estate, and then suggested that the Forge's aluminum repoussé process be implemented in the development of an elite line of art and gift items. The popularity of brass and copper for decorative accessories had been gaining impetus for several seasons, and Russel Wright's *Spun Aluminum* had recently been introduced to popular acclaim. An influential trade journal would note "the present vogue for the use of metal in the home," claiming that "perhaps never before has there been such a general public interest in decorative accessories made from such a varied assortment of metals," and specifically drawing attention to "the newer white metals . . . chromium, nickel finish, aluminum."[37] Aluminum, of course, being a relatively inexpensive metal, had a significant advantage over some of the other metals and metallic finishes, an advantage not to be scoffed at as the weight of the Great Depression grew increasingly heavy.

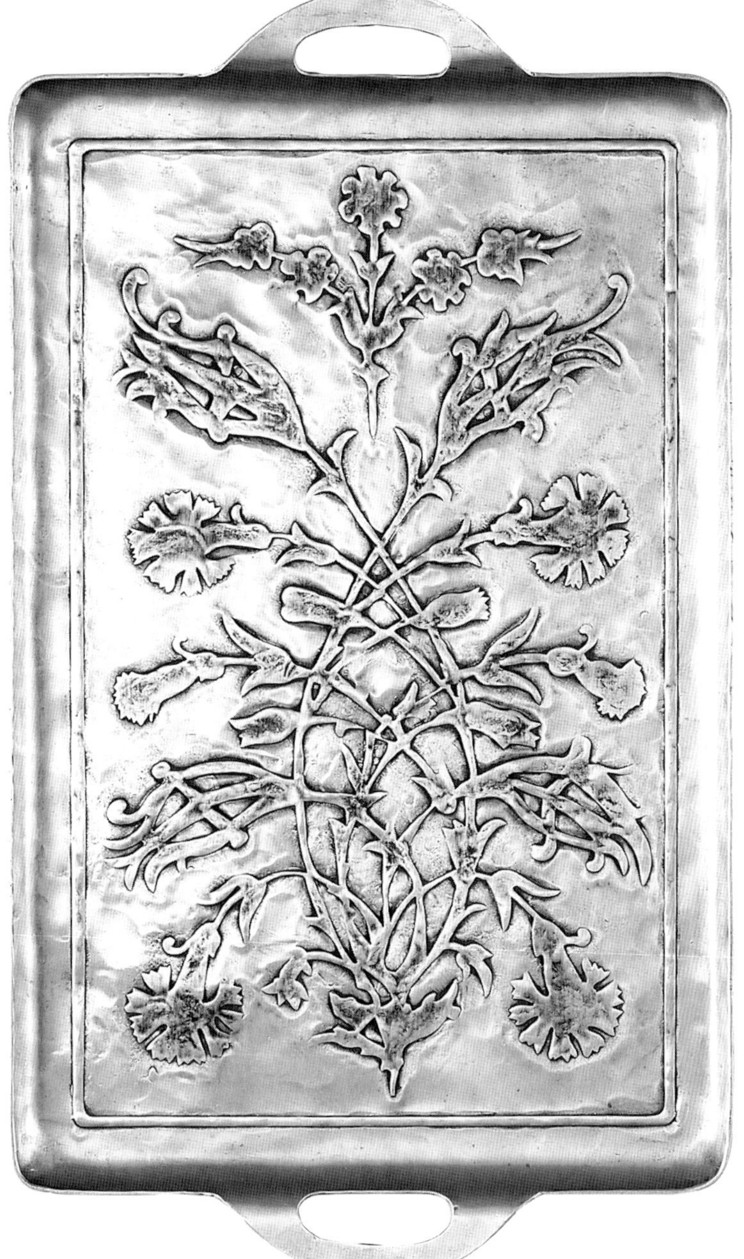

Serving tray from early die (note 35)

Before the art and giftware line was developed for Kaufmann, a different approach to embellishing the aluminum sheet had been developed. Here again, individual recollections vary. Apparently Pisoni and Rossi had taken note that, while working the

Traditional Production Process at the Wendell August Forge (note 41)

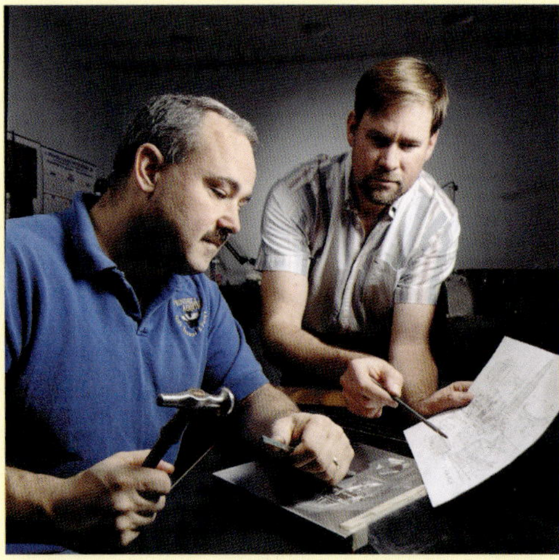

1

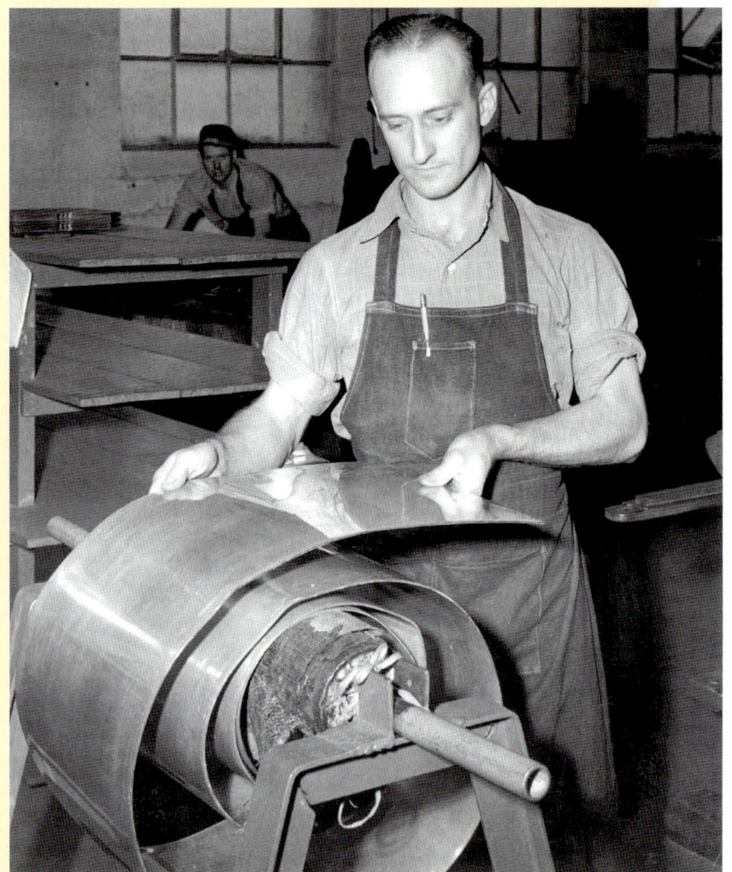

2

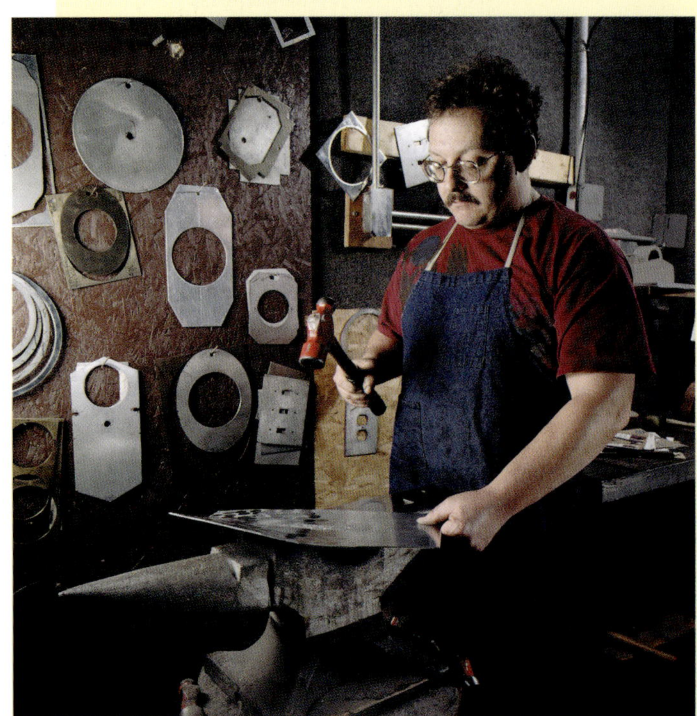

4

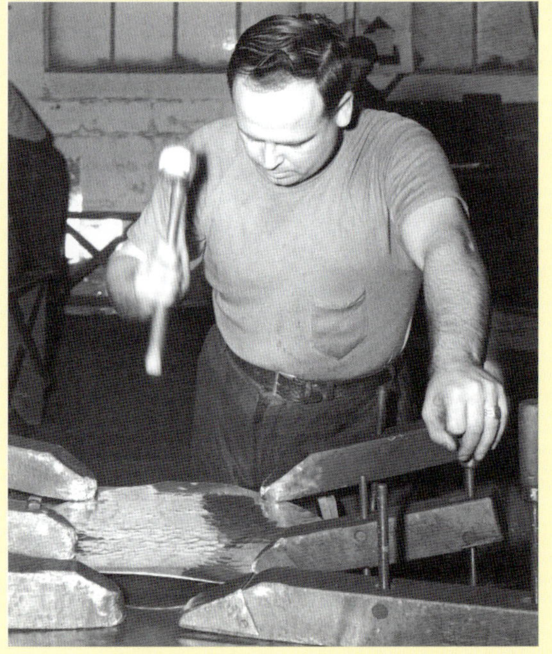

3

Clockwise:
***1** – Die engraving*
(Leonard Youngo and David Bruck, 1997)
***2** – Material selection and cutting*
(Lester Sutherland, 1940)
***3** – Repoussé (William Foster, 1940)*
***4** – Surface anvilling*
(Michael O'Mahony, 1997)

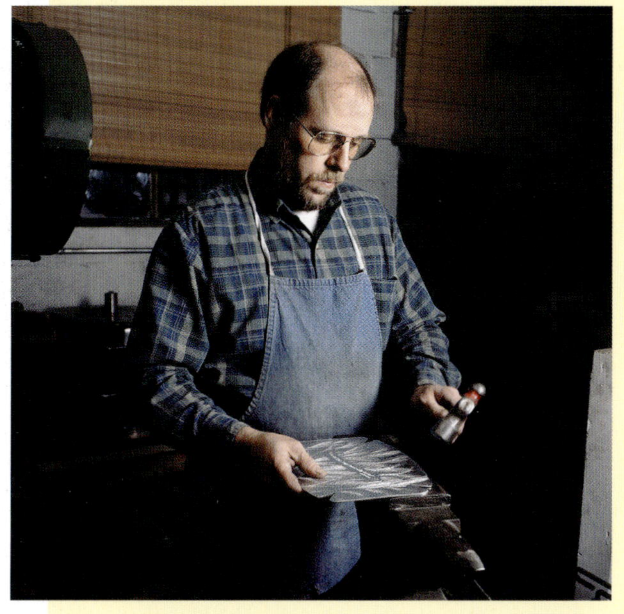

5

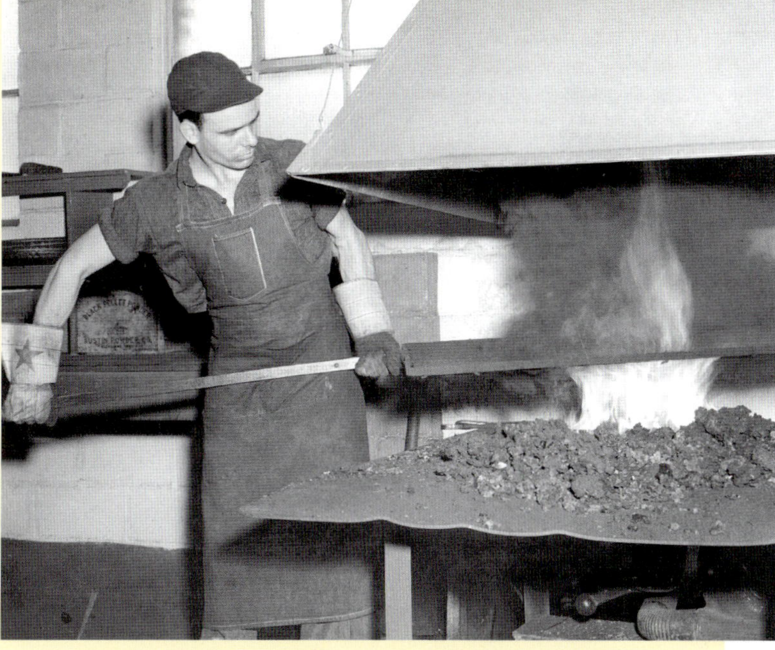

6

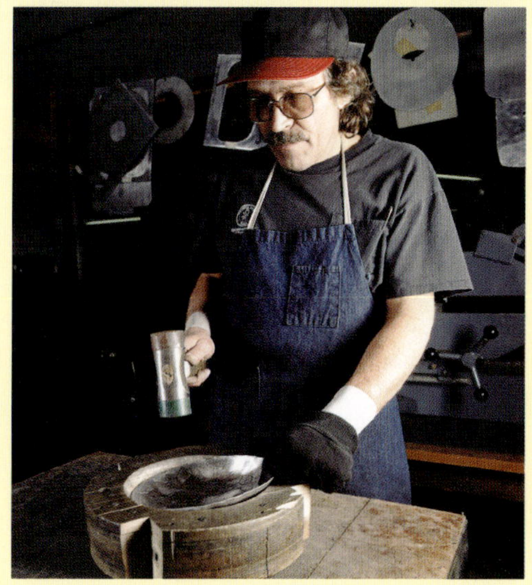

8

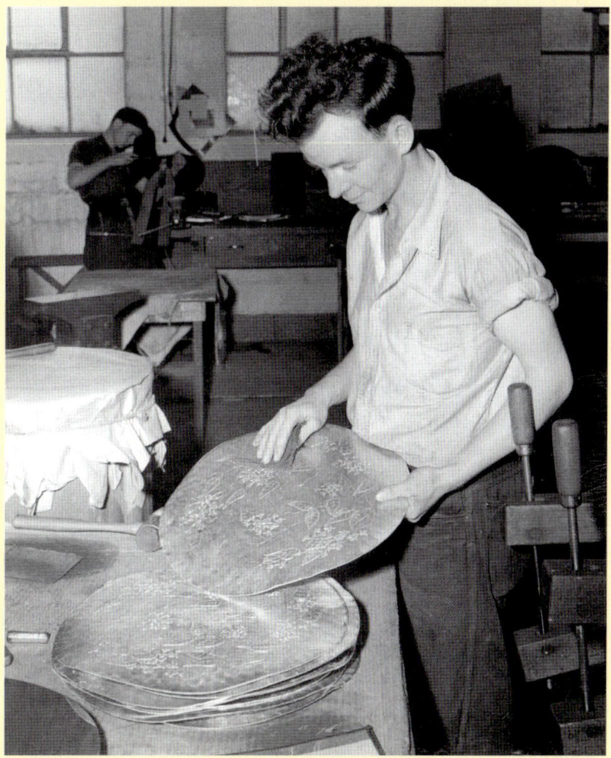

7

Clockwise:
5 *– Edging (Jeffery Brown, 1997)*
6 *– Carbon coloring (Robert Wallwork, 1940)*
7 *– Sanding and waxing (Gordon Brown, 1940)*
8 *– Forming (Ronald Winder, 1997)*

aluminum bar on the anvil, imperfections in the surface of the anvil were easily impressed in the relatively soft metal. Natale Rossi is largely credited with suggesting that the motifs simply be cut into a steel die. According to Arthur Armour, "Well, I think it was an original idea with Natale Rossi . . . He cut it [the motif] out, through a solid block of steel, so that when you drove it [the aluminum sheet] in through the solid block of steel, you had details, rather than a flat surface."[38] Natale wrote, "One morning I got to [the] shop early and I thought, what if I got a solid piece of iron about 1/2 inch thick, and cut the design in it about 1/6 of an inch deep in the iron? We could put details like leaves, not only of the outline, but we could put the veins in too."[39]

Some of the earlier motifs thus produced were relatively crude. James DePonceau recalls that several Forge metalworkers, including him, were asked to try their hand at die engraving. He relates, "Jim McCausland laid a dead grasshopper down where I was working and said to see what I could do. I just started cutting [the die] with a chisel. I didn't know you should do a drawing first. It didn't turn out very well."[40] Regardless of such difficulties, the new approach soon proved to be effective and ultimately allowed for more creatively and aesthetically rendered motifs.

Thus evolved a basic process of engraving a die, hammering sheet metal into the die, "coloring" (or "antiquing") the piece with coal tar smoke, and forming the result into an object. This approach to producing decorative metal wares, which was soon to be emulated by others, remains the basic process that is still in use at the Wendell August Forge for the creation of heirloom pieces in aluminum, bronze, pewter, sterling silver and other metals.[41]

Chapter Notes

1 Pennsylvania Writers' Project, *Pennsylvania: A Guide to the Keystone State* (New York: Oxford University Press, reprint of Dec. 1940 publication by the University of Pennsylvania).

2 Lewis D. Reddinger, *The Brockway Story, 1822-1986, A Narrative History of The Brockway Area, Jefferson County, Pennsylvania* (Brockway, PA: Brockway Area Historical Society, 1986)

3 W.J. McKnight, *Jefferson County, Pennsylvania, Her Pioneers and People*, Volume II (Chicago: J.H. Beers & Company, 1917), and James Sterrett, personal correspondence, 22 July 1997.

4 Jessie McVean Palmer, early 1900s photograph courtesy of M.C. (August) Taylor. Wendell August, 1911 photograph courtesy of Ruth Trimm and the Wendell August Forge.

5 Wendell McMinn August residence, 1101 Main Street, Brockway, PA; 1997 photograph courtesy of the Wendell August Forge.

6 Natale Rossi, "The Conception of a New Business," manuscript, 1988; courtesy of Thomas Armour and the Wendell August Forge.

7 Rossi, "Conception."

8 Sykes, Edward H., "Aluminum," *The Jewelers' Circular-Keystone*, March 1936.

9 "Wendell August Forge," *The Brockwayville Record*, 7 March 1924. All reference materials from *The Brockway(ville) Record* were transcribed from microfilm by Adam Sabatose, Brockway Area Historical Society.

10 Map of the borough of Brockway adapted from July 1978 Jefferson County Planning Commission map included in Reddinger, *The Brockway Story*; Wendell August Forge locations based on information provided by James DePonceau and Adam Sabatose, together with a site visit by the author.

11 "Wendell August Forge," *The Brockwayville Record*, 7 March 1924.

12 "Bad Fire Saturday," *The Brockway Record*, 23 July 1926.

13 *The Brockway Record*, 19 July 1929.

14 Wrought iron work on doors at the Consistory in Buffalo, New York, *circa* 1925; photograph courtesy of the Wendell August Forge.

15 Wrought iron grill work at the Latonia Theater in Oil City, Pennsylvania, *circa* late 1920s; photograph courtesy of Henry Swyers and the Brockway Area Historical Society.

16 Wrought iron grill work, with ship silhouette in brass, for the Ambassador Apartments in Pittsburgh, Pennsylvania, *circa* late 1920s; photograph courtesy of the Wendell August Forge.

17 Gust Boleen, personal interview, 17 June 1997; James DePonceau, personal correspondence, 10 July 1997; Rossi, "Conception;" Natale Rossi, in Catherine Youngo, transcription of recorded interview, 19 Nov. 1992; photographs from the files of James McCausland, courtesy of Henry Swyers and the Brockway Area Historical Society.

18 Leo M. Stevenson, personal interview, 25 July 1997.

19 Rossi, "Conception."

20 Sources include: *The A-B-C's of Aluminum* (Richmond, VA: Reynolds Metals Company, 1955); Thomas B. Canby, "Aluminum, The Magic Metal," *National Geographic*, August 1978; Charles C. Carr, *Alcoa: An Aluminum Enterprise* (New York: Rinehart & Company, Inc., 1952); C.B. Colby, *Aluminum: The Miracle Metal* (New York: Coward-McCann, Inc., 1958); June Metcalfe, *Aluminum from Mine to Sky* (New York: Whittlesey House Division of McGraw-Hill, 1947); George David Smith, *From Monopoly to Competition: The Transformations of Alcoa, 1888-1986* (Cambridge: Cambridge University Press, 1988); Edward B. Tracy, *The New World of Aluminum* (New York: Dodd, Mead & Company, 1967); Donald H. Wallace, *Market Control in the Aluminum Industry* (Cambridge, MA: Harvard University Press, 1937).

21 Smith, *From Monopoly to Competition*.

22 *Architectural Aluminum* (Pittsburgh, PA: Aluminum Company of America, 1929).

23 *Architectural Aluminum*.

24 Rossi, "Conception."

25 A.D. LeMonte, "Business Built on Aluminum Art Work," *The Youngstown Vindicator*, 9 June 1940.

26 The first experimentation has been attributed to both Ottone Pisoni and Natale Rossi. Based on the evidence accumulated to date, it is most likely that the work was done solely by Pisoni.

27 Squared spiral element (6"x11-1/2") used in the gates for the Alcoa Aluminum Research Laboratories and 17" Wendell August Forge architectural "calling card." Photography by Richard Smaltz, Smaltz Studio, West Middlesex, PA; photograph courtesy of the Wendell August Forge.

28 LeMonte, "Business Built on Aluminum."

29 Aluminum gates at the Alcoa Aluminum Research Laboratories in New Kensington, PA; photograph, dated 29 July 1930, courtesy of Henry Swyers and the Brockway Area Historical Society. Complete gate structure measures 151" (12' 7") wide and 111-1/2" (9' 3-1/2") tall (from base of ground-level posts to tops of finials); measurements courtesy of Marilyn Sullivan.

30 "Research Laboratories' New Home," *The Alcoa News*, 1930.

31 Elevator doors at Alcoa's Aluminum Research Laboratories in New Kensington, PA; photograph, dated 1930, courtesy of Henry Swyers and the Brockway Area Historical Society. Each door measures 34-3/4" (2' 10-3/4") wide and 112" (9'4") tall; measurements courtesy of Marilyn Sullivan.

32 Rossi, "Conception."

33 James DePonceau, interview, 29 June 1994.

34 James DePonceau, personal correspondence, March 1996.

35 Tray, 13-1/2"x24", made from a riveted die using some elements of the motifs from the Alcoa elevator doors; Wendell August Forge Collection. An image of this tray was used in a 1946 Alcoa advertisement featuring the Wendell August Forge, one of a sequence of advertisements to promote the creation of new firms and products using Alcoa aluminum. Photography by Richard Smaltz; photograph courtesy of the Wendell August Forge.

36 Arthur Armour, as quoted in Thomas Armour, transcription of recorded interview, 1 Feb. 1992; courtesy of Thomas Armour.

37 John Arthur Darragh, "Gold For You In The Vogue For Metal," *The Gift and Art Shop*, Oct. 1932.

38 Arthur Armour, 1 Feb. 1992.

39 Rossi, "Conception."

40 James DePonceau, personal interview, 27 July 1997.

41 Photographic sequence depicting the traditional repoussé process of the Wendell August Forge. (1) Master die engraver Leonard Youngo engraving a steel die with the assistance of master die engraver David Bruck in 1997; (2) Blacksmith and craftsman Lester "Bus" Sutherland obtaining a piece of aluminum from a coil in 1940; (3) Craftsman William "Bill" Foster hammering sheet aluminum into an engraved die to yield a repoussé motif in 1940; (4) Master craftsman Michael O'Mahony applying decorative surface hammering to flatten a piece after the repoussé process in 1997; (5) Master craftsman Jeffrey Brown finishing the edges of a piece in 1997; (6) Craftsman Robert "Bob" Wallwork coloring a piece by holding it over a coal fire in 1940 [Wallwork was a stepson of William Miller, one of the people who moved with the Wendell August Forge from Brockway to Grove City]; (7) Craftsman Gordon Brown removing excess color from and waxing a piece in 1940; (8) Master craftsman Ronald Winder hammering a piece into its final form using a wooden mold in 1997. Photography for (1), (4), (5) and (8) by Richard Smaltz; photographs courtesy of the Wendell August Forge. Photographs (2), (3) and (6), courtesy of Henry Swyers and the Brockway Area Historical Society. Photograph (7) courtesy of Mrs. Gordon Brown and the Wendell August Forge.

Chapter Two

The Golden Era: 1931-1938

Leaving Brockway, one can retrace the old state roads to Clarion and then, avoiding the bustle of the Interstate, continue on through Clarion and over the Clarion River, traveling deeper into the region where the world's first oil well was drilled. Heading southwesterly after arriving at Shippenville, a smooth curve leads into Emlenton, once home to numerous millionaires, where you can cross the Allegheny River. The roadway that climbs the west bank of the river, although remaining difficult to negotiate, bears little resemblance to the treacherous steep and sharp unpaved curves that terrified motorists and sickened passengers in the 1930s, when this portion of the route was virtually impassable when iced over from winter storms. As the distance from the Allegheny ranges and river steadily increases, the terrain gentles, and soon you have passed Barkeyville, crossed Wolf Creek, and arrived in Grove City. It was along this 90-mile route that the skilled craftsmen of Brockway's Wendell August Forge traveled for almost three hours in early 1931 to install their hand wrought aluminum grill work in the Grove City National Bank.[1]

The land purchases of the late 1700s that made possible the borough of Brockway also provided for the future borough of Grove City. Parcels of land in the area were offered to settlers, and by 1798 two permanent settlements, Slabtown and Pine Grove, had been established on Wolf Creek in the vicinity of present-day Grove City. Although bituminous coal deposits had been discovered in the area in 1850, and the arrival of the Pittsburgh, Shenango and Lake Erie Railroad in the early 1870s stimulated the opening of local mines, there were only about 200 residents of Pine Grove as late as 1875.

In 1882, the residents of Pine Grove submitted a petition to the Mercer County court to incorporate and change the name of the new borough to Grove City. The development of the borough was soon enhanced by the 1898 founding of the Carruthers-Fithian Clutch Company by John Carruthers and Edwin J. Fithian, which in the following year was incorporated as the Bessemer Gas Engine Company. In 1929, Fithian retired from his position as President of the Bessemer Gas Engine Company when it merged with the C. and E. Cooper Company of Mt. Vernon, Ohio to

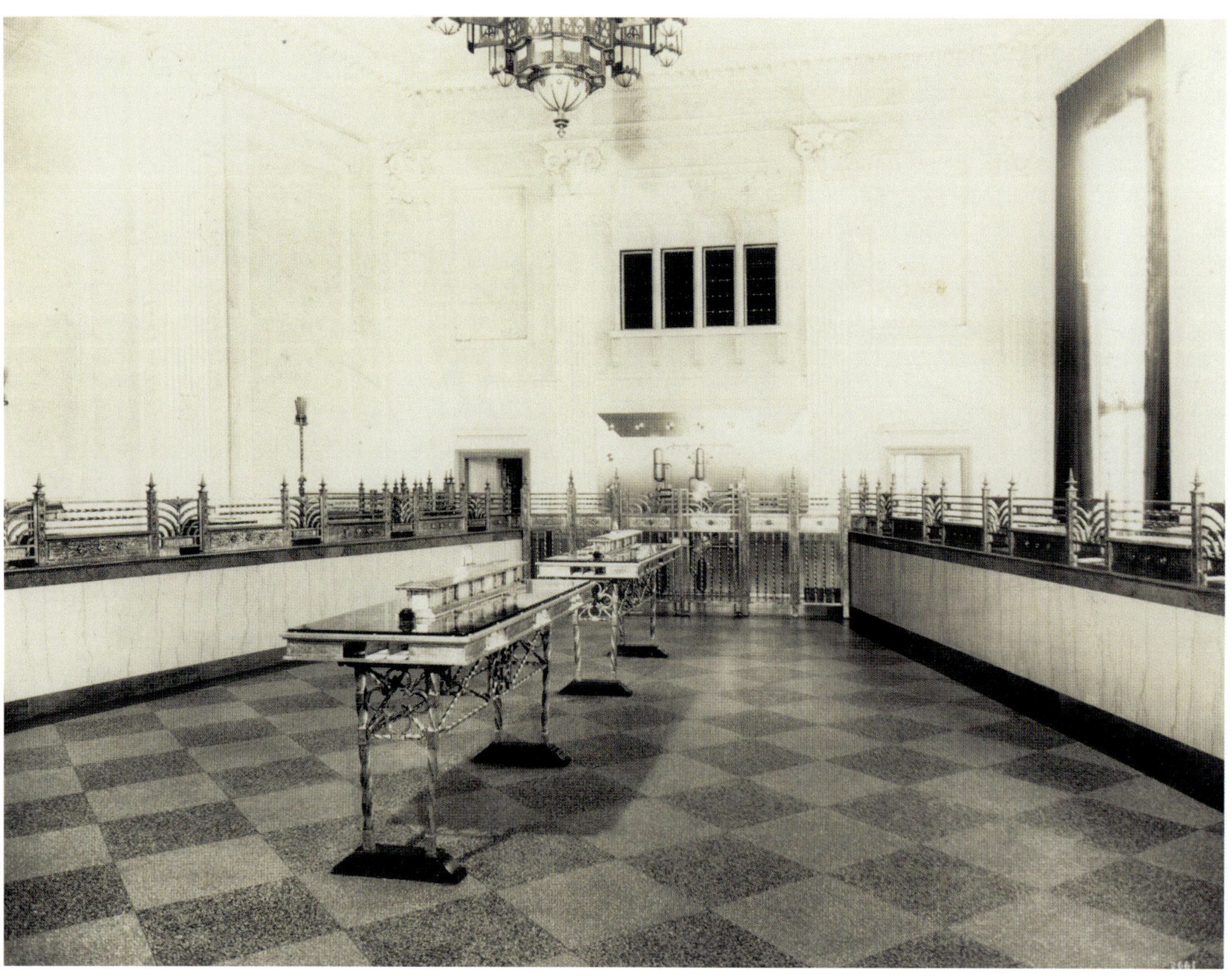

Interior of St. Clair Savings and Trust Company (note 3)

become the Cooper-Bessemer Corporation, the largest builder of gas engines and compressors in the United States.[2]

Fithian became the Chairman of the Board of Directors of the new firm, while continuing to serve as the President of the Grove City National Bank, a position that he had held since 1925. A former Pennsylvania gubernatorial candidate under the banner of the Prohibition Party, he also continued with his numerous political engagements.

Fithian's active role in banking and in politics ensured that he was aware of the new and well-received ornamental architectural aluminum work that had been completed for Alcoa's Aluminum Research Laboratories and that was to be used in the new facility for the St. Clair Savings and Trust Company in Pittsburgh.[3] Under his leadership, the services of the Wendell August Forge were engaged to provide unique and ornate aluminum counter grills for the remodeling of the Grove City National Bank Building, which reopened in mid-January of 1931.[4]

Like most of the astute bankers in the country during the Great Depression, Fithian

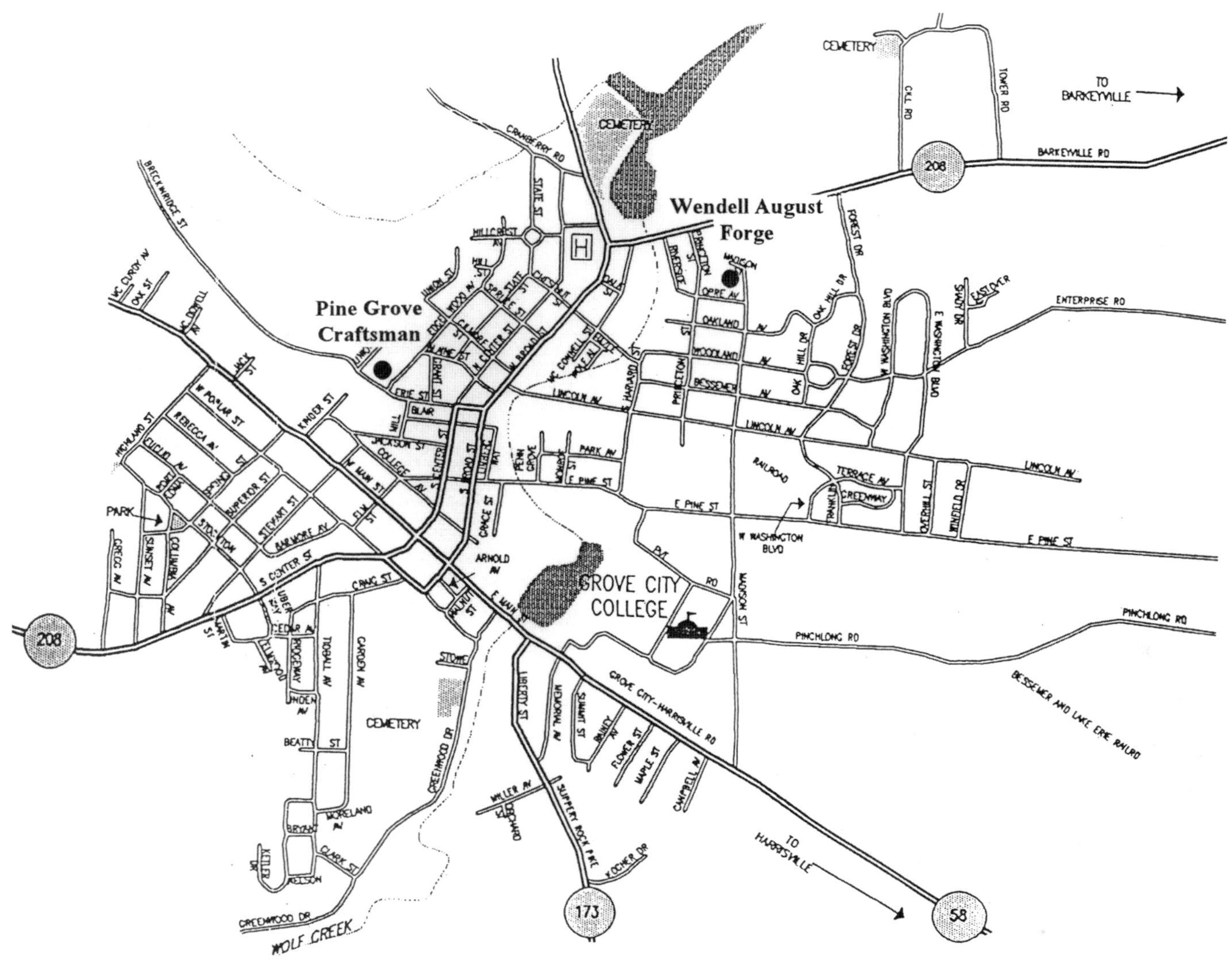

Grove City sites associated with the Wendell August Forge

was seeking new industry for his community, in order to improve the depressed economy of the borough of Grove City. He clearly perceived in the Wendell August Forge an opportunity for a new and exciting investment, an outlet for his own considerable energies, and a means of bringing new industry and economic stimulus to Grove City. Following the completion of the Forge's work for the Grove City National Bank, Wendell August was invited by the Commercial Club of Grove City to relocate the Forge. August initially moved his family to Grove City in time for the start of the Fall, 1931 school session, and "when efforts were resumed to bring his plant here, Mr. August consented to do so after Fithian acquired a financial interest as a means of bringing a new industry to Grove City."[5]

On April 16, 1932, overseen by Fithian's brother-in-law, engineer Henry Limberg, ground was broken for the construction of a concrete block building on Madison Avenue to house the Wendell August Forge when it relocated from Brockway. The firm was expected to be engaged primarily in "making the type of grill work found in the Grove City

National Bank" and "as a side line" would be manufacturing "a line of hammered aluminum novelties, such as trays, plaques, panels, and other relief work."[6] It was also anticipated that the Forge would undertake the manufacture of aluminum lawn furniture and electric lighting fixtures.

In the middle of June of 1932, the country slid into the most desperate period of the Great Depression. Northwestern Pennsylvanians struggled to survive as bituminous coal production plunged to 52 percent of its 1929 level, and the cash receipts of farmers fell to 54 percent of their 1929 level.[7] Against this backdrop, the Wendell August Forge began its final preparations for the move from Brockway to Grove City. Incorporation articles were signed on June 20, 1932 by Edwin J. and Ethel L. Fithian, Wendell M. and Jessie (Palmer) August, and Howard J. Chapin. The Fithians and the Augusts each held a 40 percent interest in the new corporation, with Chapin holding the remaining 20 percent. Edwin J. Fithian, President, Wendell M. August, Vice President and General Manager, and Howard J. Chapin, Sales Manager, would serve as the directors and officers of the closely-held firm.[8]

Early in July of 1932, with the physical plant nearly complete, the beginning of operations by the first of day of August was eagerly anticipated. Skilled workmen were to be relocated from Brockway, with additional hires coming from the Grove City populace whenever additional help might be needed.[9] Grove City was not disappointed, as on Wednesday, July 27, 1932, eight Forge employees started work. Coping as best they could to forge ornamental aluminum in a facility that was not yet complete, were James DePonceau, Benjamin Formani, brothers Robert "Les" McLaughlin and Warren McLaughlin,[10] William Miller, Ottone "Tony" Pisoni (an uncle of Benjamin Formani), and brothers Lewis "Doc" Rossi and Natale Rossi.[11] Shortly after their arrival, they would be joined by die engraver Louis Donato.[12] The ultimate employment of 18 to 20 skilled men engaged in the production of hand-hammered architectural accessories was estimated by Superintendent James McCausland.[13] Arthur J. Palmer had been engaged for the "sales of the new novelty line, which the company expects to feature," which had been "introduced by the August firm a short time ago." The high expectations for the firm were underlined by the emphasis given to Wendell August's "patent to the new process of aluminum forging," with the Grove City plant being "the only one of its kind in America."[14]

The patent claims were somewhat overstated by the press, since the patent had merely been applied for as of July 25, 1932. The applicants for the patent on a "Method of Forming Ornamental Relief Figures" were Wendell August of Grove City, James McCausland of Falls Creek, and Howard J. Chapin of Brockway, with any patent rights that might be granted being assigned to the Wendell August Forge, Inc.[15]

The first known catalog for the Wendell August Forge aluminum giftware line consisted of a single, three-fold sheet of paper, printed on both sides, in black, white, and "aluminum," with prices effective October 1, 1932.[16] The products featured included coasters, ashtrays, flower blocks, paper

The McLaughlins

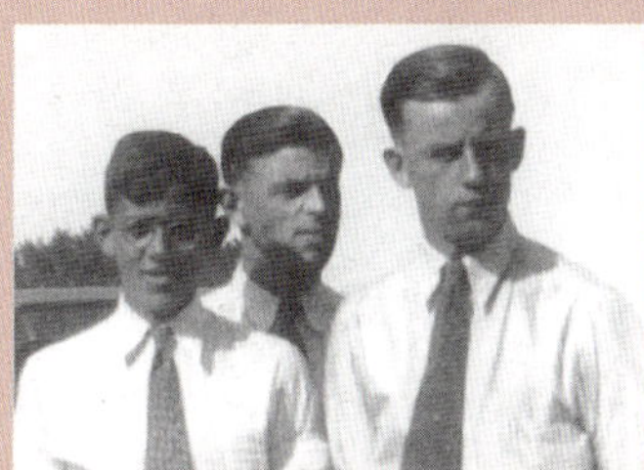

Milford, Warren and Robert McLaughlin

A cooper from Londonderry, Ireland, John McLaughlin immigrated to the United States in 1847 and in 1859 settled in Jefferson County, Pennsylvania with his wife and children. One of his grandsons, John H. McLaughlin, married Margaret Patton and established a blacksmith's shop in Falls Creek, Pennsylvania.

Two of the five children of John H. McLaughlin, Warren Patton (1897-1982) and Robert Leslie "Les" (1898-1962), joined him in learning the skilled craft of blacksmithing. When the Wendell August Forge of Brockway had a need for experienced and talented blacksmiths for its ornamental architectural work, the two McLaughlin brothers were recruited.

Warren and Les worked for the Forge while it was in Brockway, and were among the original eight craftsmen who relocated from Brockway to Grove City in 1932. They remained with the Forge until it closed down in 1942 due to World War II.

Warren's son, Milford "Buck" McLaughlin (1917-1968), began working for the Wendell August Forge in Grove City during the summer in the mid-1930s while he was in high school. When he finished school, he became a full time craftsman at the Forge. He served in the military during World War II, returning to the Forge after his discharge. In 1950, he left the Wendell August Forge when its workforce was being significantly reduced.[10]

weights, waste baskets, desk sets, trays in various shapes and sizes, several sizes of bowls, and more unusual items such as fruit knives, small tables, a *Turtle Shell Bowl* and a *Turtle Bird Bath*, and *Cat Tail Candle Sticks*.[17] The product line was clearly directed toward the luxury market. A *Moon Fish Mirror*, for example, was listed at $50.00, and a pair of *Arrow Candle Sticks* at $30.00. This was at a time when a Wendell August Forge apprentice started at 20 cents per hour and a more experienced Forge blacksmith earned only 47 cents per hour.[18]

Alcoa, eager to learn how its newest protégé was faring with this new application for aluminum, paid a visit to the Wendell August Forge at its new Grove City facility in the Fall of 1932 and reported effusively on the visit in *The Alcoa News*.[19] The Alcoa visitor noted that it was a "very difficult matter to tear yourself away from a room where a thousand and one different and fascinating articles are being fabricated from what to us is 'just another coil of 2SO [a particular grade of aluminum[20]],' " describing the Forge as "a modern aluminum forge shop where craftsmen of our day produce in handworked aluminum, decorative architectural pieces and objets d'art that rival the beauty of texture, design and patina of the products of the metal guilds of the Middle Ages which are now only to be found in museums." The future was accurately anticipated in the statement that, "it does not require an excessive stretch of the imagination to surmise that today's examples of the handiwork of those genial Grove City artisans will be the museum pieces of a later age."[21] Today, two pieces of Wendell August Forge decorative aluminum are included in the

1,971,700
METHOD OF FORMING ORNAMENTAL RELIEF FIGURES
Wendell August, Grove City, James McCausland, Falls Creek, and Howard J. Chapin, Brockway, Pa., assignors to Wendell August Forge Incorporated, Grove City, Pa., a corporation of Pennsylvania
Application July 25, 1932, Serial No. 624,640
7 Claims. (Cl. 41—24)

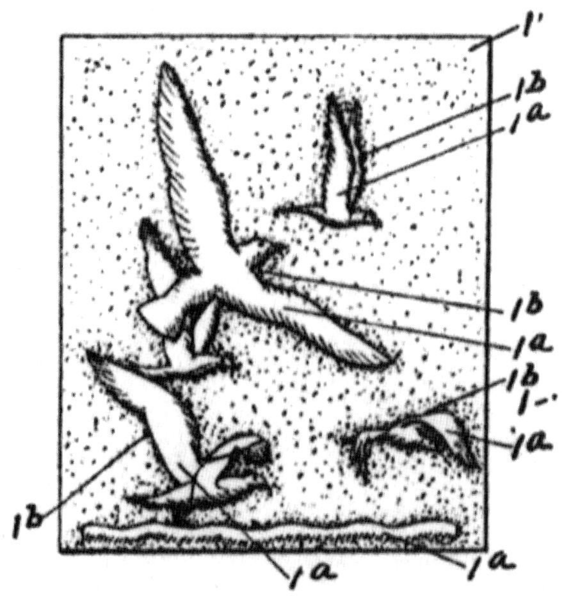

1. The method of forming ornamental plates which consists in forming a depressed portion in a pattern block; placing the plate thereupon and hammering the back of the plate forcing the material of the plate into the depressions formed in the block.

Patent abstract (note 15)

"Material World" exhibition at the National Museum of American History of the Smithsonian Institution,[22] and other pieces have been displayed in conjunction with formal art gallery as well as museum exhibitions.[23]

Certainly 1932 was a spectacular year for the Wendell August Forge. In addition to having completed the work for the St. Clair Savings and Trust Company in Pittsburgh, the Forge had provided grills for the passenger ships of the American Scantic Line, lighting fixtures for the chain of Stouffer Restaurants, architectural ornamentation for a bank in DuBois, Pennsylvania, and numerous grills, railings, and other fixtures for firms in Cleveland, Philadephia, and other cities. Since the "novelty line" (i.e., giftware) had been envisioned primarily as a means of helping to keep the plant at a stable level of operations between architectural assignments, the Forge was astounded at the enthusiastic reactions to the line that were received at the art and gift shows in Boston and New York in August and September. By November, the *Grove City Reporter-Herald* was reporting that the Forge "has lost sight of the depression in a scramble to keep up with a stream of orders for forged aluminum products that is keeping the plant busy more than 60 hours each week," elaborating that the "increased production is due almost entirely to the development of a line of gift novelties in forged aluminum which has added more than 200 accounts since August 1st."[24]

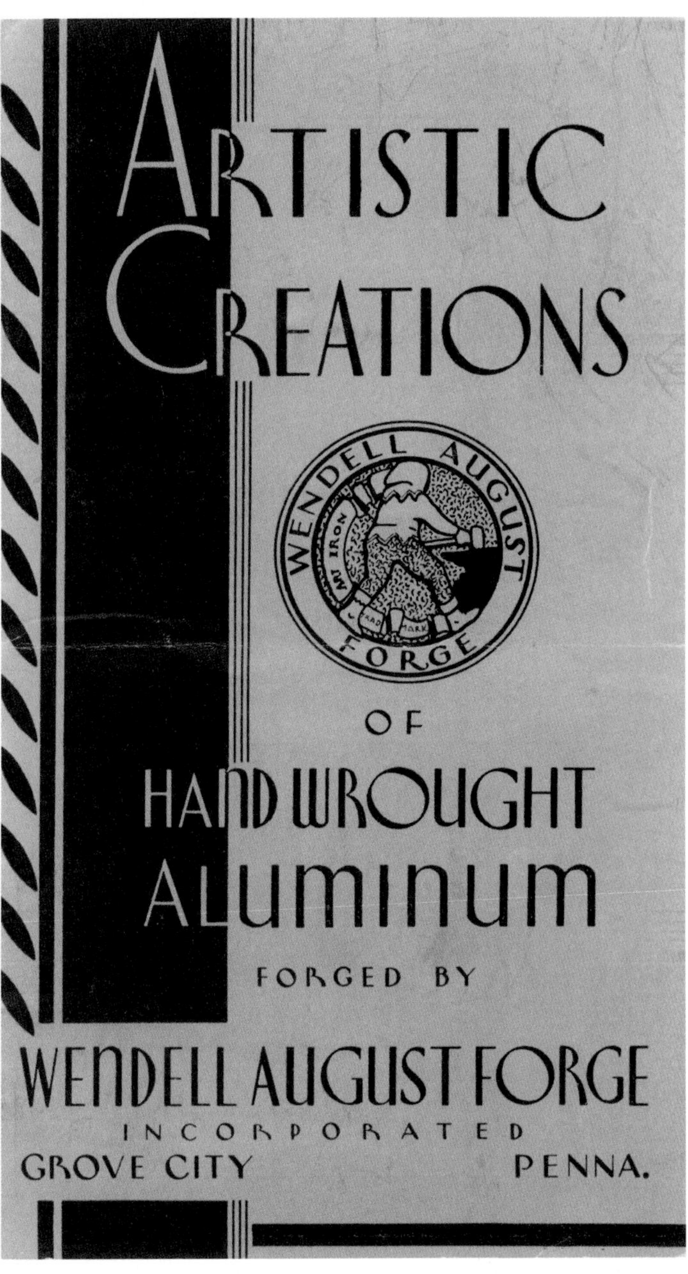

Front of 1932 brochure (note 16)

In late January of 1933, the *Grove City Reporter-Herald* carried the notice that "Arthur J. Palmer of New York City, representative of the Wendell August Forge Company, will exhibit a complete line of the plant's novelty products at the National Gift Show, to be held at the Palmer House in Chicago, Ill for two weeks, beginning Jan. 30.

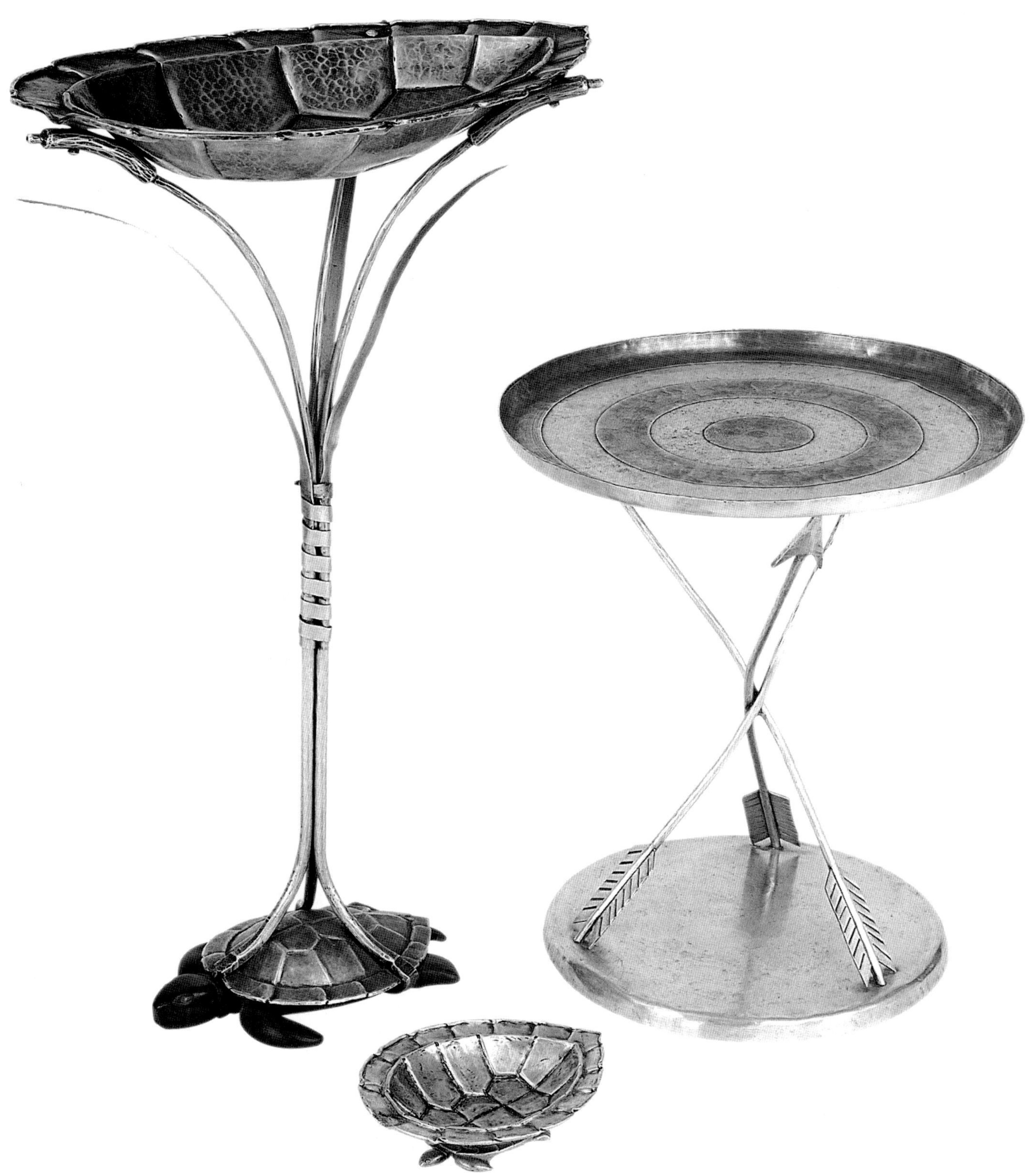

Selected items listed in 1932 brochure (note 17)

This page, left to right, Turtle Shell *bird bath.* Turtle Shell *pipe bowl,* Archery *table. Facing page,* Cat Tail *candlesticks*

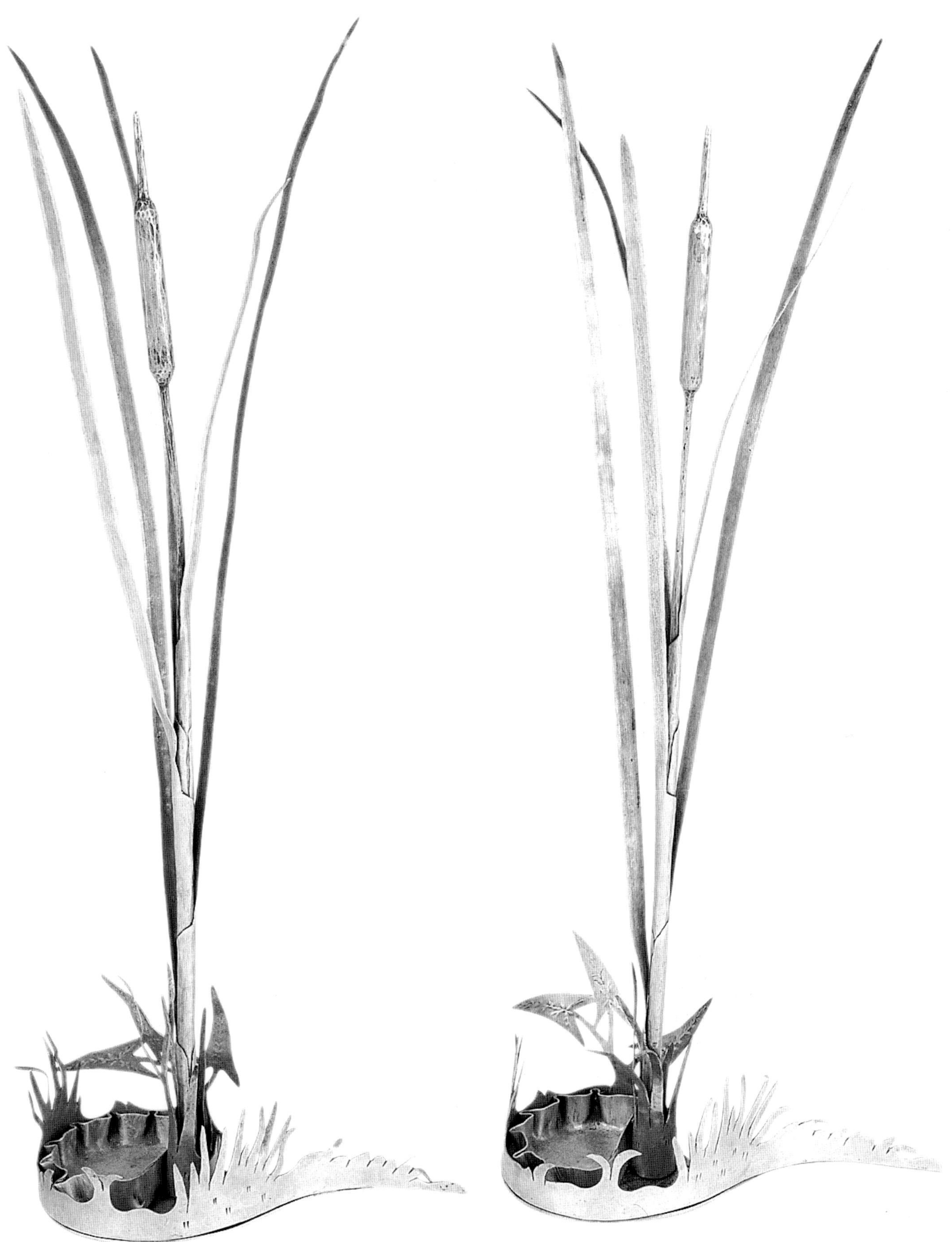

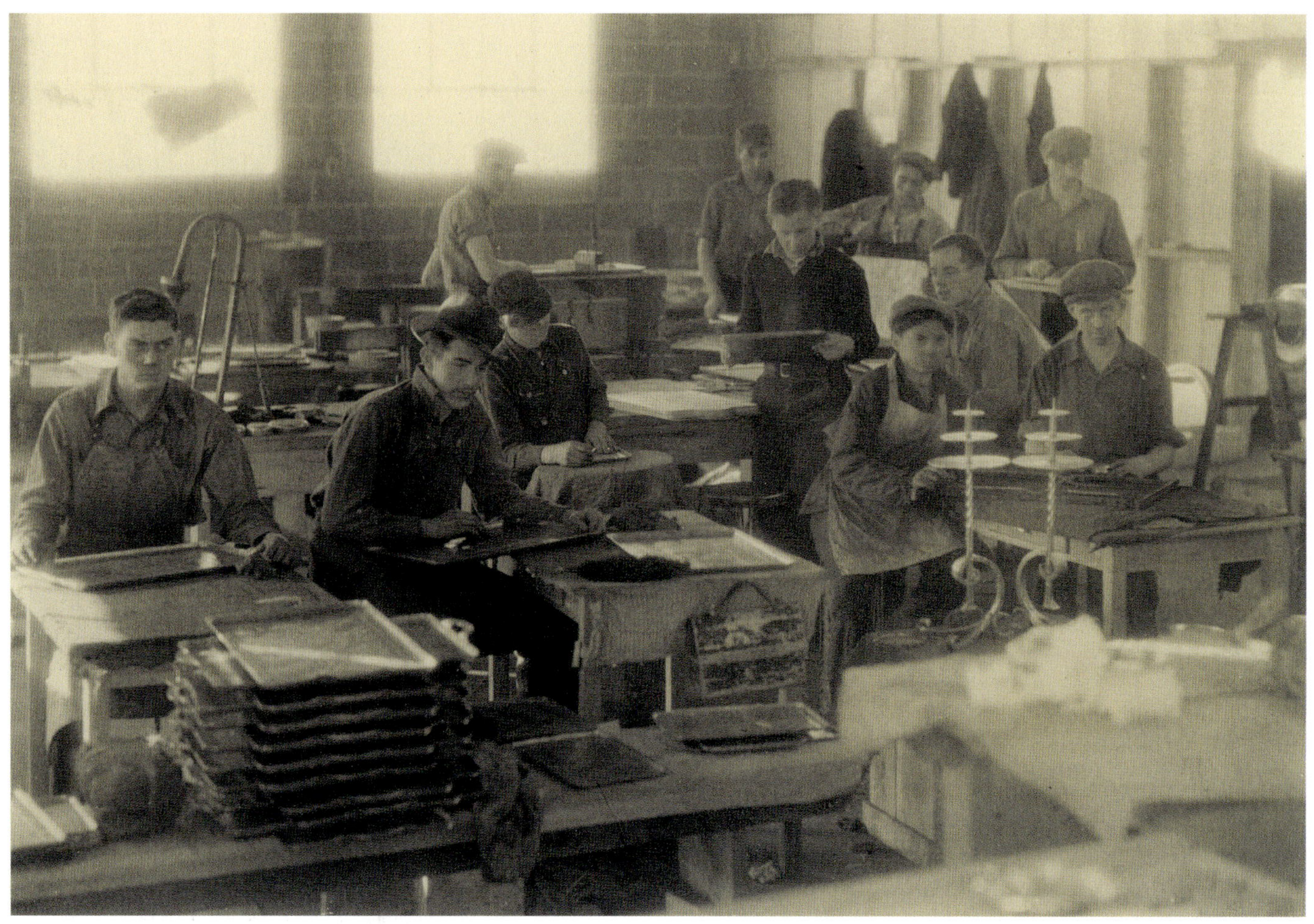

Portion of interior of Grove City facility in 1932 (note 19)

Reports from the New York Show, held recently in the Hotel Pennsylvania, are that the local company's products outdistanced all competitors, the demand being much greater than had been anticipated. A specialty is being made at present of furniture and table lamps, while additions are being made in all other lines."[25] The furniture produced appears to have been confined to chairs, small tables, smoking stands, and drink holder stands. The hand wrought aluminum chairs were certainly attractive, but their scarceness today suggests that they might not have been particularly well-received by the buyers.[26] The lamps, for which there would ultimately be more than 20 different table and floor models, were apparently more acceptable. The earlier lamps were of rather crude design in comparison to some of the other items;[27] however, more sophisticated lamps would appear by at least 1935.[28] Natale Rossi would later recall, "I think they had a designer that was very famous," continuing with the observation that the silk lamp shades were almost as expensive as the lamps themselves.[29]

The Forge's aluminum wares continued to gain additional accolades at the major gift shows in Philadelphia and Boston in February.[30] The favorable attention received at the shows, together with Alcoa's publicizing of

Louis Donato

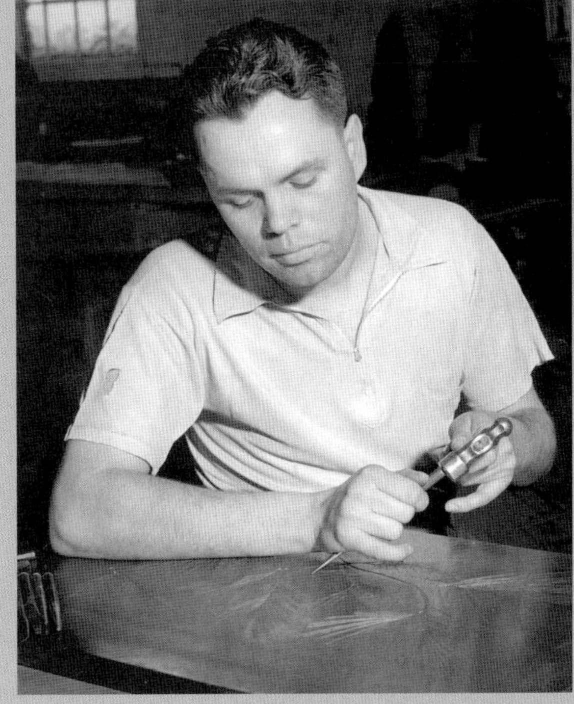

Louis William Donato (1910-1975) was born in Falls Creek, Pennsylvania, the eldest of seven children of Angelo and Edna (McFadden) Iadonato. Angelo Iadonato was an immigrant Italian stone mason and sculptor who created elaborate tombstones and architectural ornamentation such as gargoyles.

Working with his father and his maternal grandfather, carpenter W. E. McFadden of Brookville, Louis Donato's artistic talents and intellectual inclinations emerged at an early age, and by the time that he graduated from Falls Creek High School in 1928 he was honored as both salutatorian and class artist. After finishing high school, Donato continued with art classes while working with his grandfather McFadden.

In 1932, in need of an accomplished artist for the new line of aluminum art and giftwares that the Wendell August Forge was developing, James McCausland approached Angelo Iadonato about the possibility of becoming the Forge's engraver. Iadonato indicated that his son, Louis, was more talented and suggested that McCausland consider him. Louis visited McCausland with a sample of his work, a bust of Julius Caesar sculpted in a bar of soap, and was hired immediately.

When the Wendell August Forge occupied its new facility in Grove City, Louis Donato was there in the role of the Forge's primary artist and master die engraver. In 1936, he wed Grove City College art student Charlotte Virginia Spaulding.

Until the time that the Wendell August Forge closed temporarily due to World War II restrictions on consumer use of aluminum, Donato developed and drew much of the art work, as well as engraving the dies, for the Forge's countless repoussé motifs. Wendell August Forge personnel from the pre-war era recall the quiet and intense presence of Donato, either experimenting with selected engravings in bars of soap, or hunched over a steel die with hammer and chisel in hand.

Like many of his Forge colleagues, he was employed by Cooper-Bessemer in Grove City during World War II. An avid bibliophile and collector of first editions, Donato opened a news shop on North Broad Street after the war rather than returning to the Wendell August Forge. By 1949, he was a salesman and manager at Clarkson Furniture Store on South Broad Street, where he remained until his death.

Knowledge of the majority of Donato's pencil drawings, Impressionist-influenced oil paintings, and other works was lost when his widow died in 1995. With the exception of the items that were produced from the countless dies that he engraved, only a handful of his original work survives, in the possession of family members and the Wendell August Forge.[12]

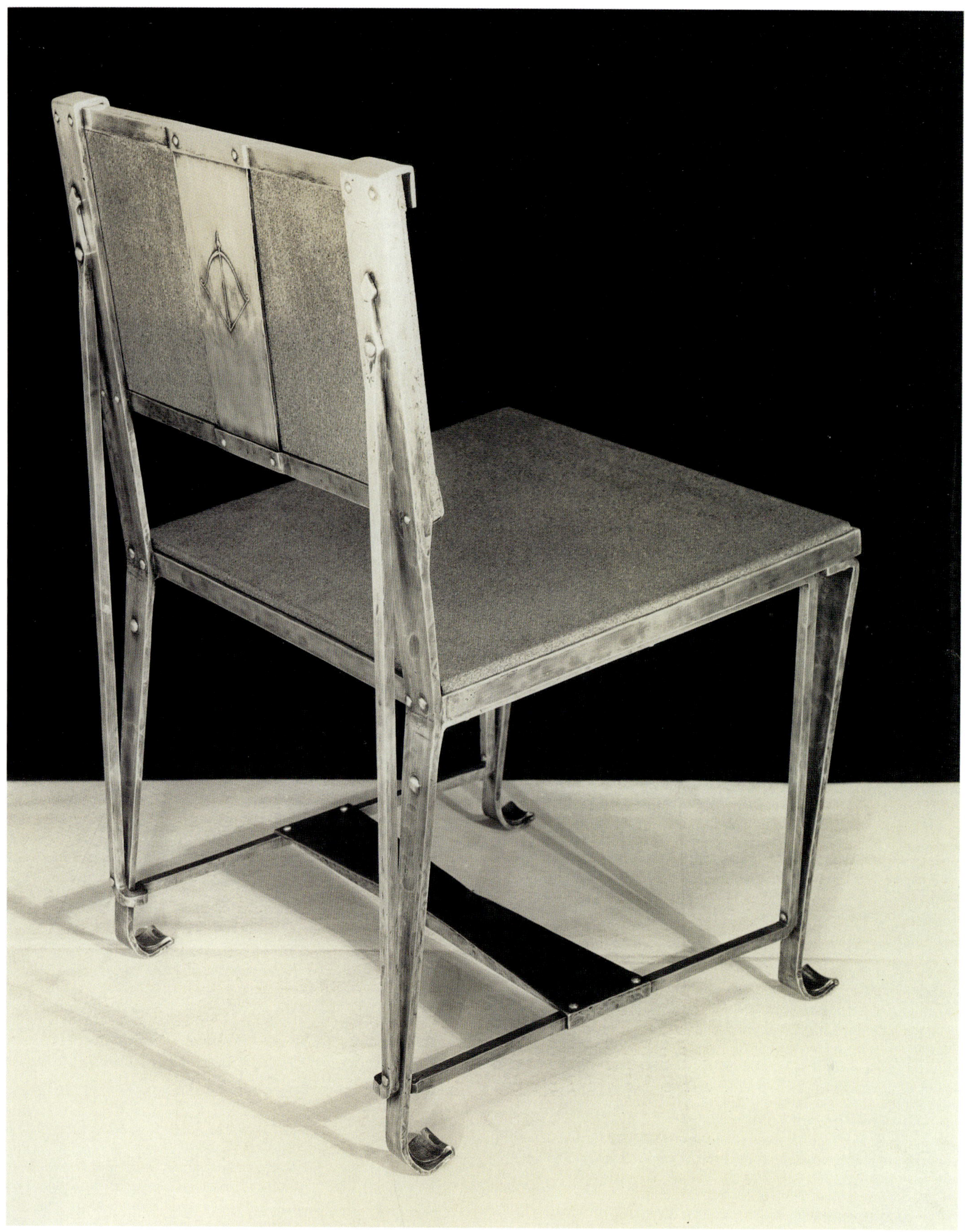

Aluminum chair with Archery *motif (note 26)*

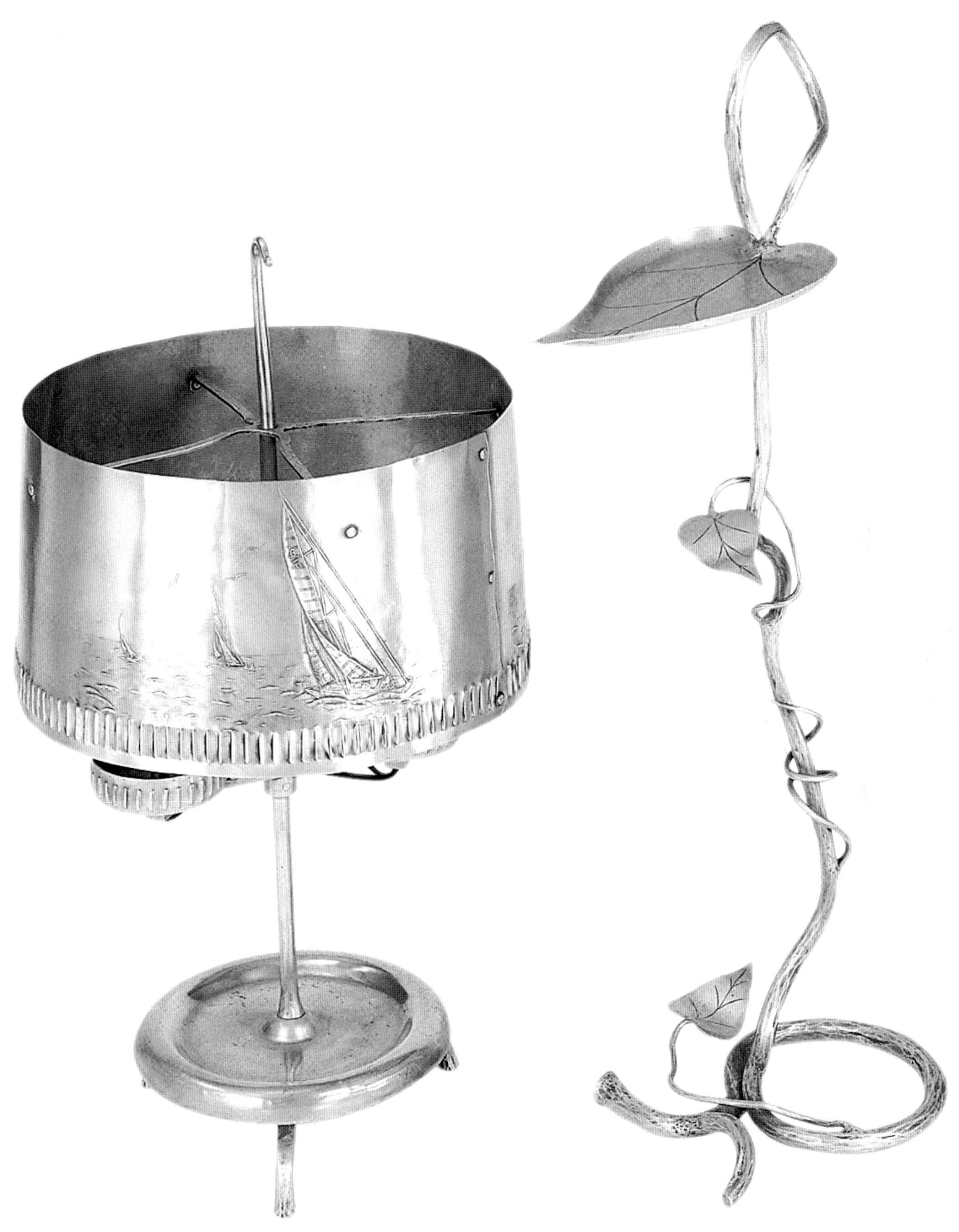

Table lamp with Sailboat *motif and vine-like drinkholder stand (note 27)*

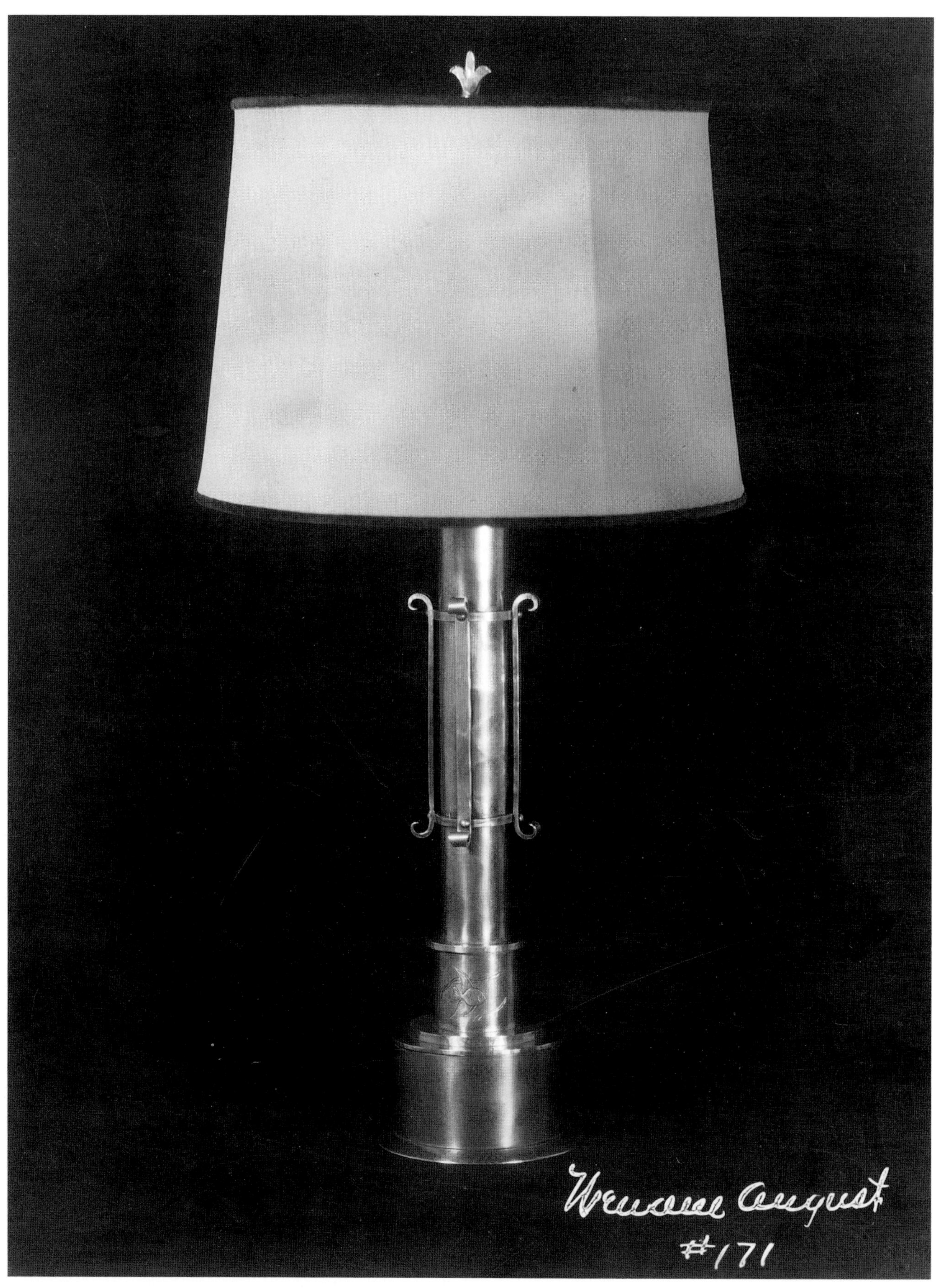

Table lamp with Tropical Fish *motif (note 28)*

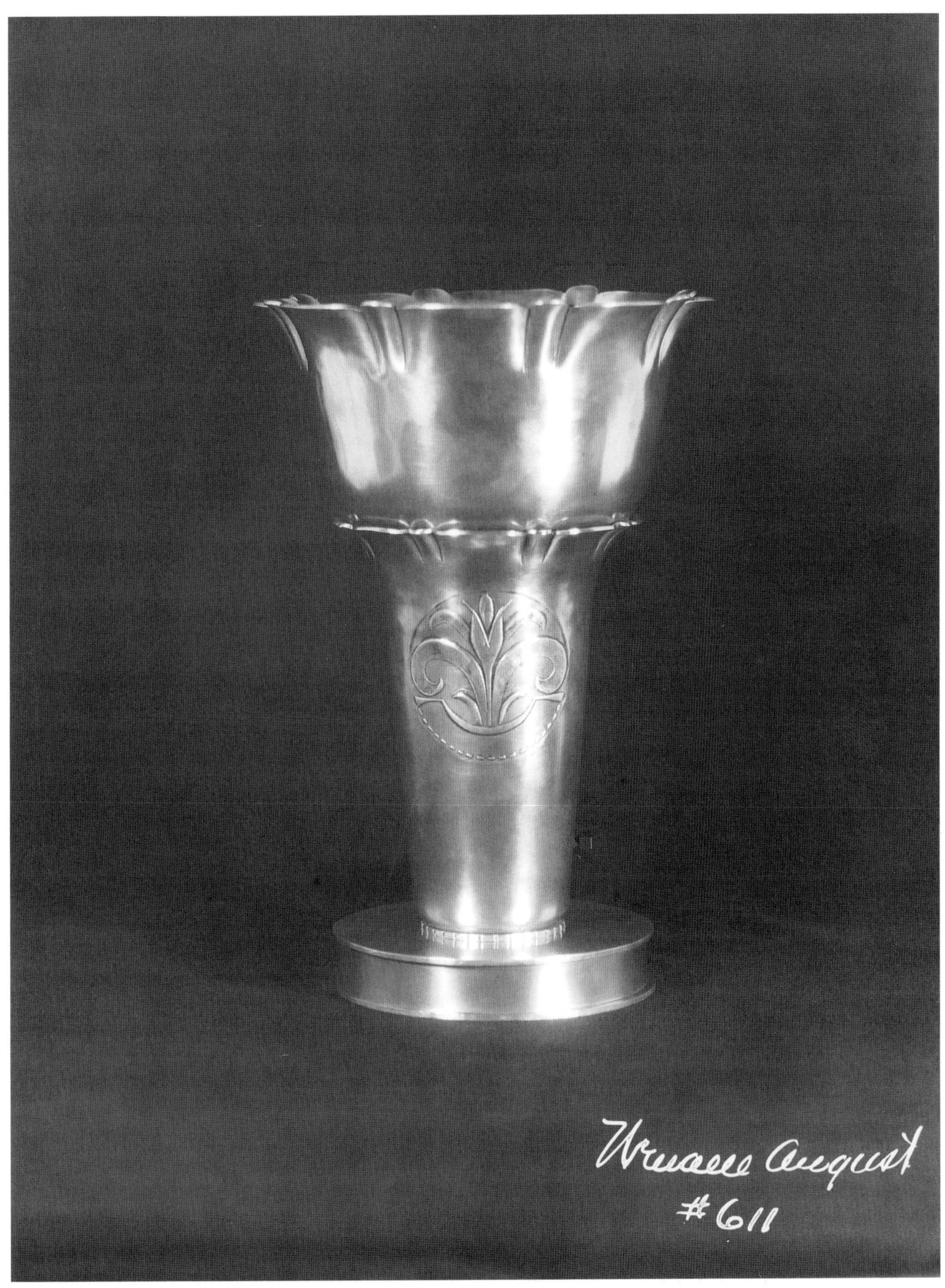

Table reflector with Grecian *motif (note 28)*

James McCausland

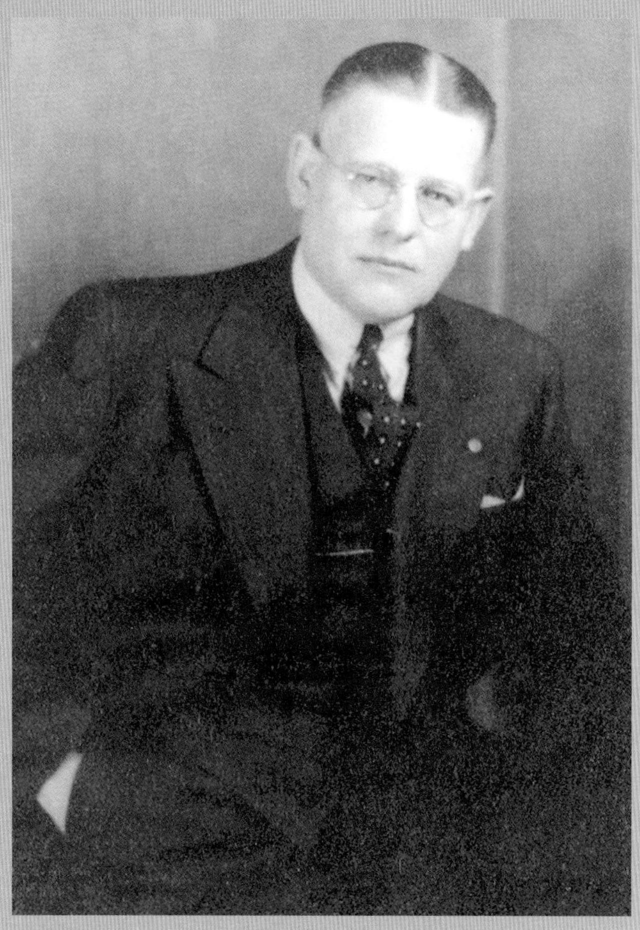

James A. McCausland (1900-1958) was born in DuBois, Pennsylvania and graduated from high school there. He attended Allegheny College in Meadville, majoring in medicine, but his education was interrupted by his service in the U.S. Army during World War I. After the war, he returned to college, earning his degree in Architecture from the University of Kansas.

In 1928, shortly after his marriage to Mary Swyers of Brockway, McCausland joined the Wendell August Forge, assuming the ornamental bronze and iron design responsibilities that had previously been contracted with the architectural firm of Howard and Hatcher. He was also responsible for supervising the operation of the Forge.

McCausland was significantly involved in the development of aluminum hand forging. The importance of his role in the evolution of the repoussé process used by the Wendell August Forge is evidenced in part by his inclusion as one of the three patent applicants — the only one who was not a principal of the firm.

Until his death, McCausland served as both the primary operations manager and the principal designer for the Wendell August Forge. A 1946 newspaper account noted that he "originated many of the popular designs," a statement reiterated by many Forge personnel of the era, who commented that McCausland created most of the designs and the drawings, and that they would then implement them. Although some of the architectural and giftwares designs can be attributed to others, it is to McCausland's design talents that the overwhelming majority of the Wendell August Forge product can be attributed during his 30-year tenure with the firm.

Less well known than his design contributions was McCausland's significant role in the process of establishing the hand forging of aluminum as an artistically worthy enterprise. Well-versed in the arts and humanities, it was McCausland who was able to articulate to prospective clientele and other audiences the aesthetic attributes of hand worked metal during the Machine Age and the appropriateness of a modern metal for such purpose. Illustrative of his approach was the eloquent speech that he penned in the late 1930's, in which he paralleled the Forge's redefinition of aluminum as appropriate for decorative art to George Bernard Shaw's Pygmalion, wherein the Forge was seen as instrumental in transforming aluminum from "the role of a domestic drudge" to "a beautiful platinum blond lady acceptable in the finest society of our land."[13]

the market possibilities of hand wrought aluminum, would serve as the epitome of the proverbial two-edged sword for the Wendell August Forge as other craftsmen and firms entered the marketplace with decorative aluminum goods. During 1933, at least four more producers of hand wrought decorative aluminum wares would begin competing with the Forge,[31] including former Forge designer Arthur Armour.[32]

In March of 1933, the country rode on a tide of hope that the woes brought on by the Great Depression would be alleviated when Franklin D. Roosevelt was inaugurated as President of the United States, declaring "This great Nation will endure as it has endured, will revive and will prosper . . . So, first of all, let me assert my firm belief that the only thing we have to fear is fear itself."[33] One of the earlier pieces of legislation implemented in conjunction with Roosevelt's New Deal was the National Industrial Recovery Act of June, 1933, overseen by the National Recovery Administration (NRA). The legislation established a wide range of industry and labor practices, including minimum wages and maximum hours for workers, the right to bargain collectively, and the outlawing of child labor. The Wendell August Forge was affected, and in August of 1933 there was a short work stoppage resulting from some confusion associated with the NRA provisions. The Forge established a 40-hour working week and a minimum employee age of 16 years, and "gave its employees pay increases averaging slightly better than 10 percent, as of August 1st, with a 12-1/2 per cent increase for apprentices."[34] Apprentices would now serve for a period of two years, at an average pay of 32-1/2 cents per hour, starting at 22-1/2 cents per hour for the first six months of the apprenticeship.

In September of 1933, the ownership of the Wendell August Forge changed. In February, Congress had passed the proposed 21st Amendment to repeal Prohibition. Grove City had a lengthy history of officially deploring the use of alcoholic beverages, and in late 1931 and throughout 1932 there was an intense, city-wide campaign to pass an ordinance to ban the sale of alcohol in the borough if the 21st Amendment was approved. Edwin J. Fithian was one of the people deeply involved in the religious-political campaign in support of the ordinance, which was passed by the borough council in April of 1933.[35] Fithian had previously been a Prohibition Party senatorial and gubernatorial candidate, and in June had been nominated as a delegate for the "dry representatives in the constitutional amendment convention balloting."[36] Within this context, it was unthinkable that Edwin J. Fithian could let his name be linked with anything that might suggest his approval of alcoholic beverages, yet a hook-handled mug produced by the Wendell August Forge was claimed by Fithian's opponents to be a beer mug.

In September, 1933, the *Grove City Reporter-Herald* printed a brief report stating in part that "In a transaction completed Friday, Dr. E. J. Fithian disposed of his interests in the Wendell August Forge, Inc., to his partner, Mr. August."[37] The *Brockway Record* was a bit more explicit in its notice, advising the reader that "Dr. Fithian and Mr. August are believed to have 'cracked up over publicity given to the

Arthur Armour

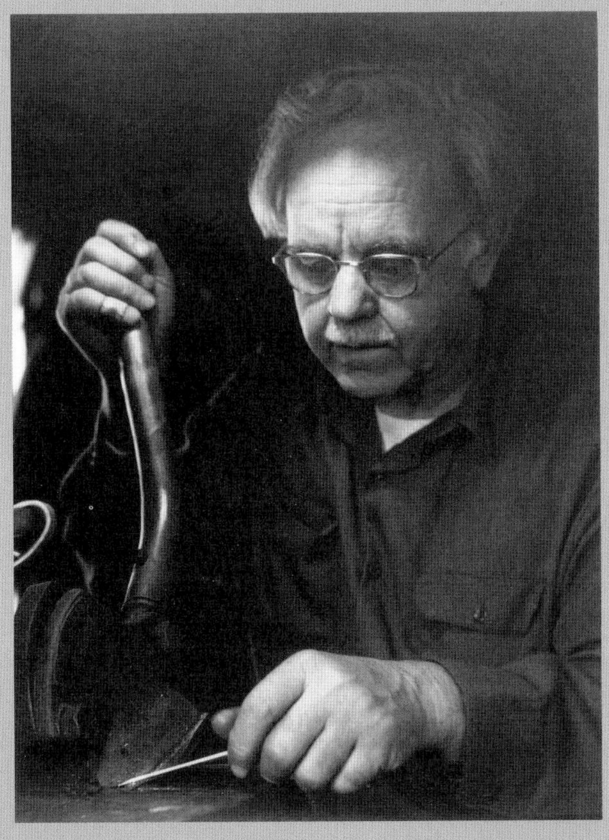

Born and raised about 45 miles northwest of Grove City in Conneaut Lake, Arthur Armour (1908-1998) earned his degree in Architecture from the Carnegie Institute of Technology in 1931, where one of his instructors was Henry Hornbostel, the architect for Alcoa's Aluminum Research Laboratories in New Kensington.

After completing his degree, Armour was with the New Castle office of W. G. Eckles & Co, the architectural firm responsible for many of the primary buildings in Grove City, including the first five buildings on the Grove City College upper campus which was begun in 1930.

In December of 1932, Armour joined the Wendell August Forge as a designer, leaving the firm after about six months. Deciding to start his own business in the creation of decorative aluminum wares, he cut his first die and fashioned some samples of his work. Catching a ride with some friends, he went to the Chicago World's Fair to explore the marketability of his goods. At the Palmer House in Chicago, he met up with A. Stanley Brussel, one of the pioneers of the gift and art industry, who was a director of the National Gift and Art Association. Brussel became Armour's manufacturers representative, describing Armour's work as "The Aristocrat of the Metal Lines; Famed for Beauty and Detail of Decoration and for Weight, Balance and Life Time Finish."

Armour, a highly-accomplished artist, adopted aluminum as his canvas. Over a period of 43 years, until he closed his shop in 1976, Armour created a distinguished sequence of handmade items in aluminum, embellishing them with about 58 different repoussé motifs that he designed and for which he engraved the dies. About 65 percent of his motifs had been developed by 1937, with all but a few making their appearances by 1941.

Armour's artistic talents are evident in not only his repoussé motifs but also the total design and production concept of each piece. In addition to developing his retail line, Armour also designed and executed a wide range of unique works, ranging from table lamps to communion sets. As stated in the vision statement that accompanied his products, the items that he created reflect "the ageless charm and beauty which have been produced by a hammer in hands of a skilled artisan."

Armour pieces have been exhibited in the American Craft Museum's Craft in the Machine Age *and in California State University's* Depression Silver: Machine Age Craft and Design in Aluminum.[32]

manufacture of beer mugs.' "[38] One can only wonder at what Fithian might think to learn that the specially-built touring car that he designed to use in his futile 1918 gubernatorial quest, now considered to be the first mobile home, is part of a collection that is displayed at the Imperial Palace in Las Vegas, Nevada, where not only alcoholic beverages but also gambling are condoned. As a result of Fithian's departure, the Wendell August Forge corporate officers became Wendell August, President, Jessie (Mrs. Wendell) August, Treasurer, and Howard J. Chapin, Secretary.

The years from 1934 through 1937 were significant ones for both Wendell August and the Wendell August Forge. Several of the people who had been integral to the success of the Forge's new ventures into hand wrought aluminum would leave the company during this period. In early 1934, James DePonceau left to work for Arthur Armour, and blacksmiths Natale Rossi (who would rejoin the Forge in early 1936)[39] and Ottone Pisoni[40] returned to Brockway to begin their own hand forged aluminum line, which they called Hand Crafted. Howard J. Chapin resigned in mid-1934 to take a position with the Veteran's Administration, leaving Wendell August as President and Jessie August as Secretary/Treasurer of the Wendell August Forge. Arthur J. Palmer, in an apparent difference of opinion with Wendell August about his sales commissions, departed in early 1935 to begin Palmer-Smith.[41]

In August of 1934, the patent that had been filed by August, McCausland, and Chapin was approved and issued, supporting all of the claims that had been made regarding "The method of forming ornamental plates which consists in forming a depressed portion in a pattern block; placing the plate thereupon and hammering the back of the plate." The elation that the applicants must have felt was soon dampened by challenges to the patent, which the U.S. Patent Office upheld. On January 15, 1935, fewer than five months after the patent issued, the Forge found it necessary to file a disclaimer, and the *Official Gazette* of February 5, 1935 carried the notice that the Forge was disclaiming "from each of claims 1, 2, and 3 any method of forming ornamental plates except a method in which the hammer blows upon the back of the plate are selectively governed with relation to the depressed portion of the pattern block."[42] With these disclaimers, the essence of the patent was negated. The Forge still had a patent, but the patent rights were now so narrowly defined as to have little value. In retrospect, it is difficult to believe that the U.S. Patent Office would have issued a patent covering a general repoussé process that had been in use for centuries.

There has been much speculation in the past regarding the exact nature of the patent or patents associated with the Wendell August Forge. Forge literature in the early 1930s carried such language as "The finish is obtained by our own process for which patent rights have been filed,"[43] leading to suppositions that the carbon coloring process might have been patented. Later in the 1930s, Forge literature included such statements as "Working patiently with hammers and polishing cloths" and "Basic patents on these methods are held by Wendell August,"[44] suggesting that

Natale Rossi

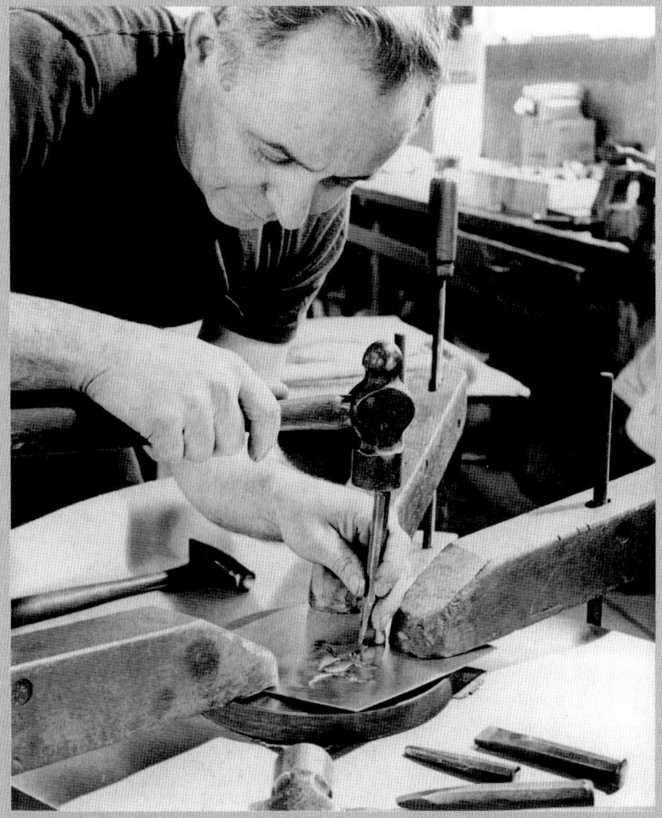

Natale Rossi (1908-1993) was born in Robecco-Sul-Naviglio, Italy, the eldest son of Pietro and Rosa Marie (Martigoni) Rossi. At the age of 7, he immigrated with his parents to Brockway, where his father found work in the region's coal mines. When his father had a heart attack, Rossi left school at the age of 16 and entered the mines to help provide income for the Rossi family. After almost three years in the mines, Rossi apprenticed himself to Ottone "Tony" Pisoni at the Wendell August Forge to learn blacksmithing.

Rossi's blacksmithing talents and creativity rapidly emerged, and during the first two decades of his long association with the Forge he was responsible for making many of its decorative architectural pieces. His most noteworthy work in this regard was his execution, by hand from aluminum sheet, of William Richard Perry's St. John the Baptist for Our Lady, Queen of the Most Holy Rosary Cathedral in Toledo, Ohio.

When the Wendell August Forge first became involved in the creation of aluminum repoussé work, it was Natale Rossi who is credited with conceiving and initially experimenting with the aesthetic and production advantages to be gained by engraving decorative repoussé motifs directly into a steel die. He was responsible for some of the earliest engraving work while the firm was in Brockway and, after World War II, was the master die engraver for the Wendell August Forge for more than 30 years.

When the consumer infatuation with Wendell August Forge hammered aluminum giftwares began its drastic decline in 1949, Rossi drew on his determination and ebullient personality to become the primary salesman for the Forge. Following the death of James McCausland, Rossi did much of the art work as well. That the Wendell August Forge survived throughout much of the 1960s and 1970s owes much to Natale Rossi.

In December of 1979, at the age of 71, Rossi retired from the Forge. Except for the interruption of World War II and a two-year period from 1934 to 1936, when he was with Ottone Pisoni in Brockway producing the Hand Crafted aluminum giftwares line and during which time he wed Irene Lucy Panada, he had been with the Forge for more than 50 years. In 1980, his craftsmanship was recognized when he was honored with a Hazlett Memorial Award for Excellence in the Arts in Pennsylvania. In the same year, in his typically indefatigable manner, Rossi also opened his own shop, By Natale, where he continued to produce hand crafted metal work until his death in 1993.[39]

Ottone Pisoni

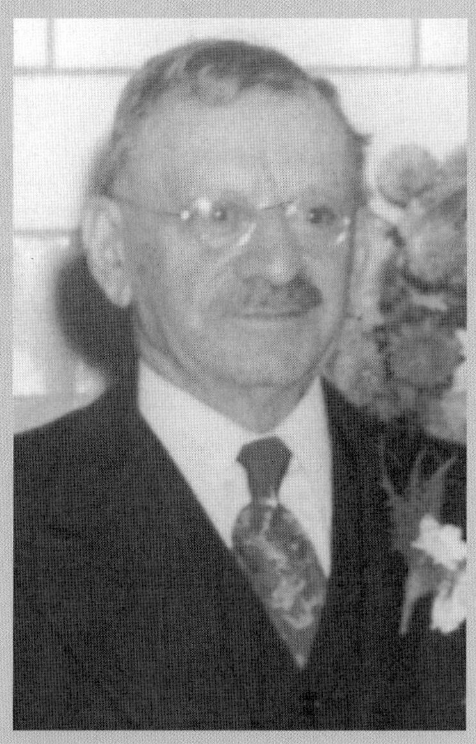

Ottone "Tony" Ignatio Pisoni (1874-1971) immigrated from Italy to the United States in 1887 at the age of 13, commencing work in the coal mines around Crenshaw, Pennsylvania shortly thereafter. He had been apprenticed to a blacksmith as a child in Italy, and in the United States his skills with a hammer quickly moved him out of the mines and into a blacksmith's shop, where he shoed mules and sharpened tools used by the farmers and miners. In 1898, he purchased a small plot of land in Brockway where he built a home, initially using one room of the ground floor for a small blacksmith shop of his own. The following year, he wed Theresa Guarnati of Brockport, Pennsylvania.

Pisoni continued to provide blacksmithing services for the coal mines, rock quarries, and local farmers until the early 1920s, when he became involved with Wendell August in the development of the Wendell August Forge. It was to Pisoni that Wendell August reportedly turned for the two additional matching wrought iron door latches that were needed to complete the new August home. When Pisoni produced the latches for a fraction of the cost of the originals, the concept of the Wendell August Forge was apparently born.

Pisoni thus became the first ornamental wrought iron blacksmith at the Wendell August Forge in 1923, and was the senior blacksmith for the Forge's numerous early wrought iron ornamental architectural installations. When the opportunity to develop the hand wrought aluminum gates for the Alcoa Aluminum Research Laboratories arose, Pisoni was the first of the Wendell August Forge blacksmiths to experiment with the hand working of aluminum.

Pisoni continued to produce ornamental hand wrought aluminum for the Wendell August Forge for about two years after it moved to Grove City, making the long trek from his home in Brockway to Grove City and to the numerous architectural sites in Pittsburgh and other cities. In early 1934, at the age of 60 years, he left the Forge to resume his own wrought iron work in Brockway and to start his own ornamental aluminum work. With Natale Rossi, Pisoni developed and made the Hand Crafted line of aluminum decorative art wares. Although Rossi left the business in about 1936, Pisoni continued with the production of Hand Crafted items until at least 1941.

Until a few months before his death at the age of almost 97, Pisoni operated his own private forge in a small building that he built next to his home. He designed and made hand wrought iron, bronze and aluminum railings, grills, lighting fixtures, candelabra and other ornamental items, while also doing repair work for miners and farmers in the Brockway area. Some of the elegant architectural ornamentation that he produced remains intact in private homes and businesses in the Brockway area.[40]

more than one patent might have existed. However, searches of U.S. Patent Office records have revealed the existence of only the single patent pertaining to the repoussé process.

Whatever disappointments that Wendell August may have had regarding the outcome of the patent had to have paled in comparison to the loss of his youngest son in 1936. The year was permanently marred when the youngest August son, Donald Dick August, was killed in a plane crash in April. "Dick, 17-year-old son of Mr. and Mrs. Wendell August of this city, met a merciful death on his first air flight, when the TWA luxury air liner crashed on a mountaintop at Uniontown, Tuesday," began the *Grove City Reporter-Herald*, continuing "His body was untouched by the flames that charred nine of the eleven victims, and it was evident that his trip home from Valley Forge Academy had ended in instant death."[45] Later in the month, the cause of the crash was still being investigated. It was suggested that the pilot had lost his bearings in the fog, sleet, wind, and rain that were prevalent during the flight, and "believed that he was over an airport a moment before he crashed in the Cheat Mountains."[46]

From 1934 through 1937, the market for Wendell August Forge aluminum giftwares continued to boom. In 1934 and early 1935, Arthur Palmer regularly frequented the major regional and national gift shows, and the product line was routinely enriched with new forms, new repoussé motifs, and more sophisticated craftsmanship.[47] Wendell August Forge art and giftwares were offered in major jewelry, gift and department stores from coast to coast, and orders were being consigned to Puerto Rico, Honolulu, Brussels, Rome, and Mexico City. Forge customers reported in the news included Mrs. John D. Rockefeller, Jr., Mayor Curley of Boston, and former Secretary of Treasury Ogden Mills.[48] Alcoa continued to promote the Wendell August Forge and its hand wrought aluminum products, clearly appealing to the luxury market with such statements as "it [forging] has seldom if ever reached a higher point than in the beautiful hand-forged grille work, plaques, trays, and a hundred other useful items of rare distinction, which are fashioned of aluminum at this most modern of forges . . . To follow the trail to the Wendell August Forge, at Grove City, Pennsylvania is to experience the thrill of discovery."[49]

The marketing of the Forge's giftwares changed markedly in mid-1935 when Arthur Palmer left the firm. Until the latter part of 1939, the Wendell August Forge aluminum products would be represented by Edward Surgeons, along with a line of bronze produced by another firm in Taunton, Massachusetts.[50] Regardless of the sales representation, it was evident that the Forge's giftware market was expanding in comparison to its other work, leading it to state as early as August of 1936 that the gift line was "the major part of the production of the Wendell August Forge."[51]

It is in the years of 1935 through 1937 that traces of the Wendell August Forge aluminum art and giftware products have been located in some important trade journals. For example, trays, ashtrays, and coasters were advertised in 1935 and 1936 in *The Jewelers' Circular-Keystone*.[52] The use of aluminum for furniture was of some interest in the design and

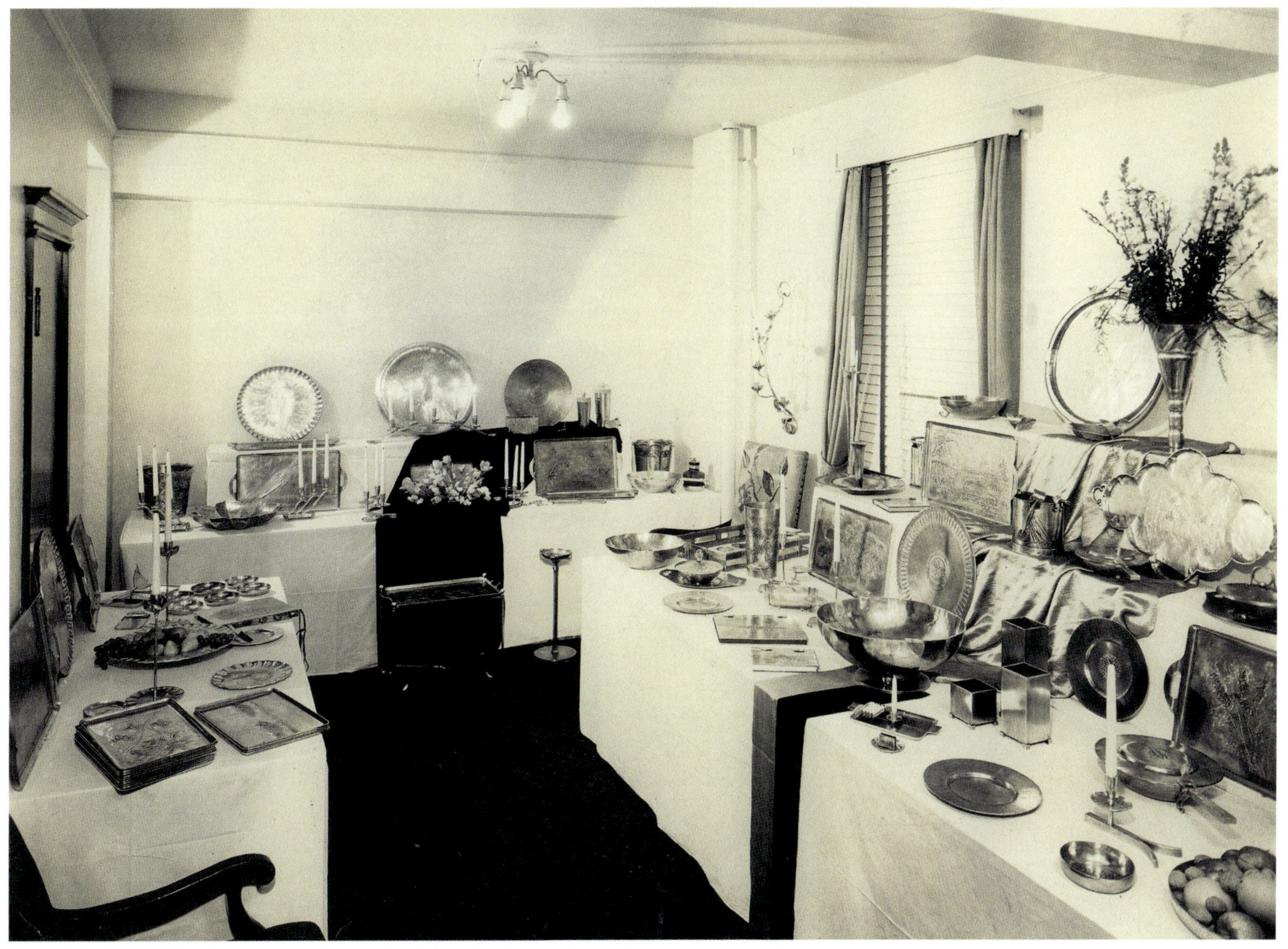

Gift show display of Wendell August Forge aluminum giftwares (note 47)

decorating communities in both the United States and parts of Europe,[53] and one of the few references to Wendell August Forge furniture that has thus far surfaced appeared in 1936.[54]

Wendell August Forge aluminum giftwares were occasionally used for awards and commemoratives, usually with appropriate citations engraved into the item that had been selected. In some cases, special dies were engraved for the creation of distinctive items for awards and business gifts, or for commemorating special events. A few such items from this time period, primarily ashtrays and coasters, have surfaced, among them being pieces recognizing Oberlin College, the Rhode Island Tercentenary, and the final completion of the Pymatuning Reservoir. Most exceptional, however, is a 1936 presentation piece commemorating a special flight of the Hindenburg.[55]

Following the 1929 around-the-world flight of the Graf Zeppelin dirigible, the public had become enamored of the possibilities for international airship travel. As one approach to increasing commercial airship travel, Hugo Eckener, who had commanded the epic Graf Zeppelin flight, arranged for a special public

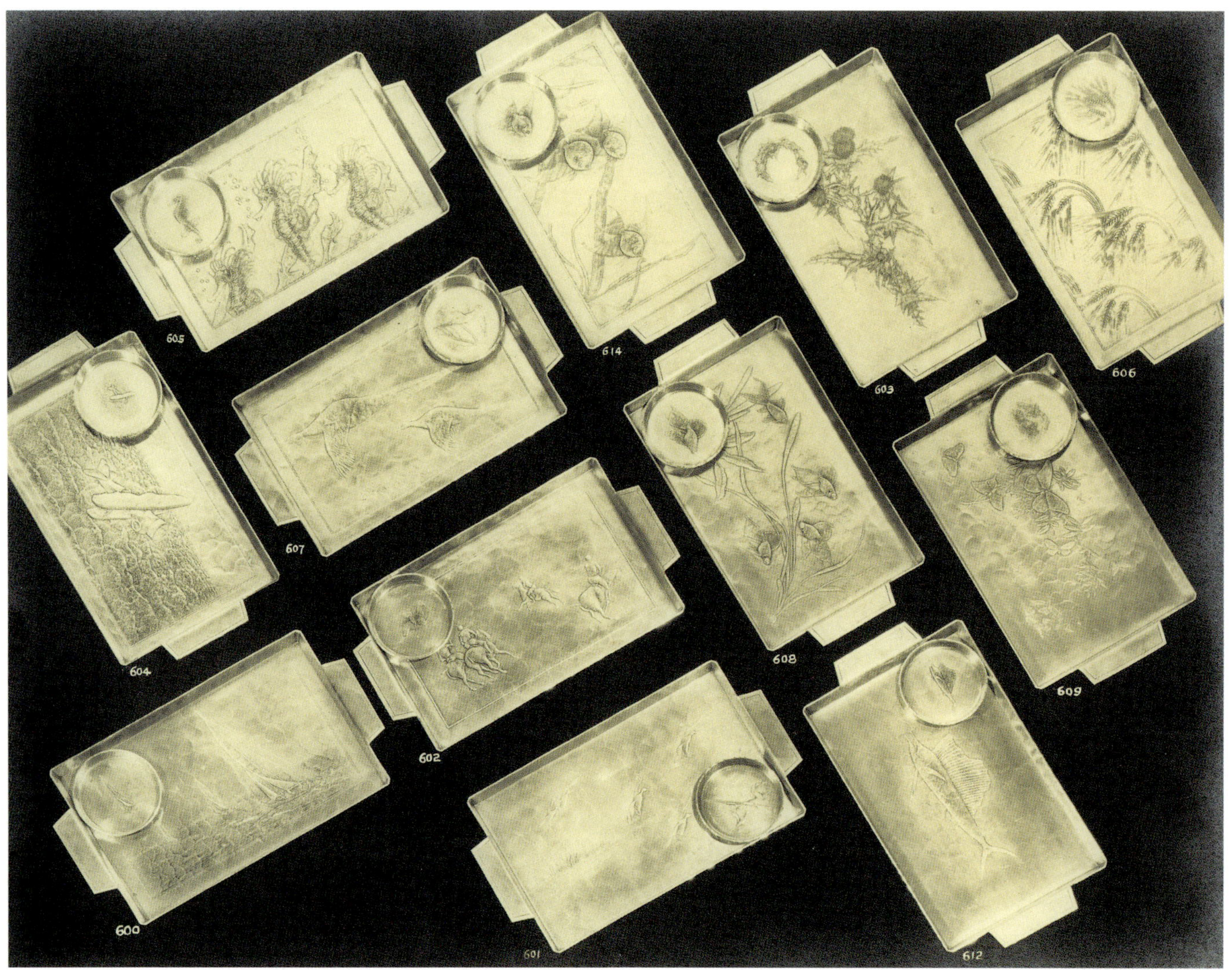

Aluminum trays and coasters (*note 52*)

relations flight of the Hindenburg airship at the end of its 1936 trans-Atlantic flight season. Dubbed "The Millionaires' Flight" by the press, the 72 guests on the flight were considered to be among the most powerful and wealthy men in the United States. The best that can be determined at this time is that the unusual commemorative pieces produced by the Wendell August Forge were presented to the invited guests. Certainly the passengers on this special flight were fortunate, because, on its first flight of the 1937 season the following May, the Hindenburg would explode at Lakehurst Naval Air Station in New York, thereby effectively ending the era of international airship travel.[56]

In July, 1937, the Forge received "the largest contract ever placed with the Wendell August Forge," the immediate effect of which was the assignment of "25 men to one order which will keep them busy for at least a year."[57] The Toastmaster Products Division of the McGraw Electric Company in Minneapolis had for some time been marketing a Toastmaster Hostess Hospitality Set each year. In early 1937, it capitalized on the popularity of aluminum

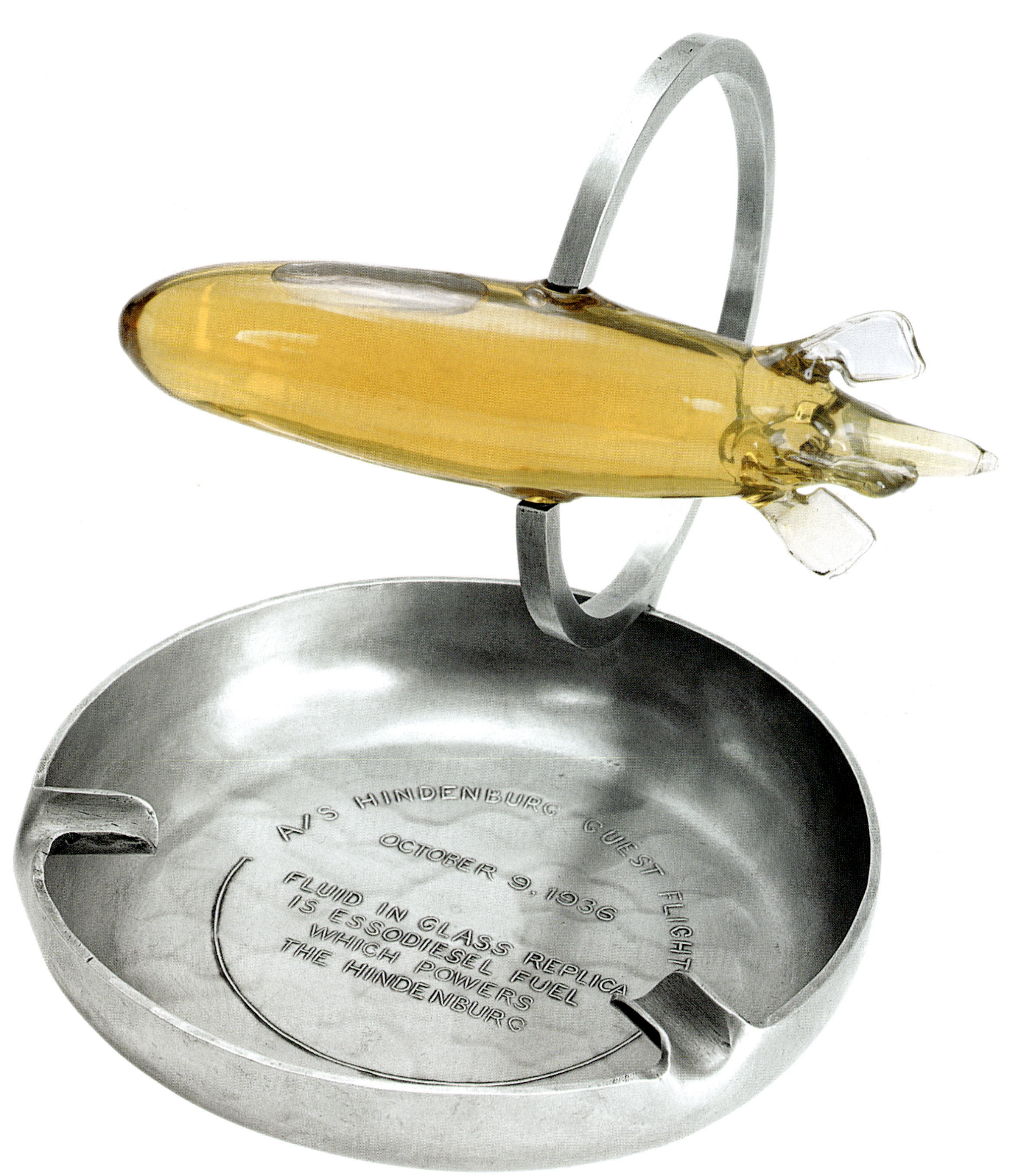

Hindenburg "Millionaires' Flight" commemorative piece (note 55)

Arthur Palmer

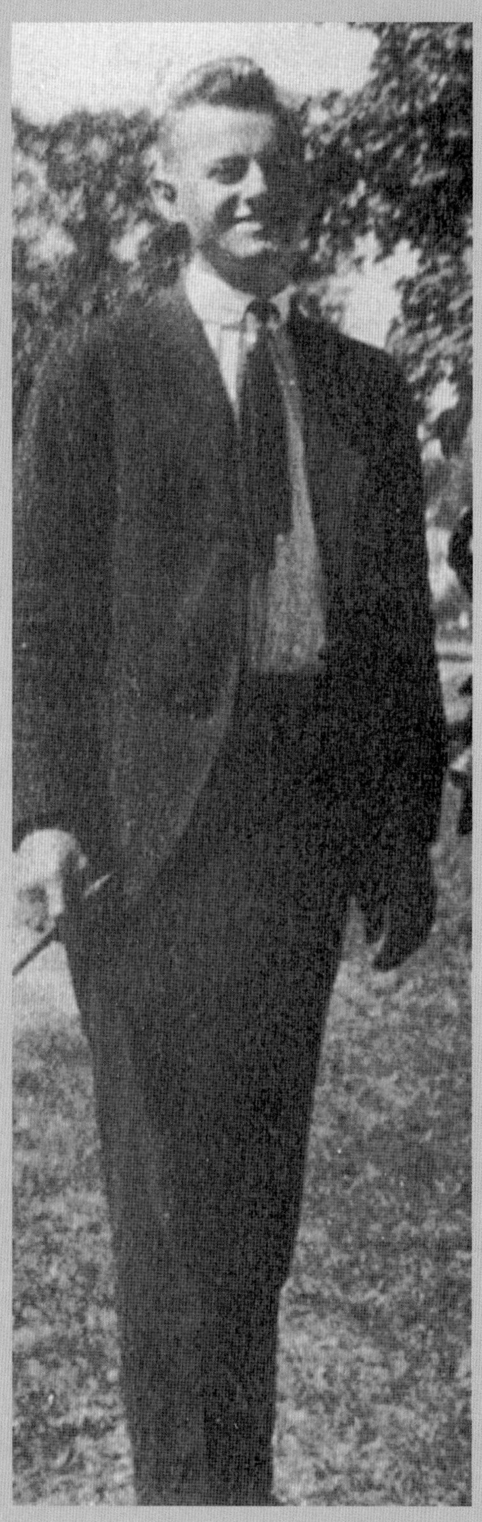

The relatively brief, but influential, foray of Arthur J. Palmer, Jr. (1897-1978) into hand wrought aluminum began with his association with the Wendell August Forge as its principal salesman for the firm's aluminum art and giftwares in 1932. Palmer brought to the position a decade of experience as a leading manufacturers representative of art and giftwares, including silver and silverplated holloware. He also possessed an acute sense of style and a well-developed sense of the preferences of the wealthy classes for which the Forge product was intended.

Palmer's influence in the design of the Forge's aluminum giftwares during the 1932-1935 time period seems to have resulted in some of the more artistically interesting works produced by the firm. His extraordinarily successful salesmanship apparently brought about his departure from the Forge in 1935, when Wendell August balked at paying salesman's commissions that exceeded the income of anyone else at the firm, including that of August himself.

With typical aplomb, Palmer recast himself into the role of industrial designer, and started his own firm in Grove City, called Palmer-Smith, to produce hand hammered aluminum giftwares. Departing from the repoussé tradition introduced by the Forge, the surfaces of most Palmer-Smith wares were sparsely chased with more than 50 different abstract decorative motifs. Most of the items relied on mass and form for their aesthetic impact, and during the latter part of the production period the sculptural impact was enhanced by the use of more than 20 cast or machined ornamental accents.

Palmer-Smith continued with the production of aluminum giftwares until the access to aluminum supplies for consumer goods was terminated with the entry of the United States into World War II. After the war, Arthur Palmer directed his attentions primarily to the custom design and merchandising of fine linens decorated with chain-stitch embroidery.

Palmer-Smith work has been exhibited in the American Craft Museum's Craft in the Machine Age *and in California State University's* Depression Silver: Machine Age Craft and Design in Aluminum.[41]

amongst the relatively affluent by contracting with the Wendell August Forge for the production of its set for 1937-1938.[58] The set, consisting of "stand, large tray, two trays (about the size of the familiar Toastmaster lap trays), a fourth small tray, and a two-slice Toastmaster toaster," was embellished with a "design, known as 'Wheat' " which used "a combination of spears and shocks of wheat."[59] The set was available for $85, and individual pieces could also be purchased. Frank Rossi was among the high school students hired during the Summer of 1937, while school was not in session, to work on the Toastmaster order. He "started out swinging a 12-pound sledge hammer for 12.5 cents per hour, forming legs for . . . coffee tables, later progressing to repoussé hammering. Using a 2.5-pound ball peen hammer at 90 strokes per minute, Frank turned out 20 trays per day."[60] Lewis "Doc" Rossi recalled, "Well, I think that was the biggest order that we ever got. We had tables, big trays for them, and small trays, but what we had the trouble with was the covers [for the toasters] . . . what we had to do was cut the slots in there . . . and then form them. And then we'd send them to the company and they'd put it on, and they'd send it back . . . it wouldn't fit right. We had a lot of trouble with Toastmaster, but not the trays. You see, to do something like this you have to have it just about perfect 'cause they're welded," adding, "I bet there's over a thousand." [61] [62]

A printed catalog for the Forge's giftwares that was produced in 1937 illustrates approximately 200 items.[63] When the various possible item sizes and the more than 60 motifs documented in the corresponding price list are accounted for, a buyer could select from more than 350 different items, ranging from trays to mail boxes to desk sets.[64] [65]

The years from 1934 through 1937 were also enriched with some of the more significant architectural work that the Wendell August Forge would execute. The Forge had had no architectural work for about six months when, in late 1934, it received a commission that would ultimately have a major impact on the Forge for the next few years. The commission was for the creation of 75 exquisitely ornate lighting fixtures for St. Bernard Church in Mount Lebanon, Pennsylvania, a major structure designed by ecclesiastical architect William Richard Perry to reflect the architecture of 12th Century France and French Catalonia.[66] The completion of this commission was followed by one for a set of large ornamental baptistry gates for the same edifice, which were finished early in 1936.[67] [68] The fixtures and gates can still be observed at what is referred to in the Pittsburgh area as "The Cathedral in the South Hills."

William Richard Perry must have been well pleased with the craftsmanship of the Forge in its execution of his designs for St. Bernard Church, because the completion of that work was followed by a commission to forge the baptistry gates for Our Lady, Queen of the Most Holy Rosary Cathedral, in Toledo, Ohio.[69] The cathedral, which Perry had designed in the Medieval Spanish Plateresque style, had been under construction for several years.[70] The completed gates, embellished with bronze rosettes, would be acclaimed as "the handsomest examples of hand-forged

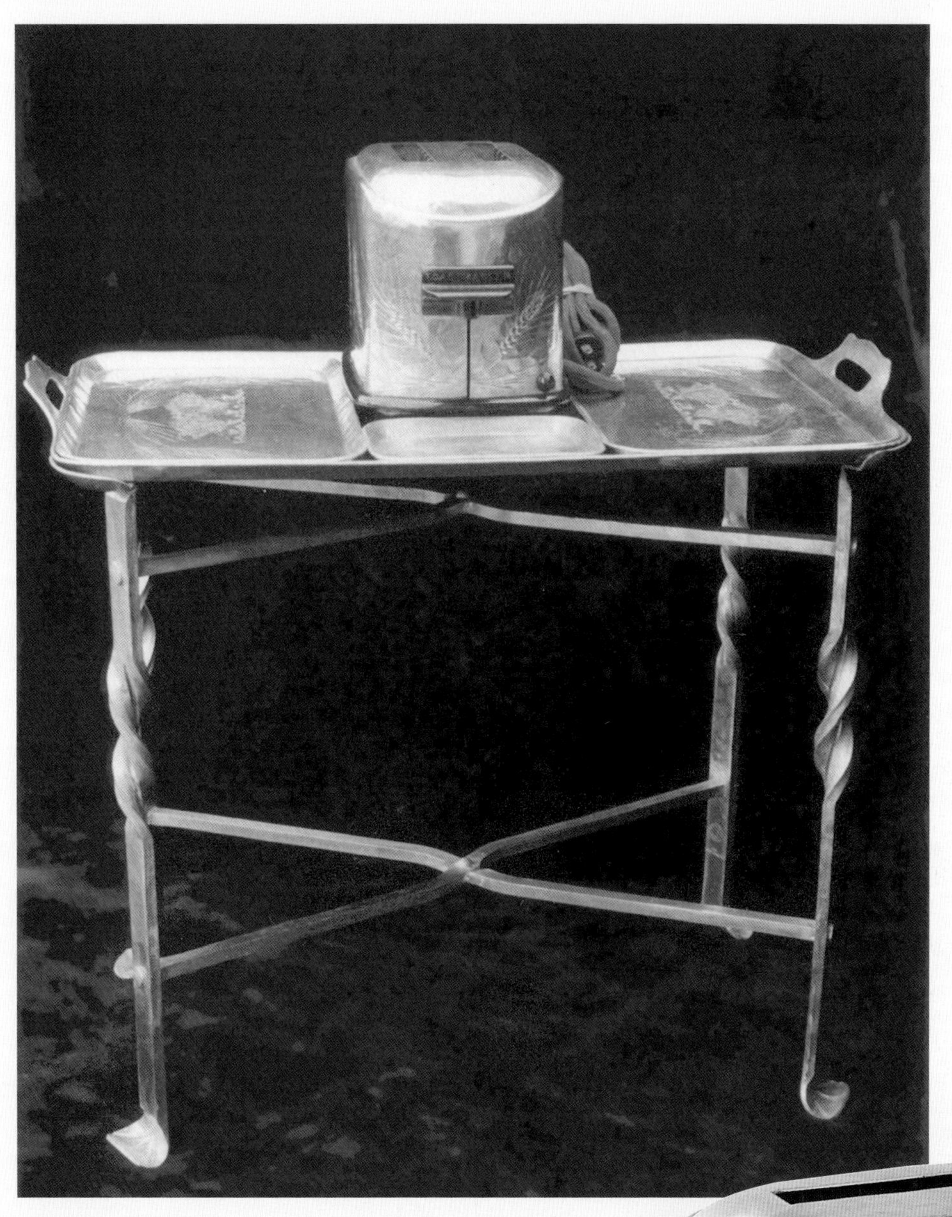

Above, Complete Toastmaster Hostess Hospitality Set (*note 58*)

Right, Toastmaster toaster (*note 61*)

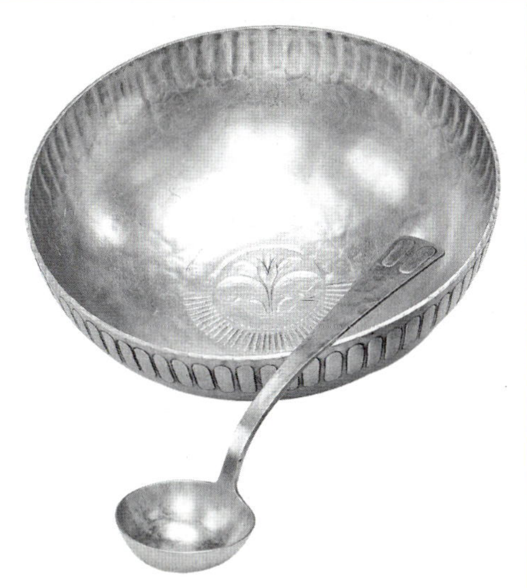

Counterclockwise, punch bowl with Grecian *motif, coffee set with* Grecian *motif, strap-hinged cigarette box with* Hobnail *motif, ice bucket with* Sail Boat *motif, and smoking stand (note 64)*

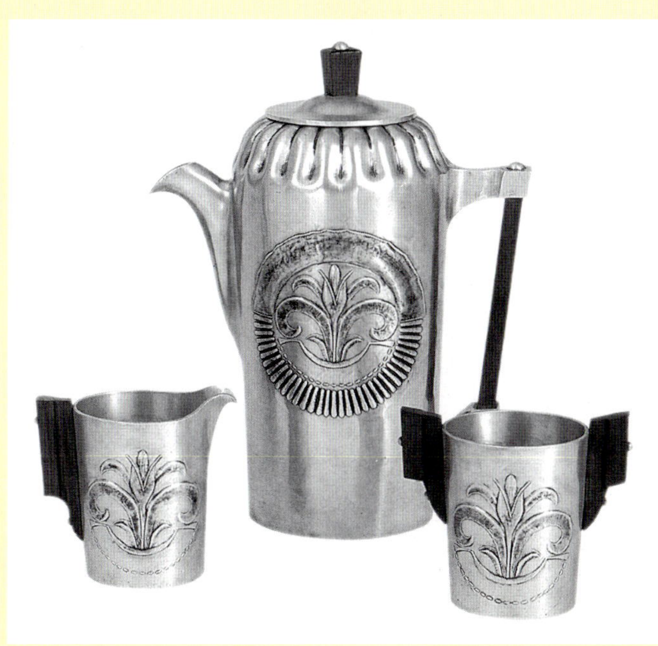

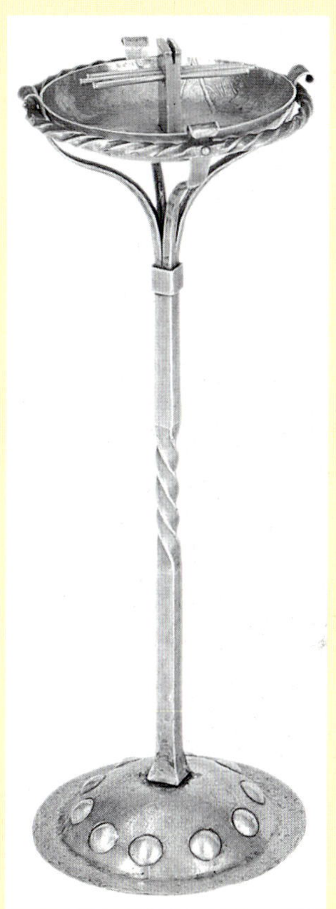

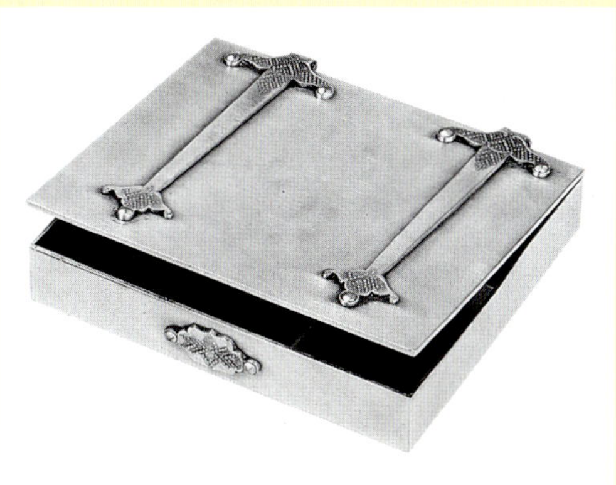

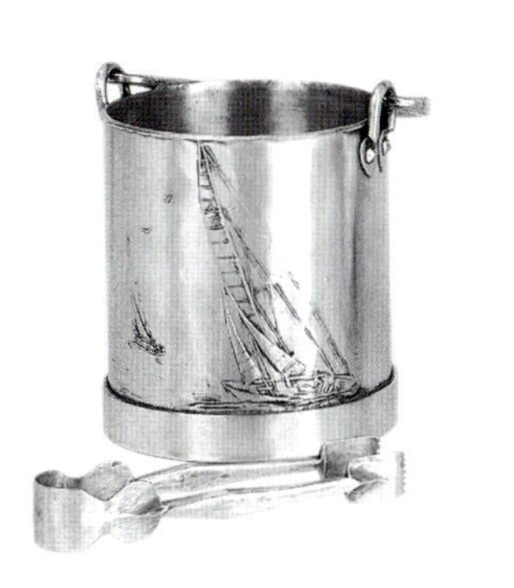

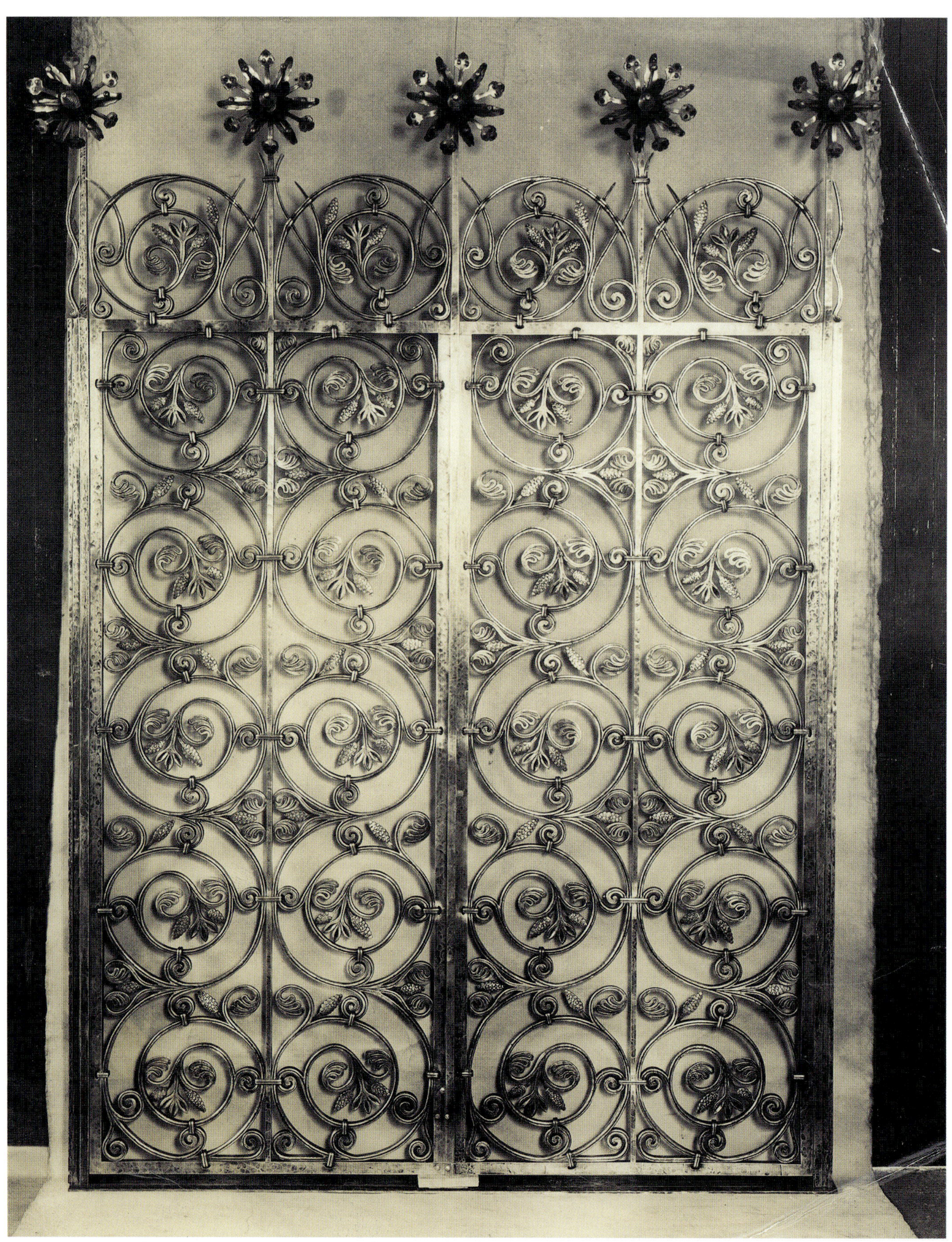

Baptistry gates for St. Bernard Church (note 67)

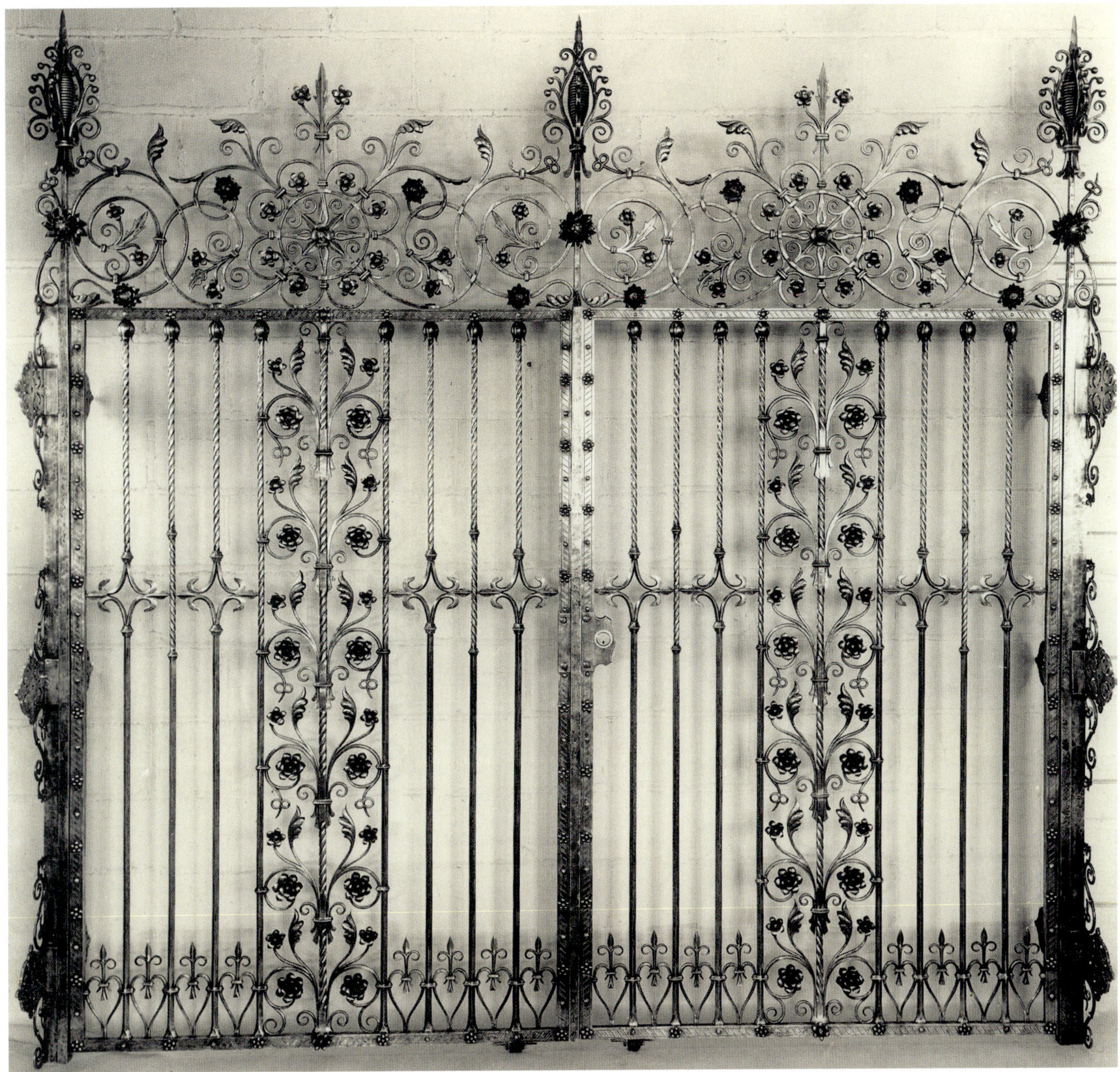

Baptistry gates for Our Lady, Queen of the Most Holy Rosary Cathedral (note 69)

aluminum ever produced in this country."[71] Although apparently not associated with these particular gates, an existing architectural sample depicts some of the techniques employed in their creation.[72]

The newspaper report on the signing of the contract for the gates included the additional notice that "Supplementary orders for this Cathedral . . . are assured the local company."[73] The supplementary orders would include massive lighting fixtures and a baptismal font cover[74] described as "the most elaborate and costly . . . ever made in America." Much of the work on the dome-shaped cover itself, using dies engraved by Louis Donato for the repoussé work, has been traced to Lester "Bus" Sutherland, and the work on the chains and apparatus for raising and lowering the cover has been traced to Lewis "Doc" Rossi.[75] The 30-inch statue of St. John the Baptist, which sits atop the cover, was made by Natale Rossi, "skilled workman who has spent 200 working days duplicating by hand, in aluminum, the drawings made by William Richard Perry, Pittsburgh architect . . .

The model from which he worked was prepared at considerable expense out of plaster-of-Paris, and critics believe that the finished aluminum figure excells [sic] the model in its faithful attention to minute details; the features, hands and feet; the lambskin robe, etc., all worked out by hand."[76] The completed baptistry stood as a monument to exceptional American hand craftsmanship.[77]

The Wendell August Forge also executed work for the Church of the Most Holy Sacrament at Greensburg, Pennsylvania.[78] The English Gothic structure, also designed by William Richard Perry, was renovated extensively in 1972, and the final disposition of the Forge's work there remains uncertain. The church, now known as Blessed Sacrament Cathedral, serves as the principal church of the Diocese of Greensburg.[79]

In 1938, the Wendell August Forge stood at what would be the pinnacle of its influence in both architectural ornamentation and the art and giftwares industry. Although the full extent of the Forge's work to this time may never be known, it had certainly completed several significant architectural commissions, its line of lamps was doing well,[80] and its giftware products were highly regarded and widely marketed.[81]

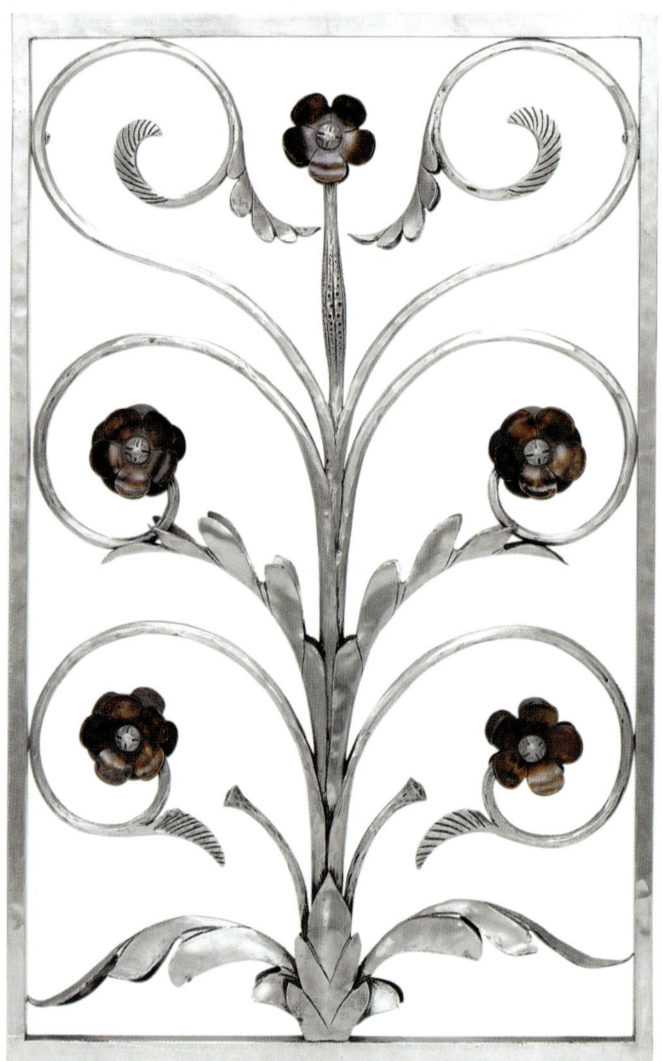
Architectural sample (note 72)

Its customer base could be expected to expand as the poverty and paralysis of the Great Depression continued to recede, driven by public works, unemployment benefits, and other governmental interventions.

It was a good time for Wendell August to combine business and pleasure in a two-month motor tour, visiting clientele in major cities of the western United States.[82]

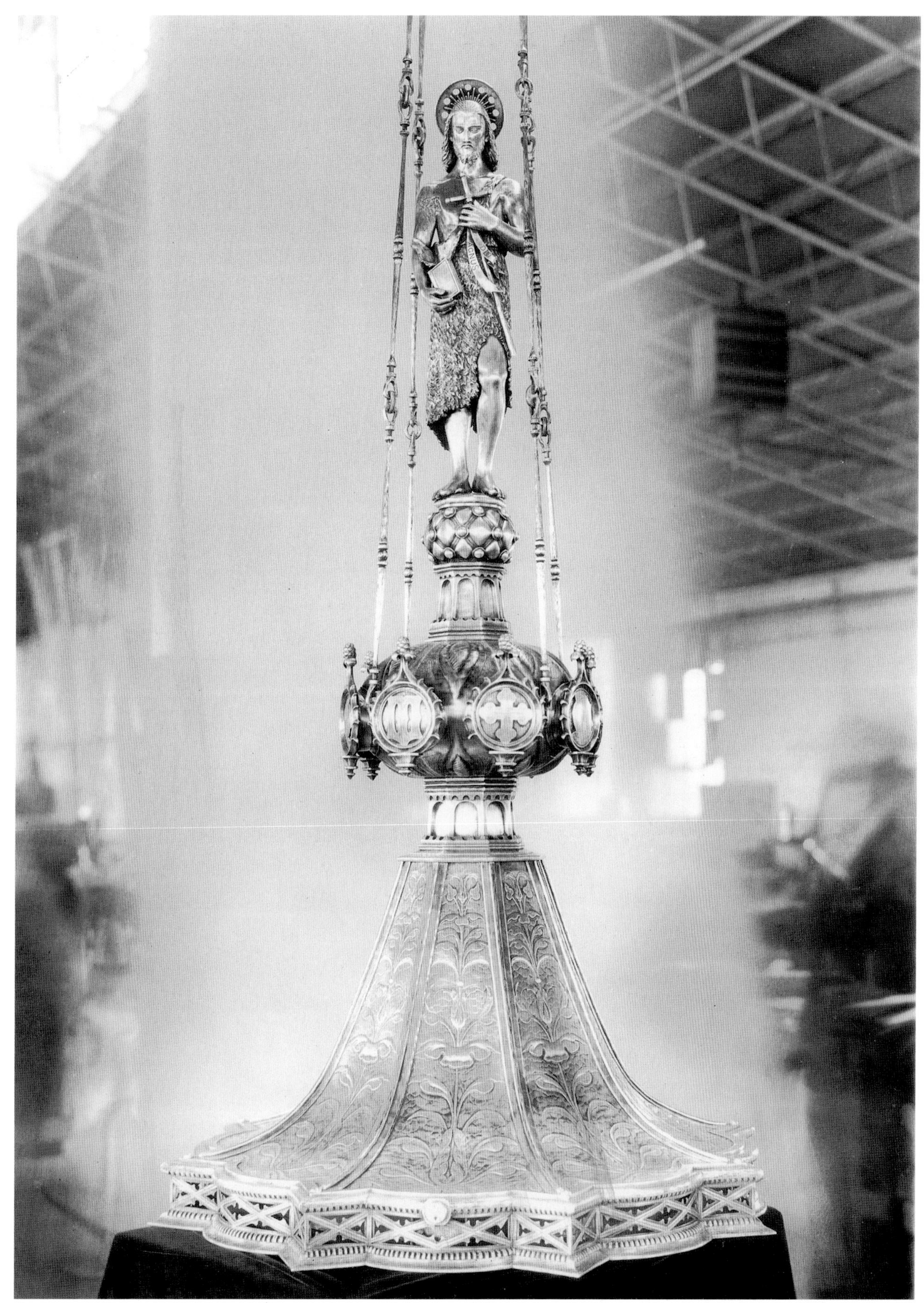

Baptismal font cover for Our Lady, Queen of the Most Holy Rosary Cathedral (note74)

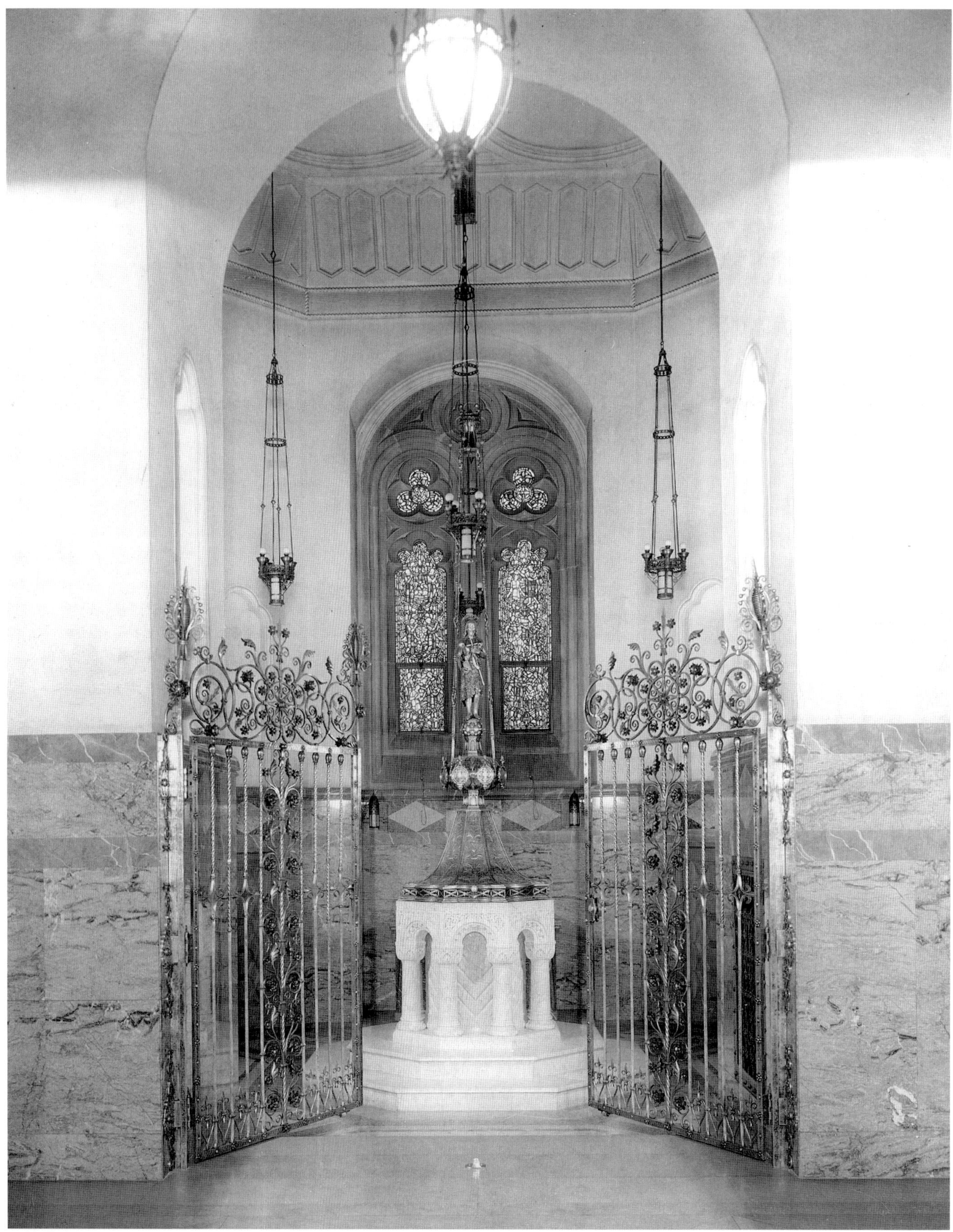

Baptistry at Our Lady, Queen of the Most Holy Rosary Cathedral, circa 1937 (note 77)

THE ORIGINAL WENDELL AUGUST FORGE HAND WROUGHT WARE

The metal is a special fine aluminum forged to a density and velvet-like smoothness simulating fine old silver. There are many other handsome pieces in our collection.

No. 147. (Below left, reading from left to right). **Match Box Cover** for Bird's Eye matches. 5 3/16"x2 3/4"x2", with elephant design...... **$2.00**

Four Compartment Cigarette Box with hinged cover. 10 1/4"x3 3/8"x1 3/8", duck design.......... **$10.00**

Two Compartment Cigarette Box with hinged cover. 5 3/8"x3 3/8"x1 3/8", duck design.......... **$7.50**

Match Box Cover for vest pocket matches. Seahorse, 1 5/8"x1 1/8"..**$ .75**

Tray. Elephant design, 9 1/2"x14".......... **$5.00**

Match Box Cover for safety matches. Sailfish design, 2 1/4"x1 1/2".... **$1.00**

(Below right): **Large Ash Tray** (Left). Sailboat design, 5 1/2" diameter **$3.00**

Small Ash Tray or Coaster (Left). Seahorse design, 3 1/2" diameter..**$ .75**

Serving Tray (Center). Pine design, 16" diameter.......... **$10.00**

Ash Trays or Coasters (Right). Yacht Wheel or Horsehead design, 4 1/2" diameter. Each.......... **$1.00**

Any one design may be had so that sets may be made of the various pieces

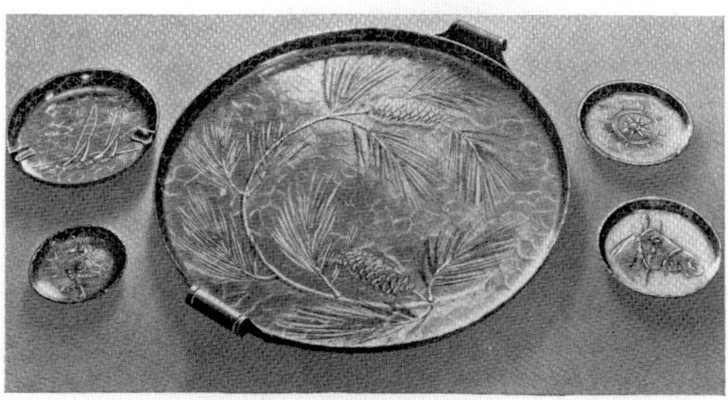

Advertisement in Hammacher Schlemmer gift catalog (*note 81*)

Chapter Notes

1 Information about the original route was provided by Frank Rossi.

2 Historical information pertaining to the development of the Grove City area is based on B. Charles Elliott, Jr., *B. Charles Elliott's Guide to the Historical and Cultural Sites of Downtown Grove City, Pa.*, pamphlet, 1995; Lillian Reeher, *Wolf Creek Legacy* (Grove City, Pa.: Allied News, 1980); *Historical Sketch of Grove City*, pamphlet, n.d. but *circa* 1897, Press of Foster, Dick & Company, Pittsburg [sic]; and *Reflections of Our Past: Celebrating 200 Years of Grove City, Pennsylvania* (Grove City, PA: Grove City Bicentennial Committee, 1998).

3 Interior of St. Clair Savings and Trust Company, *circa* 1932, photograph courtesy of Christopher Rossi.

4 "Banks are Vital Forces in Town's Growth," *Grove City Reporter-Herald*, 31 May 1933.

5 "Forged Aluminum Plant to Move Here From Brockway," *Grove City Reporter-Herald*, 19 April 1932.

6 "Forged Aluminum Plant to Move Here."

7 Based on Philip S. Klein and Ari Hoogenboom, *A History of Pennsylvania*, Second and Enlarged Edition (University Park, PA: The Pennsylvania State University Press, 1980), p 450.

8 Letters Patent pertaining to the approval of the Incorporation of Wendell August Forge by the Commonwealth of Pennsylvania on August 5, 1932; courtesy of the Wendell August Forge.

9 "Aluminum Plant to Start Work Before August 1st," *Grove City Reporter-Herald*, 8 July 1932.

10 Milford, Warren, and Robert McLaughlin, *circa* late 1930s photograph courtesy of Eleanor (Mrs. Milford) McLaughlin. Information sources include Mrs. Wayne McGinnis, personal interview, 15 July 1997; Eleanor (Mrs. Milford) McLaughlin, personal correspondence, 3 and 29 Sept. 1997; James Sterrett, personal correspondence, 22 July 1997; and W.J. McKnight, *Jefferson County, Pennsylvania, Her Pioneers and People*, Volume II (Chicago: J.H. Beers & Company, 1917).

11 Alice Marshall Hoffman, personal correspondence, 29 July 1997.

12 Louis Donato, 1940 photograph courtesy of Henry Swyers and the Brockway Area Historical Society. Information sources include Dot Donato, personal interviews, 17, 21 and 23 July 1997; Everal Donato, personal interview, 18 July 1997; Raymond Eugene and Pat Iadonato, personal interviews 18 and 23 July 1997, personal correspondence, 14 and 28 Aug. 1997, and 14 Sept. 1997; Violet Iadonato, personal interview, 18 July 1997; and Mary Lou Allen Lambert, personal interview, 21 Aug. 1997, personal correspondence, 19 and 28 Aug. 1997, 30 Sept. 1997, and 11 Nov. 1997.

13 James McCausland, 1942 photograph courtesy of the Wendell August Forge. Information sources include "Introducing One Who Needs No Introduction," *Grove City Reporter-Herald*, 26 Feb. 1946; "J.A. McCausland Services Tonight," *Grove City Reporter-Herald*, 17 Jan. 1958; James McCausland, untitled manuscript, 1939; Sterrett, 22 July 1997; Henry Swyers, personal interview, 22 July 1997; Catherine Youngo, transcription of recorded interview of Mrs. James (Mary Swyers) McCausland, 19 Nov. 1992.

14 "Eight Skilled Workmen Start Operations at Forge Plant," *Grove City Reporter-Herald*, 29 July 1932.

15 Abstract of Patent Number 1,971,700, *Official Gazette*, U. S. Patent Office, 28 Aug. 1934; patent search conducted by Bonita Campbell and William Freeman.

16 "Artistic Creations of Handwrought Aluminum Forged by Wendell August Forge Incorporated, Grove City, Penna.," brochure, 1 Oct. 1932; courtesy of the Wendell August Forge.

17 Illustrated are: #904A *Archery* coffee table, 19" diameter top, 23" high; #906 *Cat Tail* candlesticks, approximately 18" tall; *Turtle Shell* pipe bowl, 8"x10" and approximately 5" high; *Turtle Shell* bird bath, approximately 36" tall, top shell approximately 16-1/2"x23". All items from the Collection of the Wendell August Forge. Photography by Richard Smaltz, Smaltz Studio, West Middlesex, PA; photograph courtesy of the Wendell August Forge.

18 Anthony J. Pompa, personal interview, 25 July 1997, and Catherine Youngo, notes from conversation with Benjamin Formani, n.d.

19 Portion of the interior of the Grove City plant of the Wendell August Forge, showing metalworkers in the finishing area, and including a pair of #908 candlesticks listed in the 1932 brochure. The metalworkers have not been conclusively identified. The photograph is one of at least nine known to have been taken in preparation for "Skilled Workmen Forge Aluminum at Wendell August Forge, Inc.," *The Alcoa News*, 12 Dec. 1932. Photograph courtesy of the Wendell August Forge.

20 2S was the designation used by Alcoa for its commercially pure (99.0%) wrought aluminum. This designation was later changed to 1100. The suffix, O, was used to designate the softest temper of annealed wrought alloy products. 2SO was commonly recommended in Alcoa technical literature for the production of hand hammered giftware. For further information about aluminum characteristics, designations, and applications pertinent to the early 1930s see, e.g., *Aluminum in Architecture* (Pittsburgh, Pa.: Aluminum Company of America, 1932).

21 "Skilled Workmen Forge Aluminum at Wendell August Forge, Inc.," *The Alcoa News*, 12 Dec. 1932.

22 The two items on display at the Smithsonian Institution, gifts of Mrs. Jefferson Patterson, are a 4-3/4" diameter deep ash tray (#76) *circa* 1932 bearing the *Moon Fish* motif, and a 5-1/2" diameter deep ash tray (#10) *circa* 1933 having the *Dragonfly* motif. Both items bear the Wendell August Forge "bar mark" and the "PAT. APL'D FOR" mark. The pieces are misdated in the exhibit as *circa* 1940. Documentation courtesy of William Freeman.

23 A Wendell August Forge desk set from the Mitchell Wolfson, Jr. Collection, The Wolfsonian Foundation, was included in the American Craft Museum's exhibition "Craft in the Machine Age" [see Janet Kardon, ed., *Craft in the Machine Age: 1920-1945*, exhibition catalog (New York: Harry N. Abrams, Inc., 1995)] and 17 items, ranging from an architectural sample to jewelry, were included in the California State University, Northridge and Carnegie Art Museum of Oxnard exhibition, "Depression Silver: Machine Age Craft and Design in Aluminum" [see Bonita Campbell, *Depression Silver: Machine Age Craft and Design in Aluminum*, exhibition catalog, California State University, Northridge, 1995]. In 1980, Wendell August Forge items created by Natale Rossi were exhibited at the William Penn Memorial Museum in the Capitol Complex at Harrisburg, PA in honor of Rossi's receipt of a Hazlett Memorial Award for Excellence in the Arts in Pennsylvania.

24 "Wendell August Forge Rushed with Big Orders Arriving Here for Line of Aluminum Gifts," *Grove City Reporter-Herald*, 1 Nov. 1932. This article also includes in the listing of customers: "Marshall Field, Inc. and Carson-Pirie-Scott, of Chicago; J.B. Hudson Co., Minneapolis; Wm. Hengerer and Pitt Petri, the latter a large importer, of Buffalo; Hallie Bros., Cleveland; Kaufman's, Pittsburgh; Lord & Taylor, New York City; the Fred Harvey System of restaurants in the west; Bagley & Co., Jewelers, Duluth, Minn.; W.H. Block & Co., Indianapolis; J.L. Brander & Sons, Omaha, Neb.; H.W. Brown & Co., Milwaukee; Burnham & Parker, Marblehead, Mass.; A.B. Closson, Jr., Cincinnati; Daniels & Fisher, Denver; Giddings, Colorado Springs; Detroit Mantel & Tile Co., Detroit; Georgian Shop, Darien, Conn.; Greenwood Shop, Wilmington, Del.; Howe & McKendrick, Garden City, Long Island.; Hornbagen & Weldon, Marquette, Mich.; Miss Jackson's Shop, Tulsa; the Weaver Manor Shop, Newport, R.I.; and Weilles, Paducah, Ky."

25 "August Forge Co. Exhibits Gift Line in Chicago," *Grove City Reporter-Herald*, 27 Jan. 1933.

26 Wendell August Forge chair with the *Archery* motif on the back, *circa* 1933. Photograph courtesy of Henry Swyers and the Brockway Area Historical Society.

27 Illustrated are: Table lamp #101, approximately 26" tall, 13" diameter aluminum shade with *Sail Boat* motif, and 28" tall drink stand formed like a vine. Collection of the Wendell August Forge. Photography by Richard Smaltz; photograph courtesy of the Wendell August Forge.

28 Wendell August Forge lamp #171 bears a *Tropical Fish* motif and lamp #611 features the *Grecian* motif. Photographs courtesy of Henry Swyers and the Brockway Area Historical Society.

29 Natale Rossi, as quoted in Catherine Youngo, transcription of recorded interview, 19 Nov. 1992, courtesy of the Wendell August Forge.

30 "Who's Who at the Gift and Art Shows," *The Gift and Art Shop*, Feb. 1933.

31 Bonita J. Campbell, unpublished aluminum giftware products database.

32 Arthur S. Armour, *circa* 1972 photograph courtesy of Thomas Armour. Information sources include Arthur Armour and Thomas Armour, personal interviews and correspondence, 1994-1998; A. Stanley Brussel advertisements of Arthur Armour work, *The Gift and Art Buyer*, issues spanning 1934-1941 and 1946-1949, copies courtesy of Library of Congress research conducted by William Freeman; Bonita Campbell, *Depression Silver: Machine Age Craft and Design in Aluminum*, exhibition catalog, California State University, Northridge, 1995, and "An Introduction: Aluminum Decorative Arts," *Echoes*, Spring 1996.

33 Franklin D. Roosevelt, Presidential Inaugural Address, 4 March 1933.

34 "Wendell August Forge Signs NRA, Increases Wages," *Grove City Reporter-Herald*, 25 Aug. 1933.

35 "Beer Banned Here by Council Action," *Grove City Reporter-Herald*, 11 April 1933.

36 "Dr. Fithian is Nominee as Dry Representative," *Grove City Reporter-Herald*, 21 June 1933.

37 "August Buys Fithian Stock in Forge Plant, *Grove City Reporter-Herald*, 12 Sept. 1933.

38 *The Brockway Record*, 15 Sept. 1933, microfilm transcription courtesy of Adam Sabatose, Brockway Area Historical Society.

39 Natale Rossi, *circa* 1962 photograph courtesy of the Wendell August Forge. Information sources include "Grove City Man Named Officer of State K of C," *Grove City Reporter-Herald*, 7 May 1964; "Natale L. Rossi," *Allied News*, Sept. 1993; Jaci Poole, "G C Man Repairs Toledo Church Statue He Made 40 Years Ago," *Allied News*, 20 Aug. 1974; Adaline (Rossi) Reese, personal interview, 24 Sept. 1997; Christopher Rossi, personal interviews, March 1994 and 23 July 1996; Frank Rossi, personal interviews, 18 Dec. 1994, 26 June 1995, 12 June 1996, and 11 July 1997; and Irene Rossi, personal interview, 28 April 1996.

40 Ottone "Tony" Ignatio Pisoni, *circa* 1949 photograph courtesy of Lucy (Mrs. Joseph) Pisoni and Reverend Otto B. Pisoni. Information sources include Lucy (Mrs. Joseph) Pisoni, personal interview, 22 July 1997; Reverend Otto B. Pisoni, personal interviews, 19 June 1997 and 22 July 1997, personal correspondence, 24 June 1997, 20 Aug. 1997, and 16 Oct. 1997; and the St. Tobias Cemetery, Brockway, PA.

41 Arthur Palmer, *circa* 1917 photograph courtesy of Mr. and Mrs. Everett Palmer. Information sources include Bonita J. Campbell, *Arthur Palmer and Aluminum: Palmer-Smith*, research monograph, 1996, 68 pp, and "An Introduction: Aluminum Decorative Arts," *Echoes*, Spring 1996.

42 Patent Number 1,971,700, United States Patent Office. The complete patent is provided in Appendix B.

43 "Artistic Creations of Handwrought Aluminum."

44 Wendell August Forge Price List, 1 Aug. 1937, courtesy of the Wendell August Forge.

45 "Dick August is Instantly Killed as Huge Plane Crashes Into Mountain," *Grove City Reporter-Herald*, 10 April 1936.

46 "Pilot of 'Sun Racer' Felt He Was Over Airport," *Grove City Reporter-Herald*, 17 April 1936.

47 Wendell August Forge display at the National Gift Show held at the Palmer House in Chicago, Feb. 1935; photograph courtesy of the Wendell August Forge.

48 "August Buys Fithian Stock."

49 "An Ancient Art Goes Modern," *Aluminum News-Letter*, Jan. 1934.

50 Alice Marshall Hoffman.

51 Wendell August Forge Price List, 1 Aug. 1936, courtesy of the Wendell August Forge.

52 Advertisements appeared in *The Jewelers' Circular-Keystone* in August and October of 1935, and March of 1936. The photograph shown here appears in the October, 1935 advertisement. According to the data in the advertisement, depicted are: Cocktail Trays, 9-1/2"x14", $2.50 each (600-Sail Boat, 601-Ducks, 602-Polo, 603-Thistle, 604-Airplane, 605-Sea Horse, 606-Barley, 607-Pheasant, 608-Fantail Fish, 609-Butterfly, 612-Sail Fish, 614-Tropical Fish) and Coasters, 4-1/2", $6.00 doz. (0-Sail Boat, 1-Gun and Ducks, 2-Polo, 3-Thistle, 4-Airplane, 5-Sea Horse, 6-Barley, 7-Pheasant, 8-Fantail Fish, 9-Butterfly, 12-Sail Fish, 13-Duck, 14-Tropical Fish). Photograph courtesy of Henry Swyers and the Brockway Area Historical Society.

53 See, e.g., C. G. Home, ed., *Decorative Art: The Studio Year Book* (London: The Studio Limited, 1936).

54 "Aluminum's Role in Modern Home Decoration," *Furniture Index*, Feb. 1936. The article includes photographs of Wendell August Forge "coffee table, beverage cooler, ice tongs and coasters," and "occasional table, chair and beverage holder for porch and lawn use." Courtesy of Marilyn Sullivan.

55 The ashtray, 6" diameter and 6" high overall, was donated to the Collection of the Wendell August Forge by Mrs. James (Mary Swyers) McCausland. Although the piece donated by Mrs. McCausland has no Wendell August Forge mark, an identical piece bearing a Forge mark is in the Kemerer Museum of Decorative Arts in Bethlehem, PA. This second piece was donated to the Kemerer Museum by Nina Mackall, sister of Paul Mackall, who was Vice President of Sales for Bethlehem Steel and was reportedly a passenger on the flight. Information pertaining to the Kemerer piece is courtesy of Barbara Schaefer of the Kemerer Museum. Photography by Richard Smaltz; photograph courtesy of the Wendell August Forge.

56 Information pertaining to the Hindenburg flight is based on Rick Archbold, *Hindenburg: An Illustrated History* (New York: Warner Books, Inc., 1994) and Michael Macdonald Mooney, *The Hindenburg* (New York: Dodd, Mead & Company, 1972).

57 "Large Contract Taken," *The Brockway Record*, 3 July 1937.

58 Complete Toastmaster Hostess Set from the Collection of Nancy and Jack Withrow; photograph courtesy of Nancy Withrow. The stand is 19" high x 21" wide x 14" deep; the tray top for the stand is 14" x 22" (without handles); the two lap trays are each 7-3/4" x 13-1/2"; the smallest tray is 5-1/2"x6"; the toaster is approximately 6-1/2" x 8-1/2" x 9" high. Items are individually marked "For Toastmaster."

59 Featured in "Made of Metal," *The Gift and Art Buyer*, Nov. 1937.

60 Bonita J. Campbell, *Eugene Town and Frank Rossi: Hand Made Aluminum Made in California by Town*, research monograph, 1996, 38 pp.

61 Toastmaster Toaster, approximately 6-1/2" x 8-1/2"x9" high; from the Collection of the Wendell August Forge; photography by Richard Smaltz; photograph courtesy of the Wendell August Forge.

62 Lewis "Doc" Rossi, as quoted in Catherine Youngo, transcription of recorded interview, 22 March 1994, courtesy of the Wendell August Forge.

63 Wendell August Forge, Catalogue No. 5, 1 Aug. 1937.

64 Illustrated are: #277 6-1/2"x7-3/4" six-pack cigarette box with strap hinges and *Hobnail* motif; #760 20-cup coffee pot, 5" diameter and 13" tall, and one-pint cream and sugar, both 3-1/2" diameter and 4-1/2" high, all pieces with *Grecian* motif; #875 14" diameter punch bowl and ladle set with *Grecian* motif; #961 ice bucket, 6-1/4" diameter and 7" tall, with *Sail Boat* motif, and 7-5/8" long tongs; #926 smoking stand, 23" tall, and ashtray with *Ducks* motif. All items from the Collection the Wendell August Forge. Photography by Richard Smaltz; photograph courtesy of the Wendell August Forge.

65 Wendell August Forge Price List, 1 Aug. 1937.

66 A comprehensive information source pertaining to the art and architecture of the St. Bernard Church complex is Reverend Thomas R. Wilson, *St. Bernard Church* (Mt. Lebanon, PA: St. Bernard Church, 1995).

67 St. Bernard Church baptistry gates, *circa* 1936 photograph courtesy of M.C. (August) Taylor. The gates are each 33 inches wide and 8 feet tall (including the floral ornamentation at the top); measurements courtesy of the Reverend Thomas R. Wilson.

68 "August Forge to Make Gates for Toledo Cathedral," *Grove City Reporter-Herald*, 30 June 1936.

69 Our Lady, Queen of the Most Holy Rosary Cathedral baptistry gates, *circa* 1936 photograph courtesy of M.C. (August) Taylor. The gates are each 4' 2" wide and 8' 5" tall (including the ornamentation at the top); measurements courtesy of Our Lady, Queen of the Most Holy Rosary Cathedral.

70 Barbara Carter Daley, *Our Lady, Queen of the Most Holy Rosary Cathedral*, pamphlet, 1983.

71 From photograph caption in *The Alcoa News*, 19 April 1937.

72 Sample of ornamental architectural work, aluminum with bronze, 12-3/4"x20-3/4", from the Collection of the Wendell August Forge. Photography by Richard Smaltz; photograph courtesy of the Wendell August Forge.

73 "August Forge to Make Gates."

74 Our Lady, Queen of the Most Holy Rosary Cathedral baptismal font cover, *circa* 1937 photograph courtesy of Henry Swyers and the Brockway Area Historical Society. The figure of St. John the Baptist is approximately 30 inches tall; the complete baptismal font cover is approximately 7-1/2 feet tall; measurements courtesy of Our Lady, Queen of the Most Holy Rosary Cathedral.

75 Susan (Sutherland) Babb, personal correspondence, 20 Sept. 1997; Christopher and Lewis "Doc" Rossi, personal interview, March 1994.

76 "Costly Baptismal Font Made At Local Plant Sent To Cathedral In Toledo," *Grove City Reporter-Herald*, 21 Dec. 1937.

77 Our Lady, Queen of the Most Holy Rosary Cathedral, original baptistry, *circa* 1937 photograph courtesy of Henry Swyers and the Brockway Area Historical Society.

78 "Costly Baptismal Font Made At Local Plant."

79 *Blessed Sacrament Cathedral*, pamphlet, n.d., and Monsignor Robert J. Shuda, personal correspondence, 26 Sept. 1997.

80 "Do You Remember?," *Allied News*, 5 July 1983.

81 *90th Birthday Gift Catalog*, Hammacher Schlemmer, 145 East 57th Street, New York, 1938.

82 "Wendell August Leaves for Business Trip to Western Cities," *Grove City Reporter-Herald*, 19 April 1938.

Chapter Three

The War Years: 1939-1948

While the attention of most of the American public during the 1930s had been focused on the effects of the Great Depression, much of the rest of the world had been increasingly engaged in warfare. In the first half of the decade, Japan conquered Manchuria, Italian armies marched into Ethiopia, and Germany became an aggressive totalitarian state under Hitler and the Nazi party. The number and intensity of the conflicts escalated rapidly between 1936 and 1938, with the Spanish embroiled in a Civil War, Italy advancing further in Africa, Japan invading China, and Germany reoccupying the Saarland and the Rhineland, annexing Austria, and advancing into Czechoslovakia.

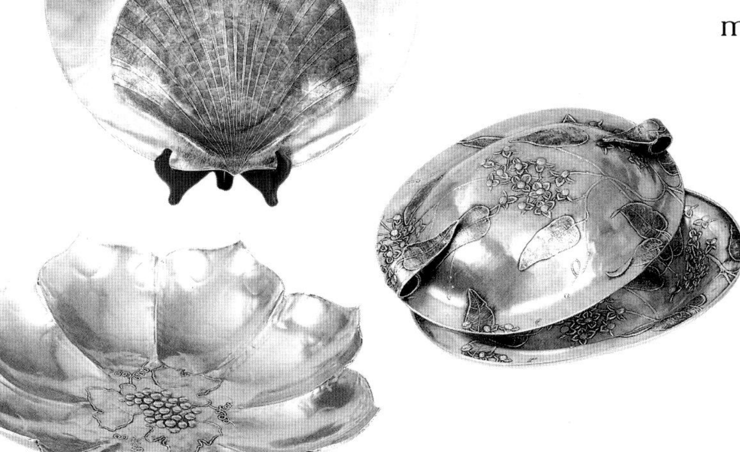

Clockwise, Hors d'oeuvres server with Shell *motif, covered bon bon with* Bittersweet *motif, star bowl with* Grape *motif (note 3)*

As much of the globe plunged deeper into war, the United States was continuing to emerge from the Great Depression. By 1939, President Franklin D. Roosevelt's New Deal was firmly in place and prosperity was once again within reach of much of the populace. The restoration of consumer purchasing power was stimulating the demand for giftwares, and the disruptions in the supply of imported goods caused by the distant wars meant that the demand must be met by increases in domestic production. For example, the consumer penchant for hand forged aluminum had not yet been sated, and the absence of foreign competition contributed to the doubling of the aluminum giftware product lines on the market during 1939 alone.[1]

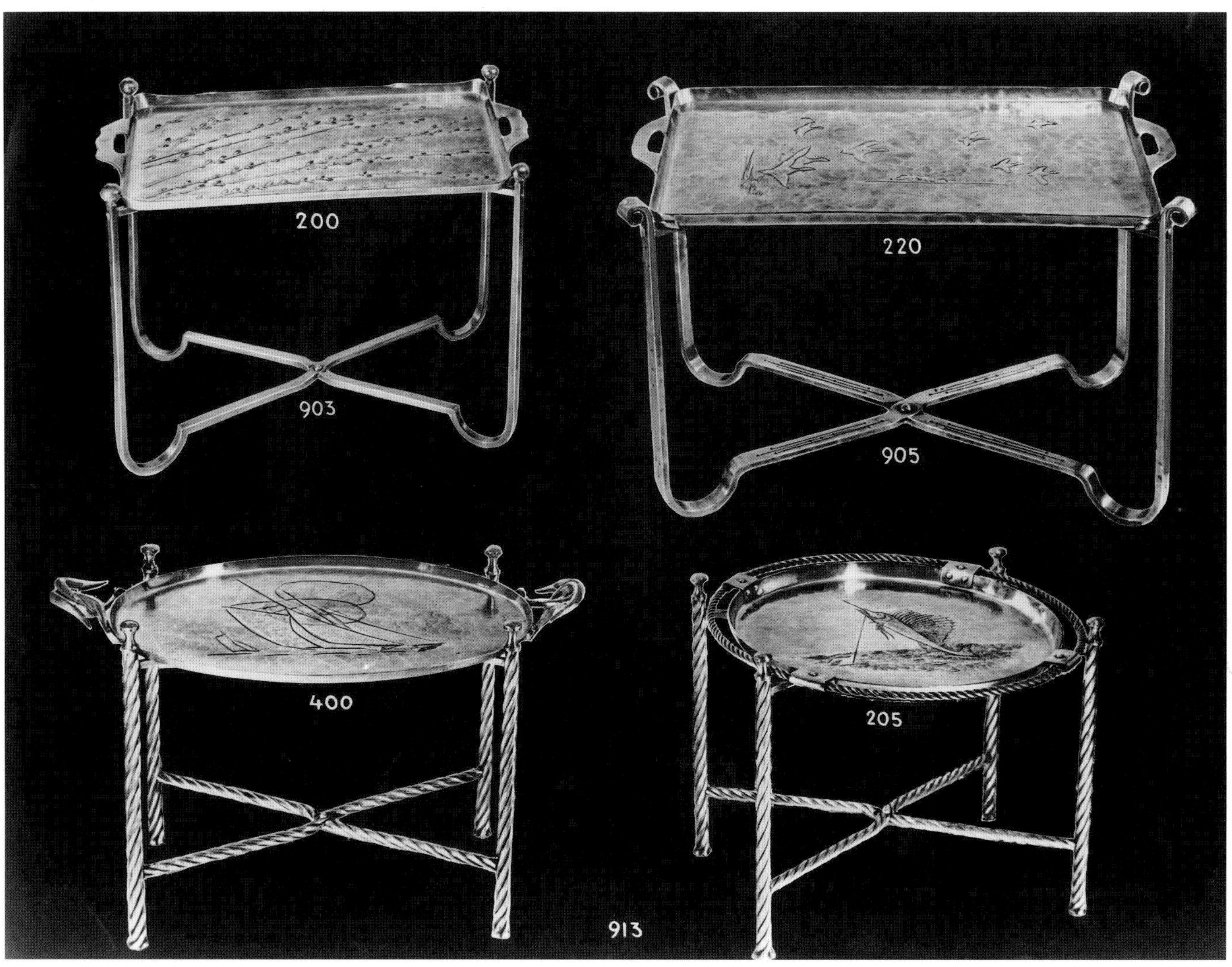

Four styles and sizes of butler tables, with trays bearing Pussywillow, Ducks, Yacht *and* Sailfish *motifs (note 4)*

In addition, the opening of the 1939 New York World's Fair was rapidly approaching, providing a wealth of merchandising opportunities for firms across the country. Themed to the "World of Tomorrow," the Fair was at that time the largest international exposition ever undertaken, and its Board of Design was dominated by luminaries of the newly-emerging field of industrial design. The Fair's emphasis on decorative arts for the home contributed to the boom in the design, production, and sales of giftwares and allied products.[2]

In Grove City, the Wendell August Forge prepared for the 1939 season with not only its own World's Fair offerings featuring the Trylon and Perisphere logo, but also the introduction of the first few pieces of its redesigned line.[3] For this line, the Forge parted from many of its conventionally-formed bowls and trays, introducing more stylish forms that were either angular and in keeping with geometric Art Deco, or sleek and curvilinear consistent with the streamlined approach that was in vogue. A portent of the post-war product, some of the bowls had "tool-marked" (notched) edges.

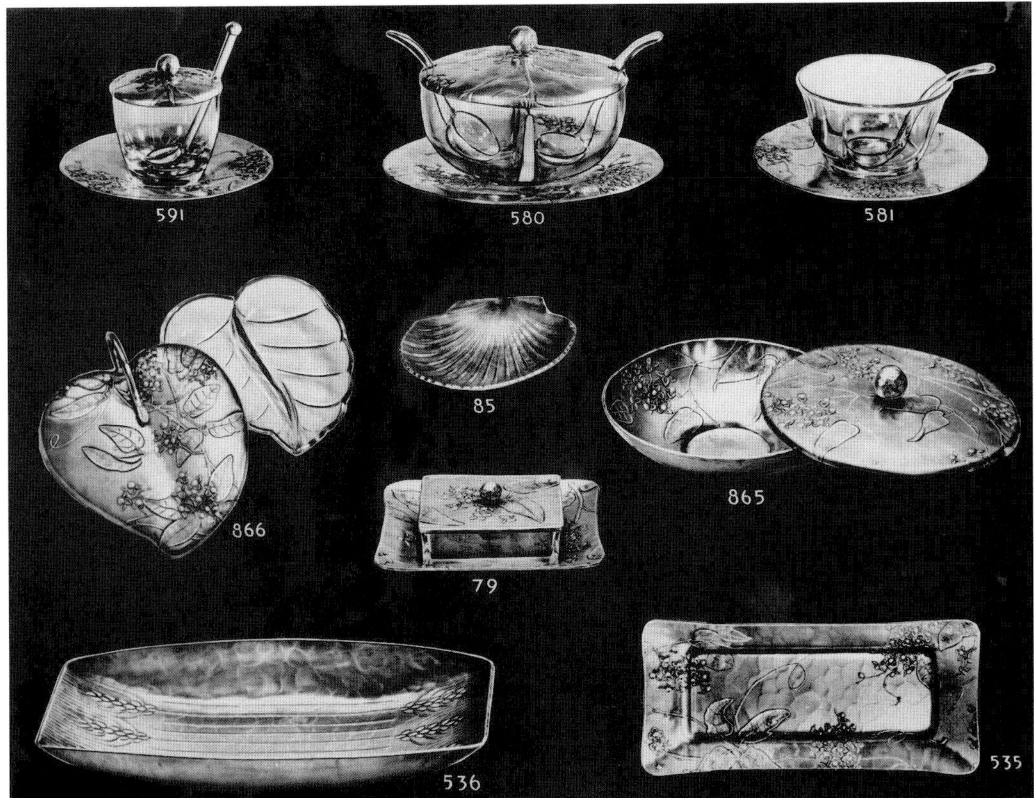

Small Shell *bon bon, celery tray with* Wheat *motif, and an array of pieces with the* Bittersweet *motif (note 5)*

Most of the items that had disclosed the wrought iron roots of the firm, such as massive candelabra and lamps, were being removed from the product line, an exception being the continuation of the butler-type tables that had been featured with modest variations in style since at least 1932.[4] Only in these tables could one discern vestiges of the ornamental work of just a few years before.

Most of the new forms were also embellished with new repoussé motifs, such as *Bittersweet*, which were allover designs that covered most of the surface of an engraved die and the pieces that were made from it.[5] These motifs provided significant production advantages, since a single large engraved die could be used to produce a large variety of pieces. With few exceptions, the earlier motifs featuring naturalistic and relatively realistic images of fauna, such as seahorses and elephants, and sporting themes, such as yachting and polo, were retired from the product line. At the same time, the number of motifs offered was drastically reduced from more than 60 to fewer than 30.

In conjunction with the more sophisticated forms and motifs, the creative use of California pottery with some items evoked a more sophisticated image for the Wendell August Forge giftwares line. The rich combination of crackle-glazed La Mirada[6] pieces and Wendell August Forge repoussé aluminum work provided ample evidence that domestic industry could produce wares that were at least as desirable as the imports that were no longer available.[7] [8] A few of the new items, particularly the vases, were very time-consuming to make. According to Frank Rossi, "I remember those [pieces with pottery] before the war. Those ducks, they were just terrible to cut out; took a lot of time." [9] [10]

The design genesis of the new line was primarily attributable to James McCausland. It was in keeping with the past practice of the Wendell August Forge for McCausland to do the design work, and the available evidence, including a tissue drawing of the *Apple Blossoms* motif, points to McCausland as the creative force behind the new line. "Jim would do the design, then Louis would cut it,"[11] said Frank Rossi, later adding that "Donato, he cut all the dies before the war. He sat back in the corner by himself, hunched over a die all day. He used to cut earrings in a bar of soap first to see what they would look like before he cut the steel. He was a better cutter than Natale."[12] According to Joseph Leone, "Lou Donato did all the dies; his shoulders were all humped from working over those dies all the time."[13]

At the same time, a more aggressive approach to marketing was undertaken by the Forge. Janis-Tarter, Greeman Inc., a highly-regarded manufacturers' representative firm in the art and giftwares industry, was engaged to market the line. The new business relationship was announced in a leading trade journal, with a dramatic, white-on-black, stylized rendition of the Alcoa Aluminum Research Laboratories gates dominating a full page, noting that "The open gates shown below illustrate the first use of hand forged aluminum. Made by Wendell August Forge, they were the forerunner of the art of forging aluminum into smart gift items."[14]

In addition, James McCausland traversed the region with a romantically-inspired speech, extolling the virtues of aluminum as a material worthy of use for the creation of art. Comparing the role of the Wendell August Forge to that of *Pygmalion* linguist Henry Higgins, and aluminum to Liza Doolittle, McCausland wrote, "We decided to find a new metal . . . and to become first in the Forging of it rather than to continue the competition in Iron. We naturally turned to stainless steel, then new; also to several of the new nickle [sic] alloys. All these were rejected for some quality that made fabrication uncertain and costly. So we really assumed the role of the linguist when we first caught sight of our heroine aluminum, blond-interesting-but certainly not beautiful in her surroundings." Following an eloquent description of the experiments pursued, McCausland's essay continued "there was the question of how she should be ornamented. At first it was the general opinion she would have to keep in the reserved field – only silhouette for design and that simply done. Later it was accepted that fancy costume was flattering and the designs were increased to the ones she wears today. By a carbonizing process we added the contrasting shadows which tend to further accent her beauty. Finally after several years she has been acceptable . . . where ever the smart crowd was gathering. Then there appeared the new role – that of hand wrought Jewelry. Now she was not only to be admitted to the room where the party was held, but she was to mingle intimately with the individuals, even to attempt, through the shedding of her own personality to accent the beauty and grace of her wearer."[15]

The image that McCausland created certainly hit home in northwestern Pennsylvania. "We got invited to all the weddings," said Frank Rossi, "Oil City,

Franklin, all the other towns. We could use the shop to make things on our own time if we paid for the aluminum."[16] Joseph Leone noted that "Everybody knew if they invited somebody from the Forge, they'd get some Wendell August. I made lots of different things, not just the regular stuff, for weddings."[17]

In September of 1939, German troops invaded Poland, touching off World War II. For a majority of the United States, the European conflict brought economic blessings, as war material production increased and the availability of imported goods dwindled. "The war in Europe," according to the *Grove City Reporter Herald*, "offers salesmen of Pennsylvania anthracite and bituminous the opportunity to capture markets lost in recent years to foreign coal, because the war is now taking them out of world trade."[18] By the end of the year, headlines such as "Cooper-Bessemer Directors Meet, Declare Dividends," crediting a backlog of orders three times the size it had been at the beginning of the year, were becoming common.[19]

From a commercial perspective, the Western Hemisphere had been effectively isolated from the rest of the globe by war.

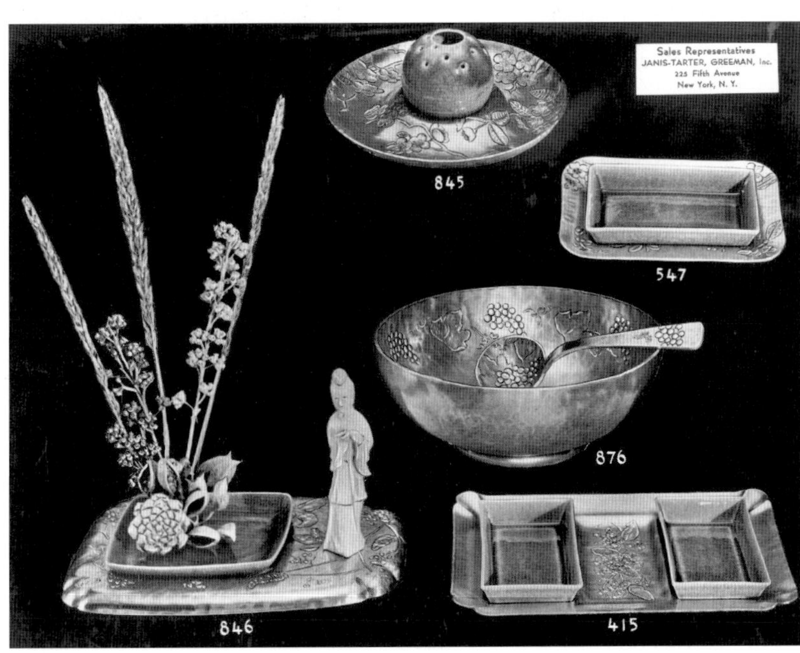

Punch Bowl with Grape *motif, and serving and floral arranging pieces with La Mirada pottery (note 7)*

Although domestic production was soaring to meet the demands of both consumers and the foreign munitions buyers, concerns about the potential need for and requirements of national defense became prevalent. Trade journals articulated these concerns. In January of 1940, an article in one such journal began, "Doubt and uncertainty four months ago pervaded the giftwares industry. Germany, Austria, Czechoslovakia and Poland, the first three of them dominant producers of gift merchandise, had disappeared one after the other as sources. Could other nations supply the same or substitute commodities? Could blockades and counter-blockades plug all overseas shipments? Could American factories produce wares as good in design, workmanship and price as foreign-made?"[20] In a similar vein, one of the announcements associated with the Chicago Gift Show began, "War and rumors of war fill the air as the 1940 Chicago summer market period approaches, and to the usual worries of the buyer is added a whole new set of perplexities brought to the front by the successive shutting off of imports from country after country, and, most recently by the

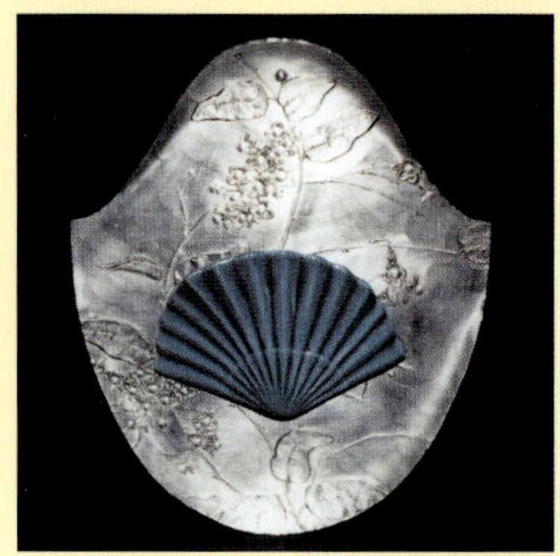

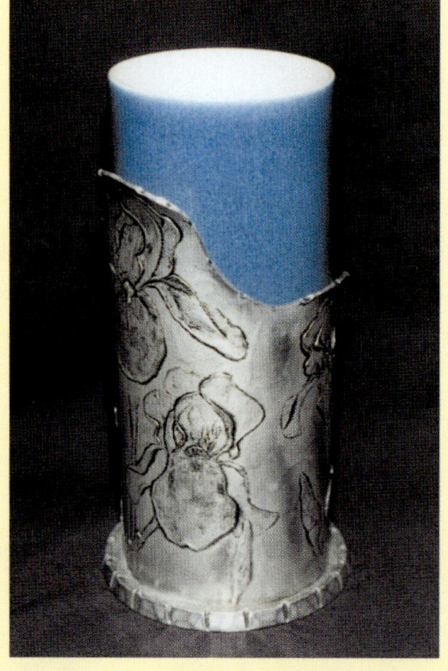

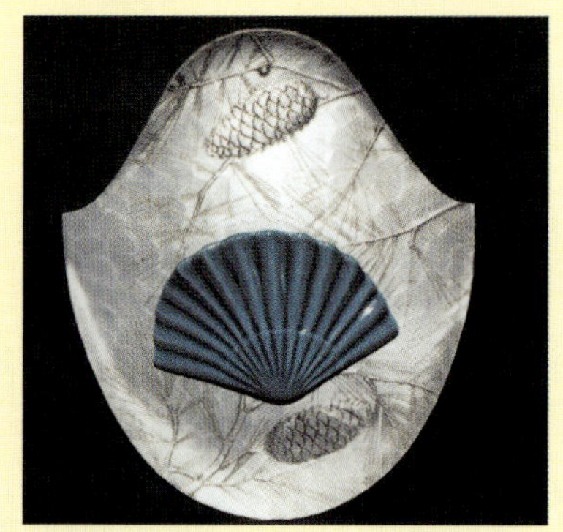

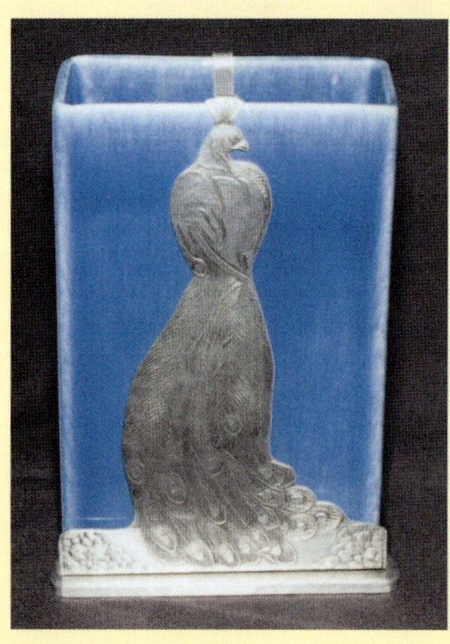

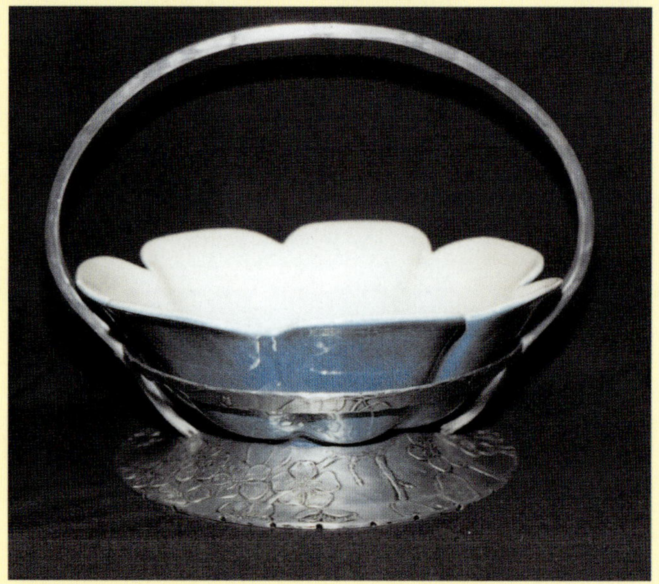

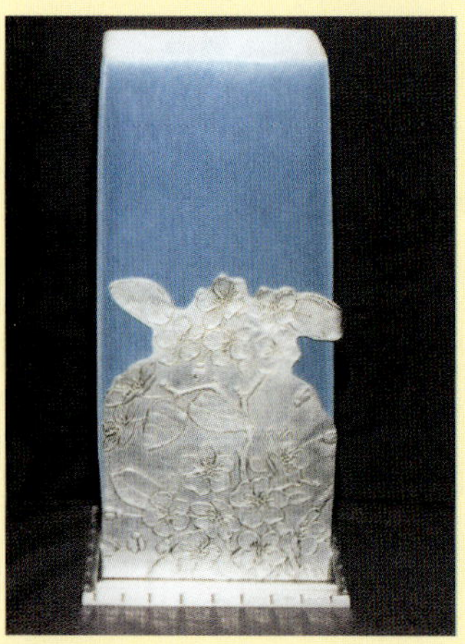

Aluminum with La Mirada Pottery (note 8)

Top, left to right, Wall pocket with Bittersweet *motif, vase with* Iris *motif, wall pocket with* Pine *motif*

Center, left to right, Vase with Peacock *motif, basket with* Apple Blossom *motif*

Left, Vase with Apple Blossom *motif*

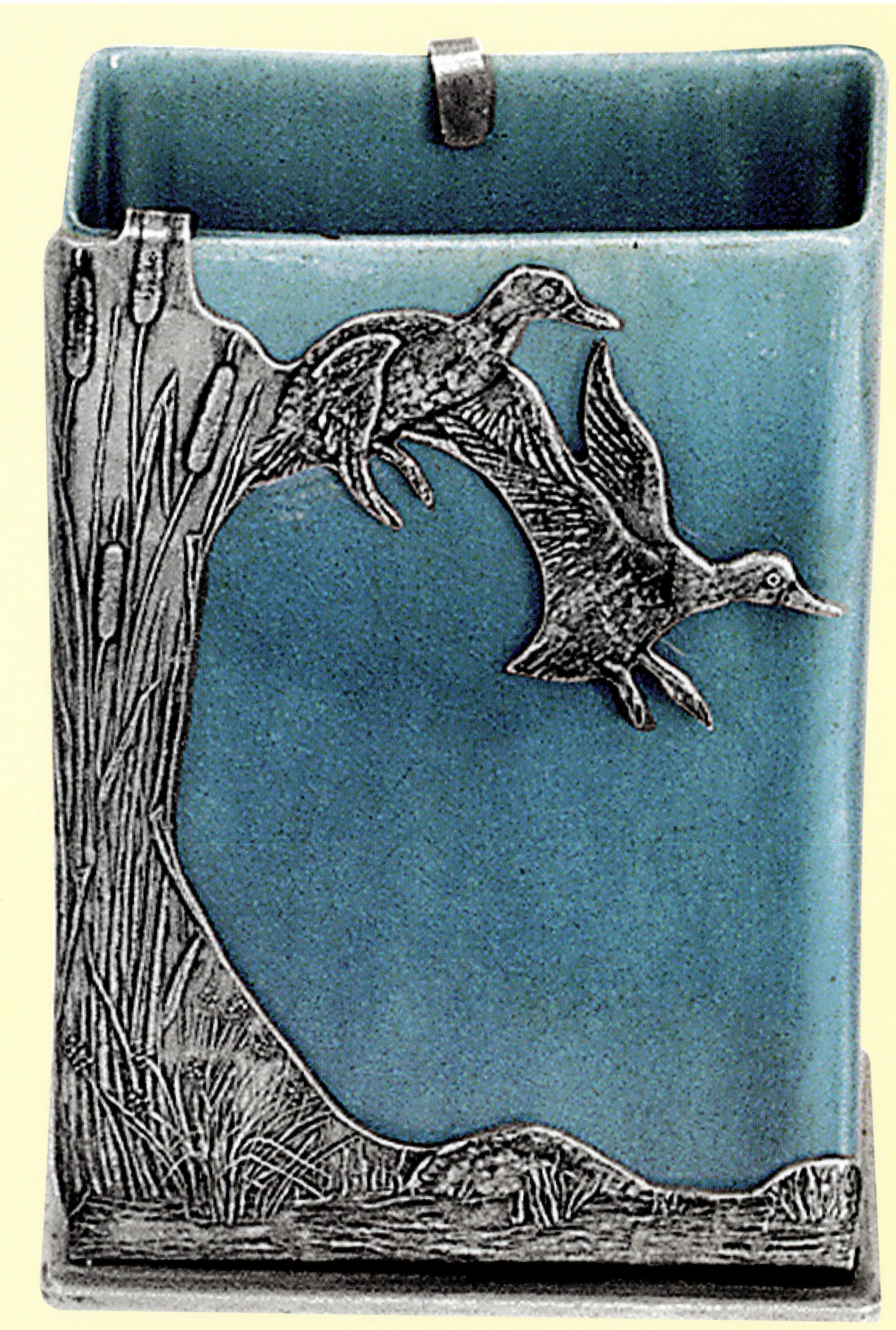

Aluminum and La Mirada Pottery vase with Ducks *motif* *(note 9)*

implications of the national preparedness campaign, which is certain to affect many a commodity line including our own, as manufacturing facilities and supplies of raw materials are drafted for the needs of national defense."[21]

The response of domestic metalware manufacturers to the paucity of imported goods was apparently more than acceptable to the giftwares buyers. By the Summer of 1940, industry observers were noting that "outstanding activity in metalware," including aluminum, was being reported by jewelers, and that no branch of the industry was showing "more creative ability of designer, higher craftsmanship of manufacture, or consumer appeal than high grade metalware."[22] The demand for decorative aluminum art and giftwares certainly did not abate, and additional manufacturers, including such relatively large firms as Continental Silver Company and Farberware, entered the field.[23]

The Wendell August Forge was foremost among the firms that responded to the challenge of meeting consumer demand with quality products. The workforce at the Forge continued to expand until it reached a peak of almost 50 full-time skilled metalworkers, with temporary increases for the heavy Summer and Fall giftware production periods.[24] "Grove City Plant Booms" read the subtitle of a 1940 newspaper article about the Forge, in which the sound of hammers was likened to a "clattering melody."[25]

Likening the sound of hammers to a "clattering melody" might yield good newspaper copy, but for the metalworkers, it was deafening noise. "Hammering all day with a ball peen hammer was *hard* labor, and the noise ruined your hearing. Nobody I know from there didn't lose some hearing,"[26] said Anthony "Tony" Pompa. Joseph Leone remembered, "It made a real racket, all that hammering. None of us hear so good." "Eight hours a day with a 2-1/2 pound ball peen hammer was *work*," recalled Frank Rossi, continuing "For awhile there, Gordon Brown was doing repoussé, those 9 by 14 inch trays, he hammered 40 trays a day. Jim Breese got them from Gordon and flattened them, then Tony Pompa formed them, I sanded, and Joe Leone did finishing."[27] Rossi also noted that Jim Breese "did the most work" in the tray group, "he would straighten out each piece, shear the edges, band saw the handles, and file all sharp edges."[28] "We used rawhide hammers for flattening them after the design was put on. Had to leave the hammer in water overnight to soften the rawhide 'cause otherwise it would leave marks," added Anthony Pompa.[29] The creation of James McCausland's beautiful, blonde heroine was not without considerable cost to the people responsible for her production.

Against this backdrop, Robert August completed his degree program in economics and engineering at Yale University in 1940 and returned home to Grove City to start a new business and, later in the year, wed Mary Boquine, daughter of R. Gilbert and Nellie (Sullivan) Boquine of Dunkirk, New York.[30] On his return in June, he teamed up with Jack Carruthers to form Pine Grove Craftsman (also referred to as the Pinegrove Company) to begin the production of a modest new aluminum giftware line.[31] The new company's officers

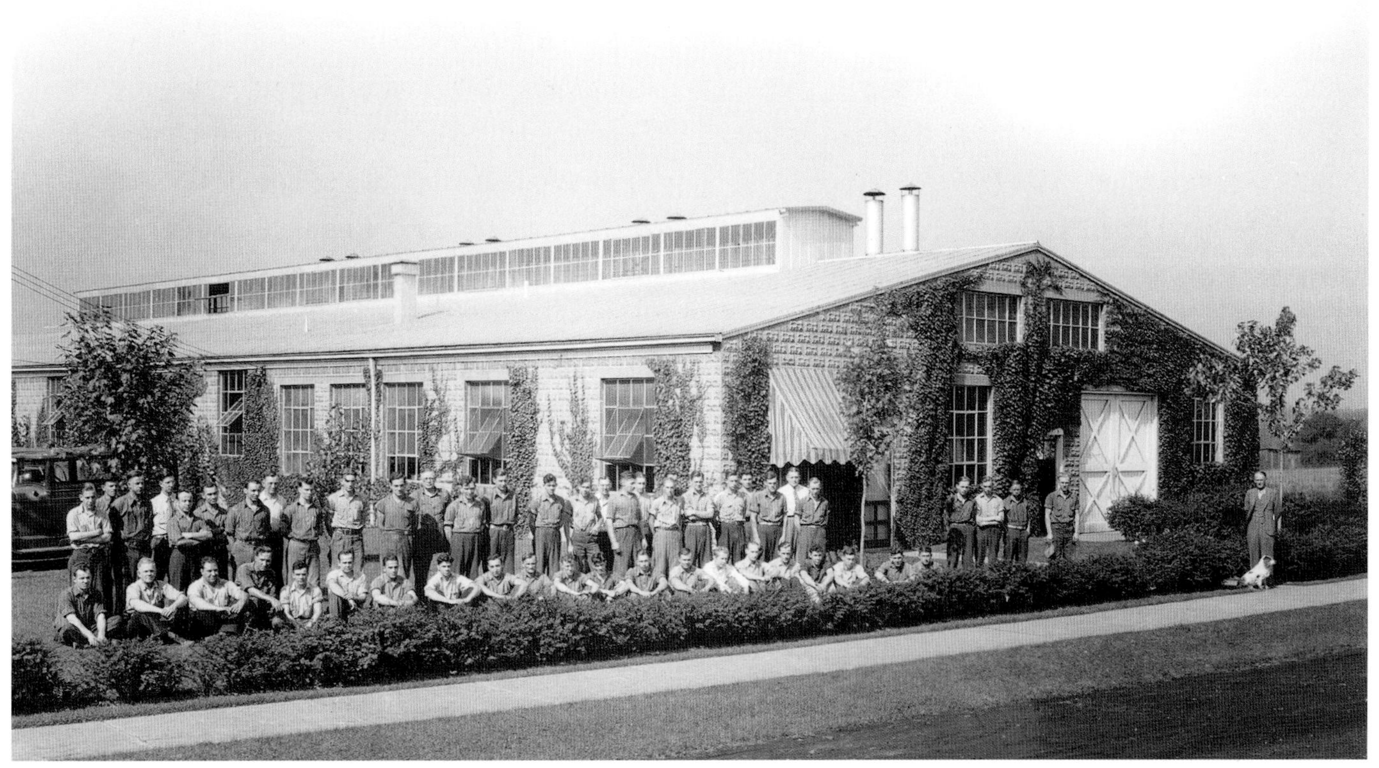

Wendell August Forge workforce at the Grove City facility in 1940 (note 24)

were Carruthers as President, Robert August as Secretary, with Wendell August (who invested in the new firm) as Treasurer. "It was kind of funny," remembered Carruthers, "Bob came over one day to our house, right across the alley, and he had the checkbook that was Pine Grove Craftsman, and there was a place for signature for president and secretary. And we had no election, so Bob said do you want to be president, so I said why don't we flip a coin. So that's how I was president of Pine Grove Craftsman."[32]

Locating at 401 Erie, they intended to produce a machine-made line, but until the equipment arrived, some items were produced partially by hand. The new firm used skilled craftsmen from the Wendell August Forge, including Natale Rossi, who engraved the dies[33] and indicated that "we cut dies real shallow and then they rolled it out."[34] According to Carruthers, "We had a mat roller which we got from the *Baltimore Sun*, which was a big machine that we put the dies on and then we put the aluminum on top of the dies and ran it through the big rollers and pressed the design into the aluminum," continuing, "We had the forming press which broke your arm every time we used it . . . We had a die and we stuck [a piece] in there and this plunger would come down . . . and push it, bend it right into the die. Then we'd have to trim it and buff the edges around. We had a couple of buffing machines and fellows running those."[35] Among the craftsmen who worked for Pine Grove Craftsman were Gordon Brown and Tom Trepasso.[36]

Although no printed documentation pertaining to the new product line has been located, the composite of numerous recollections and artifacts suggests that the initial pieces that had hammered motifs were marked "Wendell"[37] and those that had rolled motifs were marked "Pinegrove".[38] There appear to have been no more than 15 items, consisting primarily of ashtrays, coasters,

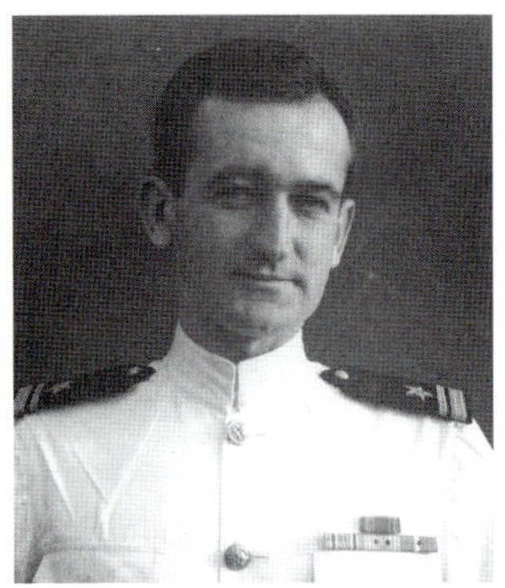
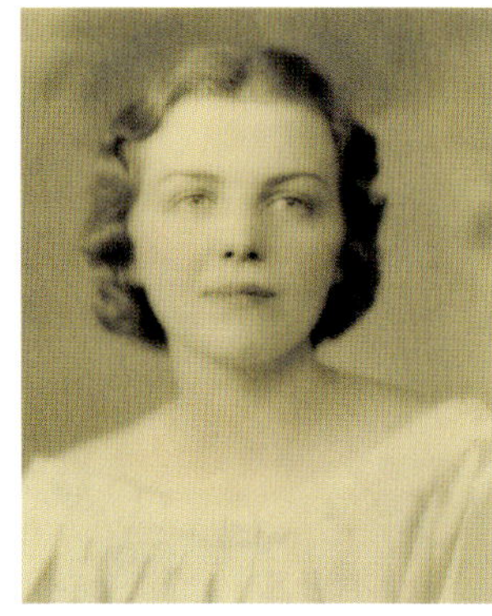

Robert August and Mary Boquine (note 30)

baskets, bowls, and trays, bearing one of at least six shallow, stylized motifs. No formal names for these motifs have yet emerged; however, the motifs that appear on pieces that have been found by collectors include the equivalent of pine cone, sail boat, hummingbird(s) with flowers, flying geese, bittersweet, and a bonsai-style conifer.[39] According to Jack Carruthers, "Salesmen that Gus [Wendell August] had also pushed our stuff but then we went around. I took a trip over to Sharon, Akron, Cleveland, just visiting jewelry stores and department stores and things like that."[40]

However promising the beginning of Pine Grove Craftsman may have seemed in June of 1940, its future had begun to dim within four months. "Lack of Aluminum Cuts Plane Output" screeched a September, 1940 *New York Times* headline, reporting that "half of the airplane manufacturers were . . . being forced to curtail production because of inability to obtain aluminum."[41] By April of 1941, Jack Carruthers had joined the military, and in July of 1941, Pine Grove Craftsman closed its doors for the last time.[42]

The Wendell August Forge, along with the rest of the decorative aluminum industry, was not faring much better. For example, by August of 1941, one company's line was advertised as available only in pewter, and by September of 1941, another decorative aluminum manufacturer was advertising simply "metalware," with no hint of what the metal might be.[43] Anticipating the intensive aluminum drives that would consume some decorative aluminum wares during World War II, *The Saturday Evening Post* carried cartoons emphasizing the need to conserve and collect aluminum

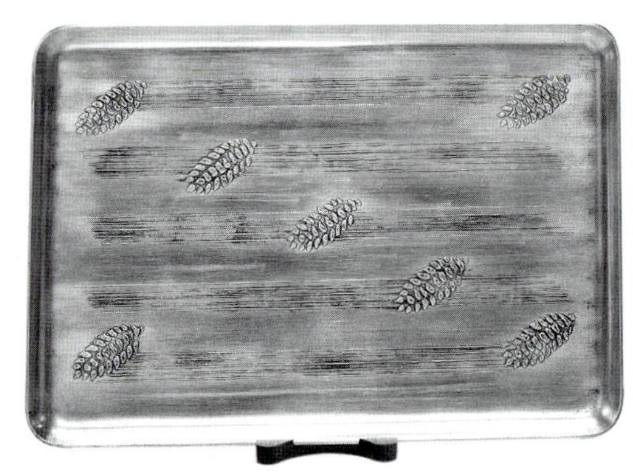
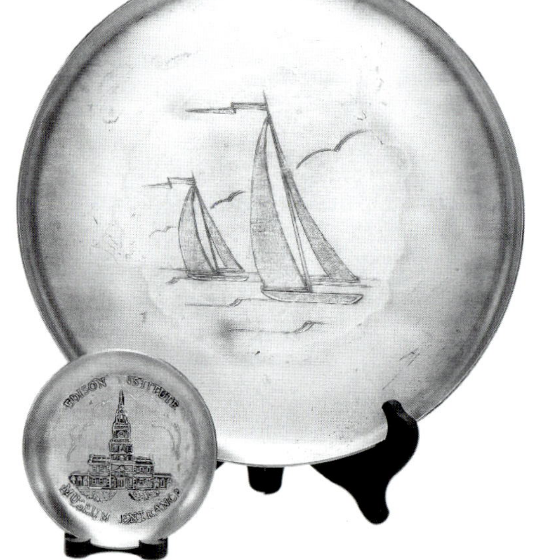

Pine Grove Craftsman pieces marked Wendell (note 37)

for the National Defense Program.[44] Access to aluminum supplies for consumer products was completely terminated in December of 1941, with the declaration of a state of war with the Axis powers by President Franklin D. Roosevelt. Not only were the lines of supply for aluminum cut off completely, but unprocessed aluminum that was stored in inventory was also confiscated within weeks. Unable to acquire aluminum, the Wendell August Forge attempted to produce its wares in pewter, but by early 1942, pewter was also impossible to obtain.[45] "We couldn't get any aluminum but Mrs. Roberts could get pewter," said Natale Rossi, "And then they went to the show and somebody bought it all and we couldn't get any more pewter. We couldn't make any more."[46]

With no metals available for consumer product goods, aluminum giftware manufacturers either closed down or converted to the production of war materials. The Wendell August Forge had not been successful in acquiring war production contracts, and in April of 1942 the Forge closed its doors.[47] Many of the Forge craftsmen reported for military service, and most of the others found employment at Cooper-Bessemer in Grove City, which was expanding its production capacity to support the war effort. Several months earlier, Robert August had already taken a position as a plant and field superintendent with a firm that manufactured and erected prefabricated houses.[48]

In August of 1942, after his concerted efforts to obtain war materials subcontracts failed, Wendell August was finally able to obtain a contract to house and feed military personnel who were being trained at Cooper-Bessemer.[49] According to the *Brockway Record*, "The forge building of steel and concrete, 160 by 160 feet, is to be turned into a dormitory with 100 bunks. The idle forge will be removed at once and the building altered to provide the sleeping accommodations (as well as three class rooms, offices, a projection room, and two rooms for petty officers). In addition, Mr. August will erect a mess hall 32 by 64 feet complete with kitchen and serving facilities on land south of his plant. A sub-contract to supply meals has been

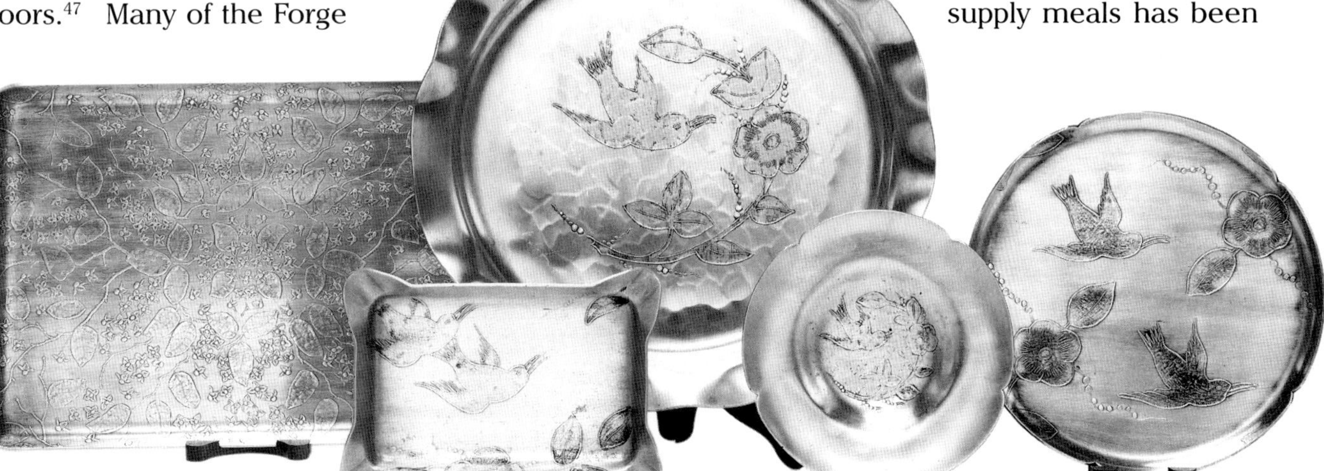

Pine Grove Craftsman pieces marked Pinegrove (note 38)

awarded to C.E. Frampton, of the diner, it is said. Navy and Coast Guard men being trained in Diesel engine operation at the Cooper Bessemer plant will occupy the barracks. For the past year they have been housed in private homes."[50]

From early 1942 until early 1945, it was as though the entire globe was ablaze. The first real signs of hope for the Allies came with the Allied invasion of Italy in early 1944. On D-Day, June 6, 1944, the Allied invasions of Western Europe commenced with landings on the coast of Normandy. With the beginning of these invasions, domestic manufacturers in the United States began looking toward conversion back to peacetime production. By mid-1944, Japan's strength in the Pacific began to ebb, forcing it into a defensive policy, and by late 1944, Germany began losing control of the Eastern Front to Russia. During the last half of 1944, advertisements began to appear in the United States for aluminum art and giftwares that were "scheduled for production soon after critical materials are released" and would be "available soon after the war is concluded."[51] Limited quantities of aluminum giftwares became available in time for the year-end holidays, with advertisments carrying such qualified statements as "Everyone cannot be supplied today; proportions of limited merchandise may seem unfair — but

Edward Deniker

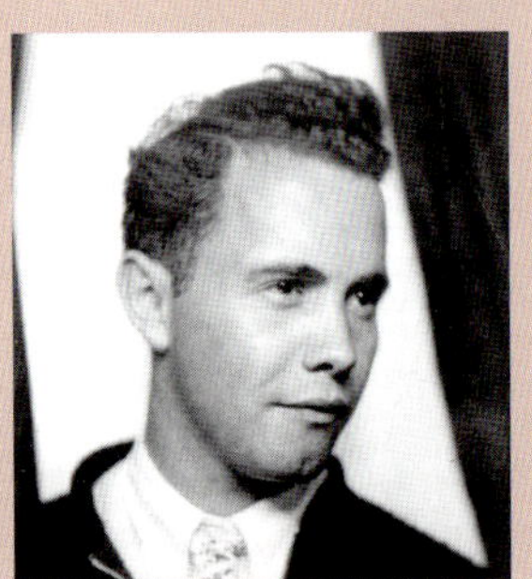

His father's serious injury in a mining accident forced Edward B. Deniker (1913-1983) to leave high school and seek work to help support the family. Deniker's premature departure from school ultimately led to his employment with the Wendell August Forge from about 1932 to 1934, where he gained his initial training as an aluminum artisan. He then joined Arthur Armour and gained additional expertise at Armour's shop from 1934 through 1941.

After World War II, when aluminum once again became available for consumer goods, Deniker initiated his own line of hand-hammered aluminum giftware, which he produced from 1946 until about 1958 at his home in Briscoe Springs. The items that he created were sold primarily throughout western Pennsylvania and eastern Ohio.

Reflecting his gentle manner and love of nature, most of Deniker's work was embellished with delicately-cut repoussé motifs, such as Dogwood with Butterfly *and* Two Ducks with Cattails. *Additional repoussé motifs (e.g.,* Apple Blossom*) were used extensively for cuff bracelets, some of which (e.g.,* Mouse*) were developed for children. When the State of Pennsylvania ceased issuing front license plates for automobiles, Deniker fashioned decorative plates featuring the Pennsylvania Keystone and his* Ruffed Grouse *state bird motif to take their place.*

Deniker also created numerous unique hand-hammered works, including an aquarium, cabinet latches, photograph frames, and a communion set for the Holy Trinity Lutheran Church in Grove City. The communion set, which is still in use, is embellished with repoussé created from a die engraved by Arthur Armour.[54]

we are doing the best we can under circumstances entirely beyond our control" and "Thanks to a recent aluminum release by the W.P.B. [War Production Board], we hope to make *some* . . . items . . . available to our regular outlets *very soon* — possibly even in time for the holiday season."[52] Clearly the backlog in consumer demand, accompanied by the unexpended income from the war years, could expect to find an outlet with domestic manufacturers.

In April, 1945, the sudden death of President Franklin D. Roosevelt shocked the United States and its Allies, but grief over the loss of the President was rapidly buried with the exultation over the final collapse of Germany in May of 1945. With the ability to focus assaults in the Pacific, the war in the Pacific was at long last formally concluded in September, 1945. Although some of the firms engaged in the production of aluminum art and giftwares prior to World War II did not continue in that market after the war, the majority of them reentered the market, and then were joined by an equivalent number of new entrants by the end of 1945.[53] Among the new entrants were former Wendell August Forge metalworkers Edward Deniker,[54] James DePonceau,[55] and, working together, Frank Rossi and Eugene Town.[56] The popularity of metal wares prior to the war combined with their virtual absence on the market during the war to yield a high level of consumer demand for anything made of metal. Incomes during the war years had been high, and the highly restricted access to consumer goods since the latter part of 1941 had ensured that the population had significant savings to expend. Because of the devastation of much of the world, imported metalwares were virtually nonexistent, necessitating that the demand be met by domestic industry. In addition, the mystique associated with aluminum had once again been heightened by the prominent and crucial role that it had played during World War II. The result was an enormous surge of demand for domestically-produced decorative aluminum giftwares.

Wendell August found himself in a difficult situation. Although large quantities of some competing aluminum giftwares had begun appearing in the consumer market by mid-1945,[57] the Forge's production facility was still being used as a barracks, and the most recent lease period was not scheduled to expire until July 1,1946. In addition, Wendell was now 60 years old and in failing health. Since 1937 he had suffered from ulcers, for which he had had major surgery in September of 1944.[58] Both Bob and Mac August, his sons, were serving in the U.S. Navy. By late August of 1945, Wendell was desperate for additional help to get back into full production. Robert and his father petitioned for Robert's release from active duty in order to run the firm, arguing that the Wendell August Forge had orders in hand sufficient to provide employment for many more men and help the local economy, but that Wendell's poor health precluded him from doing so.[59]

Until it could be moved back into its facility, the Wendell August Forge made use of the building that Pine Grove Craftsman had occupied before the war. As former Forge metalworkers were released from war contract work, particularly with nearby Cooper-

James DePonceau

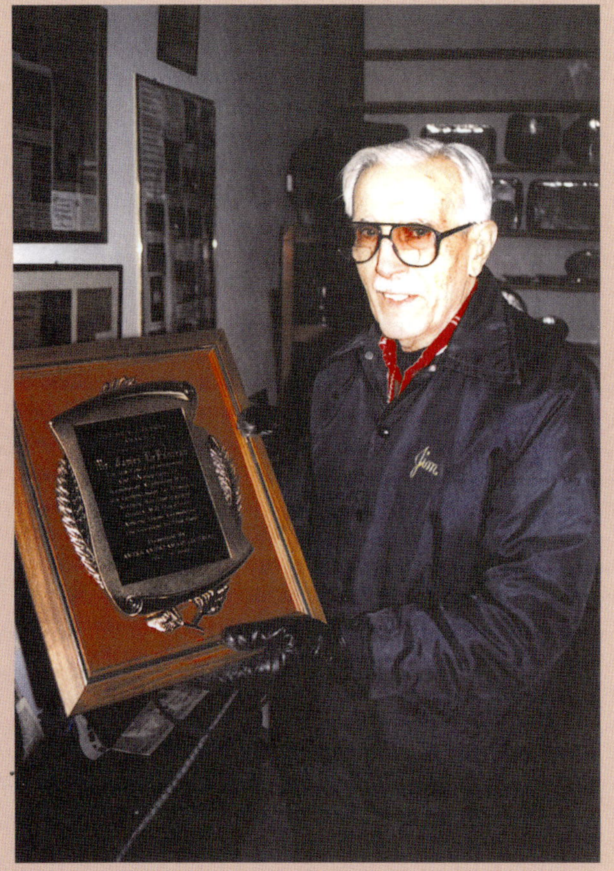

Completing high school in three years at the age of 17, James V. DePonceau (1912-present) found his first job at the Wendell August Forge in Brockway in 1929, while the firm was still engaged in the production of ornamental wrought iron. He was with the Forge when it initiated the development of the hand wrought aluminum industry with its work for Alcoa's Aluminum Research Laboratories, and carbon-colored the decorative motifs on the elevator doors.

DePonceau went with the Forge when it was relocated to Grove City in 1932 and continued there until, in January of 1934, he left to join Arthur Armour. At Armour's, DePonceau was responsible for carbon-coloring all of the pieces made in the shop. He also made or shaped many of the pieces. He remained at Armour's until the United States entered World War II.

As the end of the war approached, DePonceau began preparations for his own business, and on V-J Day in 1945, produced the first items for his own line of hand crafted decorative aluminum, which was carried in gift shops and department stores in several cities in the northeastern part of the United States. He produced his work in Grove City until 1952, when he relocated to Chautauqua, New York. In Chautauqua, he continued to produce decorative art aluminum until 1955, when he launched a completely new career.

At the age of 43, DePonceau undertook a college degree program, earning his B.S. in 1960. He started teaching Special Education courses, while at the same time pursuing his M.S. degree, which he received in 1964. He remained in Special Education, actively engaged in teaching and fund raising for Special Education programs, particularly the Chautauqua County Resource Center, until his retirement in 1977.

DePonceau continues to produce hand crafted decorative art aluminum wares at the DePonceau Aluminum Forge in Chautauqua. Among the many items that he continues to carry in his product line are bowls, candle holders, dinner bells, trays, and waste baskets. His die engravings are primarily of popular floral motifs, such as Daisy, Dogwood, Fuchsia and Primrose. DePonceau has also acquired many of the engraved dies that had belonged to Clayton Sheasley and John Wesley Findley, thus helping to preserve the artistic heritage of the hand wrought aluminum industry.

For his contributions to decorative art aluminum, DePonceau was honored with a Lifetime Achievement Award in 1994. His son, Joseph DePonceau, carries on the tradition at DePonceau Aluminum Craft in Rochester, New York; his grandson, Robert, is also learning the art.[55]

Made in California by Town

Eugene Town (1910-present) and Frank Rossi (1922-present) developed their skills with hand hammered aluminum wares as craftsmen at the Wendell August Forge. Eugene, a brother-in-law of Natale Rossi, had worked for the Forge from 1933 to 1939. Frank Rossi, a younger brother of Natale, had worked there from 1937 to 1942 and from late 1945 to early 1947.

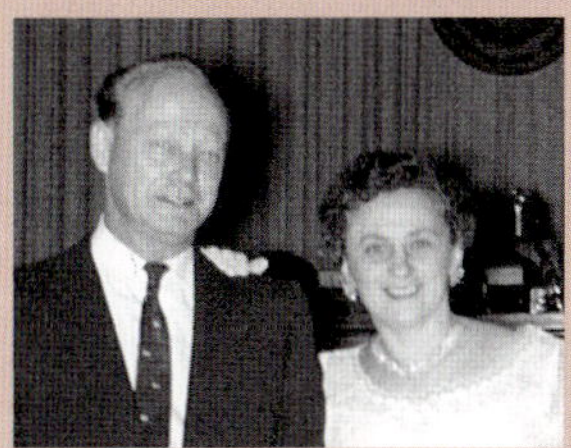

Eugene and Josephine Town

In mid-1946, Eugene Town initiated a hand-hammered aluminum business in Bell Gardens, California. At that time, Eugene had little die engraving experience, and so the dies for the Apple Blossom *and* Pine Cone *motifs were engraved by Natale Rossi in Pennsylvania and shipped to Eugene Town in California. By early 1947, the line had met with sufficient success that assistance was needed, at which time Frank Rossi left the Wendell August Forge to join Eugene Town in Azusa, California.*

Town and Rossi produced a wide range of decorative items bearing the Apple Blossom *and* Pine Cone *motifs, such as album covers, bowls, candle holders, cigarette boxes, napkin holders, tables, trays and waste baskets. They also produced special order advertising items (primarily coasters) and tourist items (primarily trays). Most of the dies for these items were engraved by Eugene Town, of particular interest being* Mt. Hood, Mt. Rainier, Mt. Whitney, San Francisco Turntable Cablecar, *and the California missions* San Carlos del Rio Carmelo, San Francisco de Asis Dolores *and* Santa Barbara. *Town and Rossi ceased production in mid-1949, when aluminum supplies in Southern California became erratic as a result of the Korean War.*

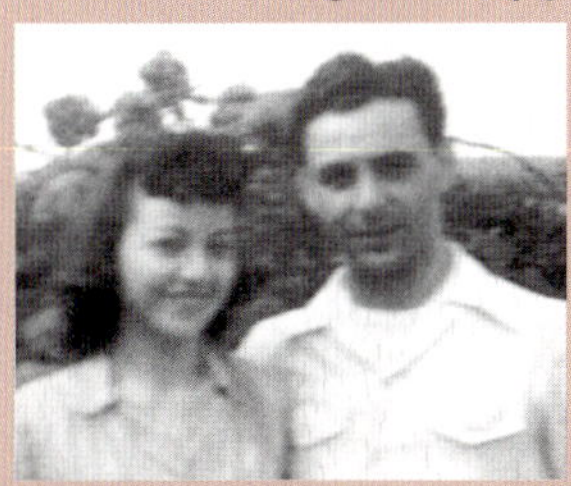

Frank and Lenora Rossi

Work from the Town and Rossi collaboration was exhibited in California State University's Depression Silver: Machine Age Craft and Design in Aluminum.[56]

Bessemer, they returned to work at the Wendell August Forge in its temporary facility, as did many of those in the armed services. Anthony Pompa was among those who, when he was discharged from the service, returned to the Forge. He recalled, "We were at the place over on Edgewood and Erie and I got $1.25 an hour."[60] By October of 1945, when Frank Rossi returned from Europe to the Forge after being discharged from the Army, Bob August had returned to Grove City.[61] The military was closing down as many of its war-time operations as rapidly as possible, and by the end of February, 1946, the plant on Madison Avenue had been vacated by the Navy and Coast Guard. The lease had been terminated before its expiration date, and the Forge could finally begin to convert its facility back to its original use.

By April of 1946, the Wendell August Forge was back in full production at its Madison Avenue plant, albeit with some changes. The ownership participation of Jessie (Mrs. Wendell) August had been severed, and the officers were now Wendell M. August (President), Robert E. August (Treasurer), and Wendell M. August, Jr. (Secretary and Sales Manager).[62] Die engraver Louis Donato did not rejoin the firm, and Natale Rossi took over the engraving responsibilities. Frank

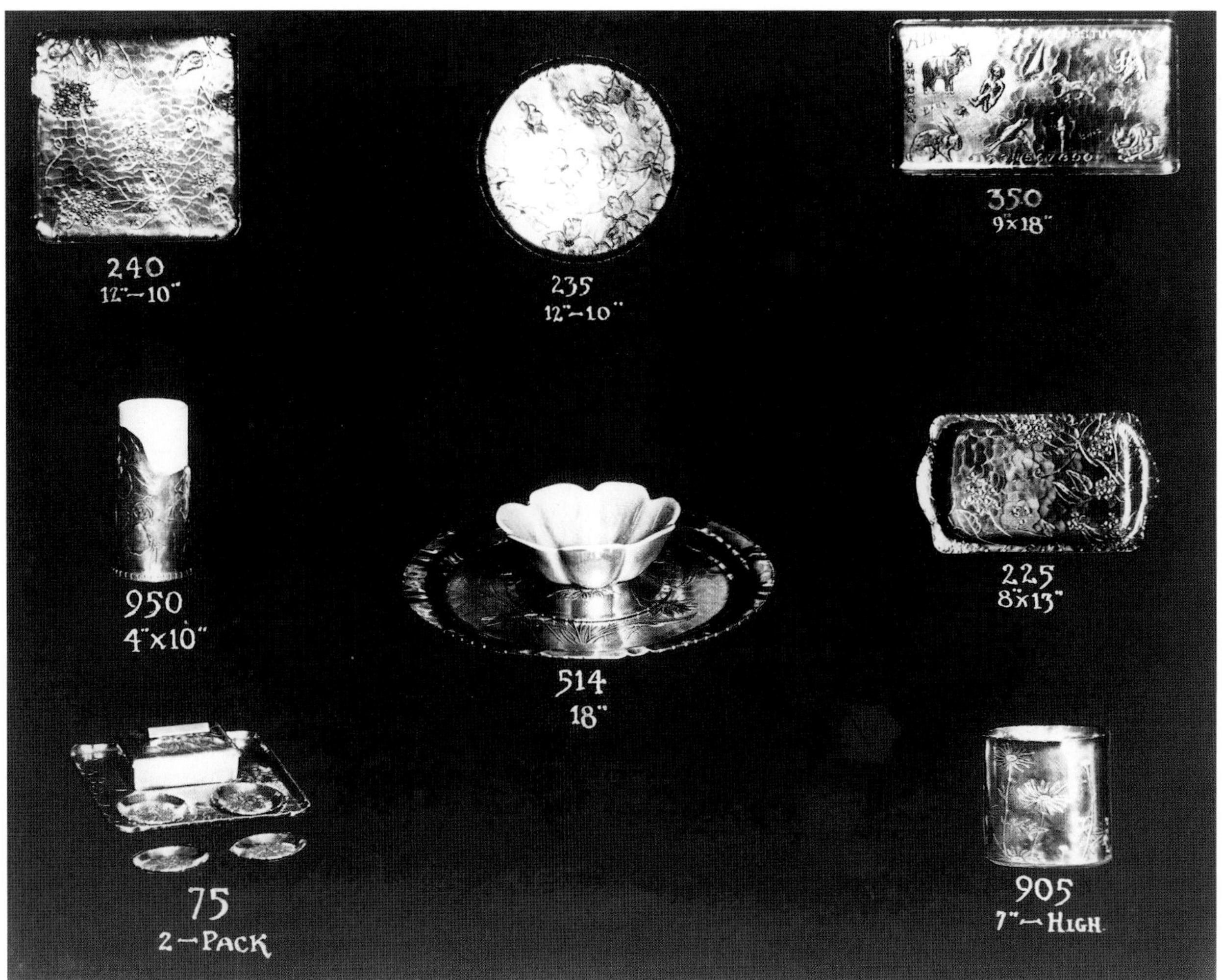

Sample of 1947 Wendell August Forge aluminum giftwares (note 65)

Rossi noted, "After the war, Natale started cutting dies for the commercial stuff."[63] Joseph Ziccardi recalled, "What I heard was that Mr. August told him [Donato] that he didn't need any more dies, that he had enough."[64]

The post-war product was essentially the same as it had been in the early 1940s, its most distinctive difference being the prevalence of tool-marked (notched) edges on the majority of the items rather than just a few. Although a handful of new items incorporating La Mirada pottery pieces were introduced, most of the forms of the Wendell August Forge giftware items remained the same, and the number of motifs available was restricted to fewer than 15.[65]

With demand exceeding the supply of virtually everything, including Wendell August Forge aluminum giftwares, allocations to buyers were required. "Metal items were scarce," wrote Natale Rossi, "so he [Wendell August] would send his customers a quota . . . business was so good."[66] Although the Wendell August Forge seems to have undertaken no advertising, it received national coverage through Alcoa's 1946 advertising series encouraging new businesses and products in aluminum.[67] This series, with its reference to the Forge's architectural work, may have helped to stimulate some of the immediate post-war architectural commissions such as

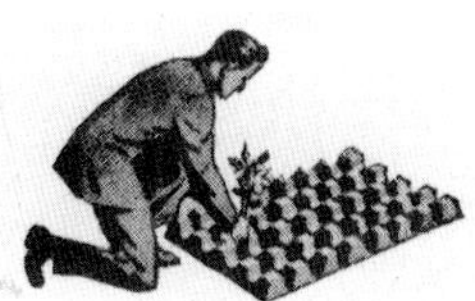

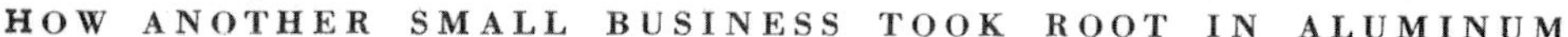

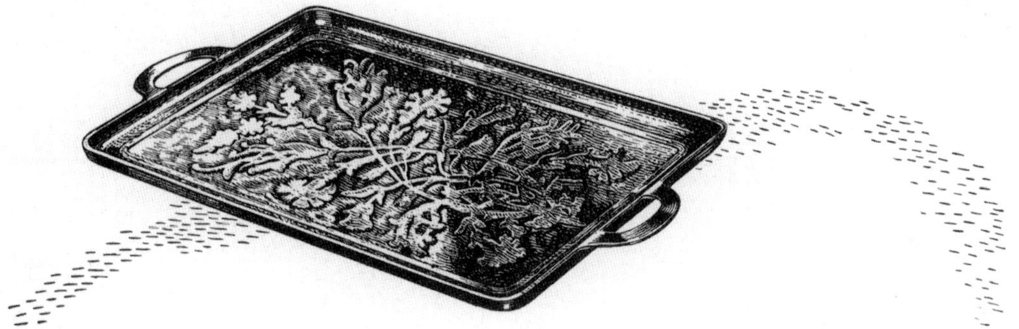

The Gate Begat a Tray;

The Tray Begat a Business

You couldn't exactly say the architect launched the business.

Nor two blacksmiths who made some aluminum gates.

Nor the department store executive who sparked to an idea.

And we of Alcoa were merely building a building.

But today in Grove City, Pa., there's a thriving company that owes its existence to the part each of these played.

The architect's plans for the Aluminum Research Laboratories' new building, in 1929, called for imposing entrance gates made of cast aluminum. Wendell August, blacksmith shop owner, asked for a chance to bid but on the basis of *hand forging* aluminum bars. He got the job and put his two smiths to making the gates.

When the architect saw their beautiful handcraft he suggested hammering out several aluminum serving trays, in repoussé, as souvenirs for some of our executives. Delighted . . . and sensing a new, marketable product . . . they showed their trays to a department store president.

He sparked. An order followed . . . and Wendell August Forge Inc., was started. This was, we believe, the first company in the aluminum giftware business. Since then, many others have gone into it and done well.

Not just in *making* things of aluminum . . . a broad field in which Alcoa offers the richest fund of knowledge in the world . . . but also in *marketing* aluminum products, Alcoa stands ready to assist any manufacturer, large or small, who has a *practical* application for aluminum and needs help to get going. Consult your nearby Alcoa sales office or write ALUMINUM COMPANY OF AMERICA, 2109 Gulf Building, Pittsburgh 19, Pennsylvania.

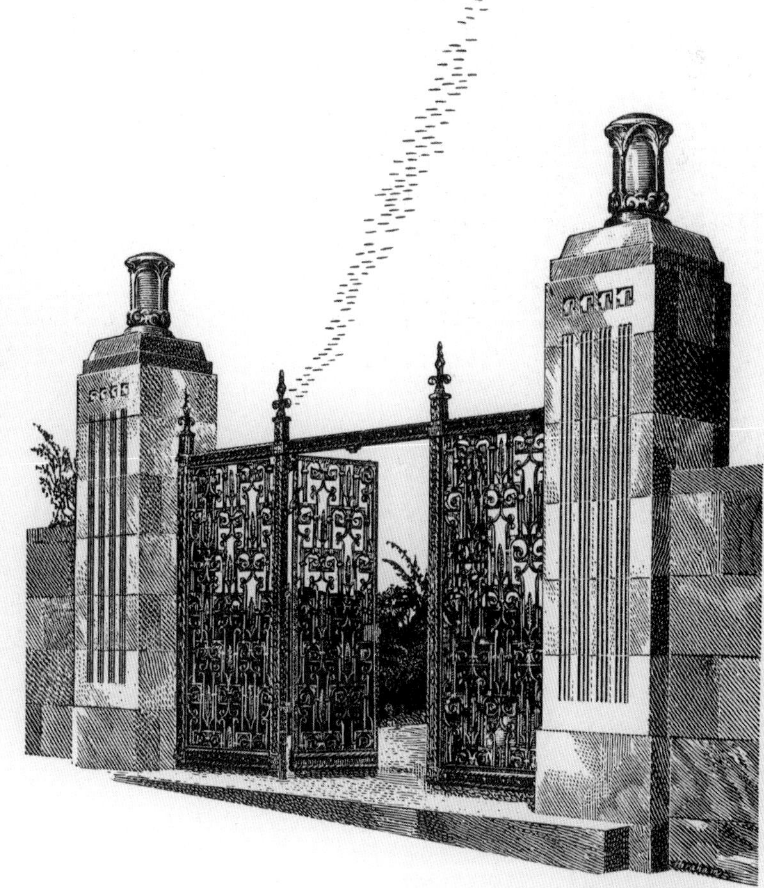

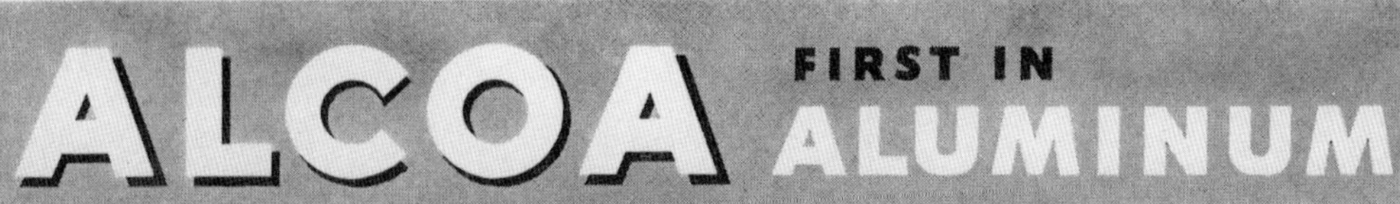

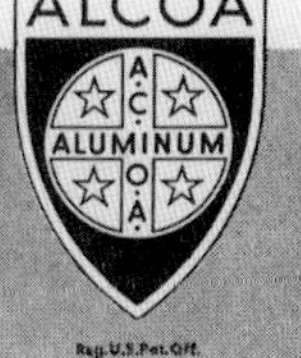

1946 Alcoa advertisement featuring the Wendell August Forge (note 67)

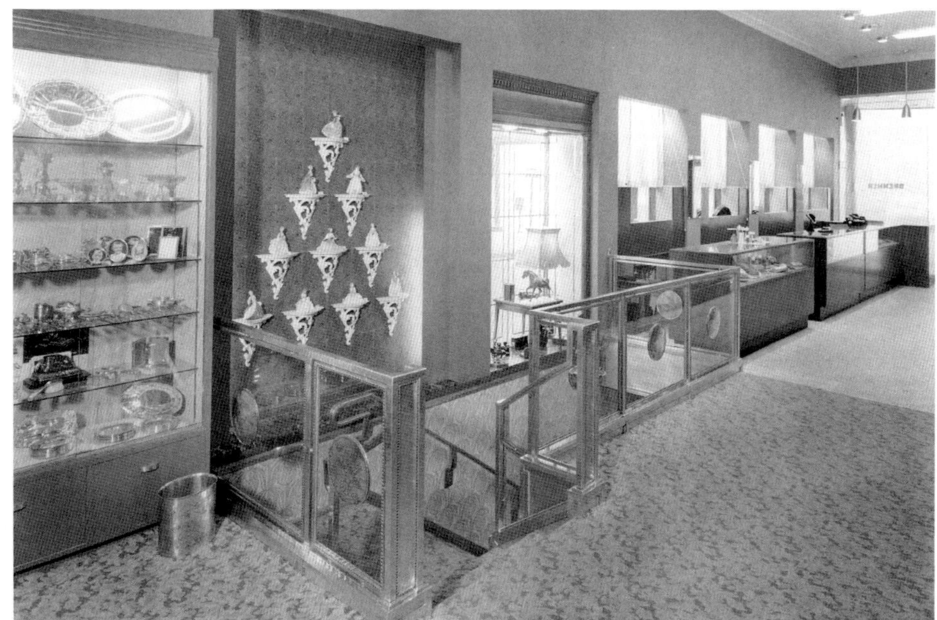

Left, Aluminum railing at Brenner Jewelers
(note 68)

Below, Selection of aluminum jewelry
(note 70)

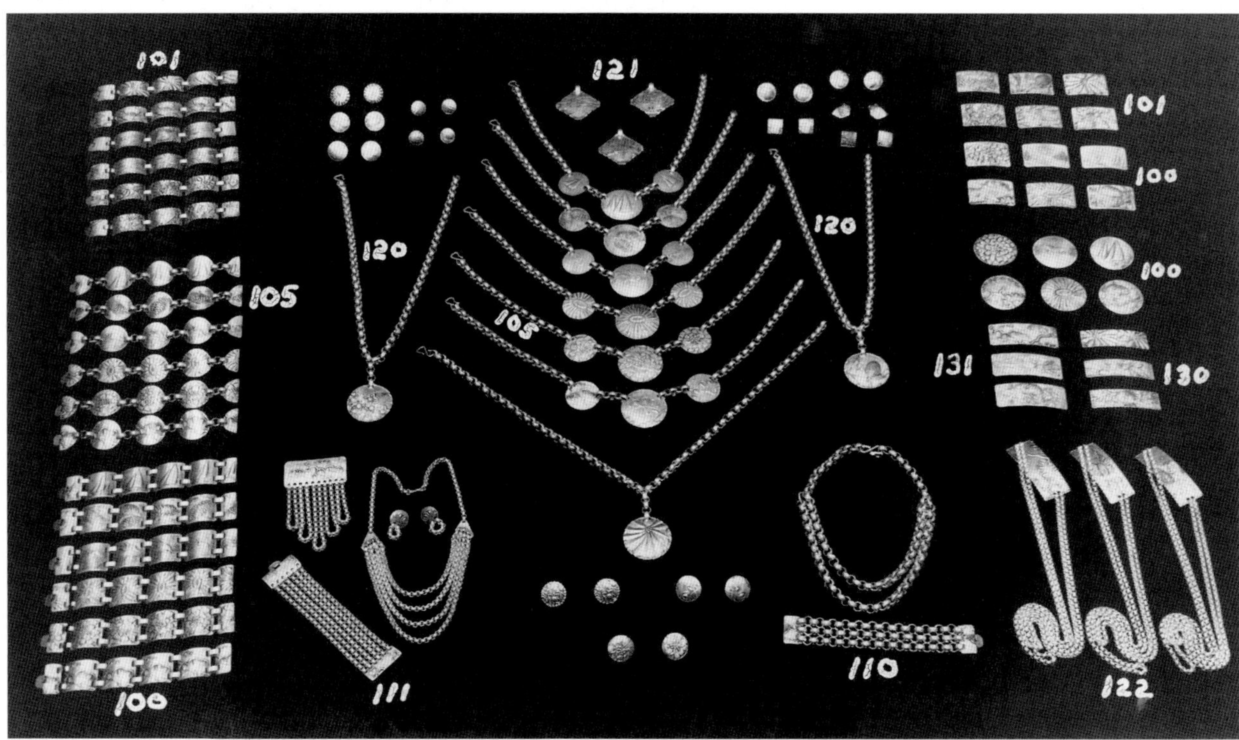

railings for Brenner Jewelers in Youngstown, Ohio[68] and "interior work for 20 banks in 1946 and 1947 alone," which included "railings, gates, wickets, check desks, and lighting fixtures."[69]

Jewelry was in high demand.[70] [71] "For a couple of years before the war, we didn't make much jewelry," remembered Frank Rossi, "but after the war we made a lot. Bill Boyd used to make jewelry, and I made it after the war. Had a little hot plate to warm the pieces on before they [the findings] were soldered. We got the chains on big spools from some company; we'd cut the chain off the spool like it was thread."[72]

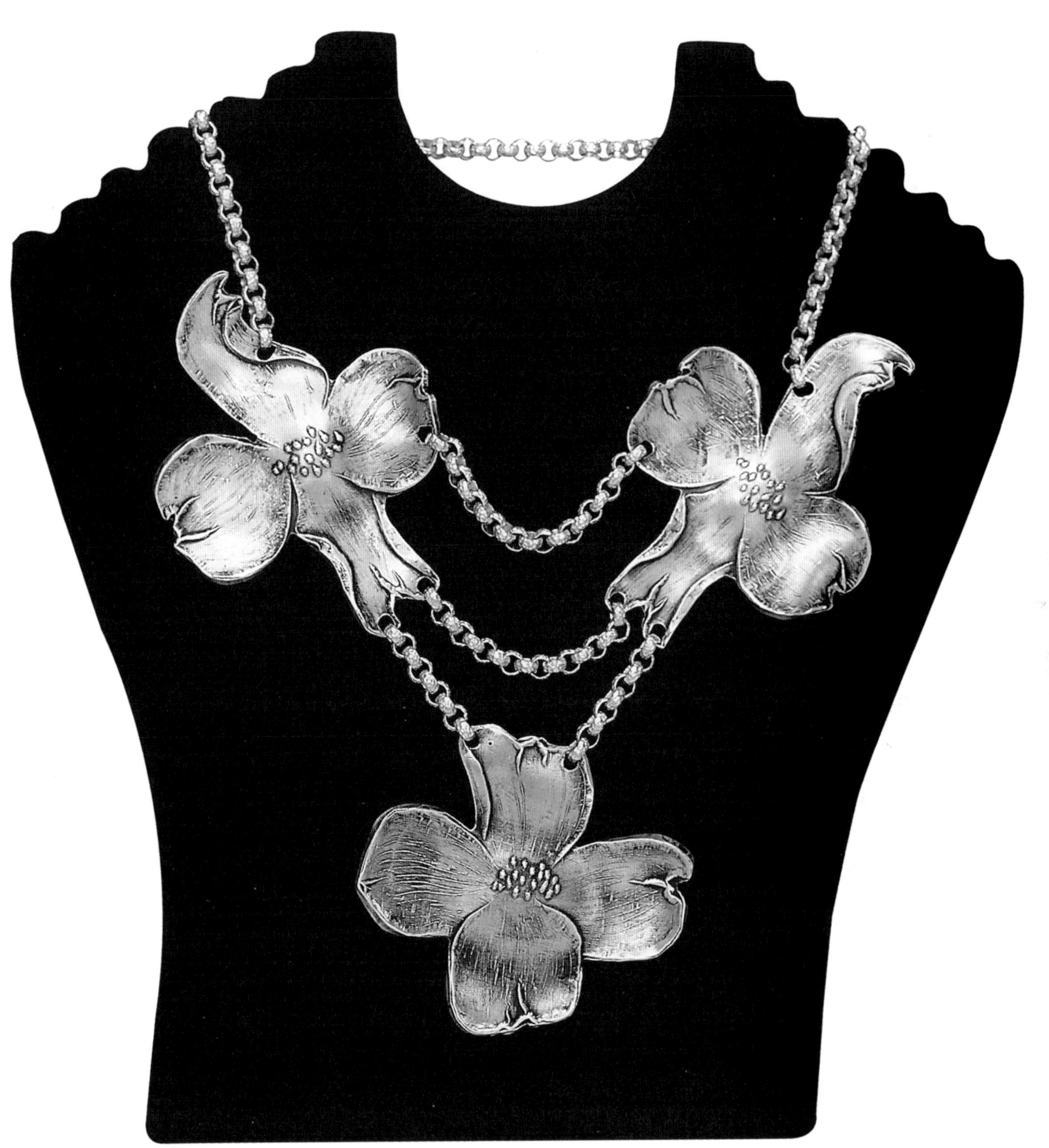

Elaborate Dogwood *necklace (note 71)*

At least three somewhat unusual jewelry-related ventures seem to have taken place in the 1946-1948 time period. One of these ventures involved the production of hand forged aluminum brooches and belts, marked "Platanelle," for H. Pomerantz, Inc. To date, the motifs on the pieces reported by collectors have been traced to dies that were engraved at the Wendell August Forge.[73] A second venture, possibly associated with Pomerantz, involved the production of hand made aluminum buttons for dresses.[74] The third venture involved pieces marked "Hawaiian Patterns." Collectors have reported belts, pins, earrings, and brooches, as well as coasters and trays, bearing motifs of bamboo, hibiscus, and tropical fish.[75]

In addition to assisting his father with the Wendell August Forge, Robert August also started his own firm, REA Metal Products, Inc., in 1946. The company was located next door to the Forge, and its corporate officers were Robert E. August (President), Wendell M. August, Jr. (Secretary), and Wendell M. August (Treasurer).[76] Its primary product was a set of coin trays with a stand for holding coin rolls at teller stations in banks. Each tray was anodized in a color consistent with the colors of the paper wrappers that were used for different coin denominations.[77]

Another product line of REA Metals Products, Inc. consisted of three items (two vase sizes and one planter), readily identified by the floral medallion placed on each one, for use by florists.[78] According to Joseph Ziccardi, "The ones with the riveted medallion, that was shortly after the war. Ralph Trepasso and Fred Todarello worked on them over in the REA building."[79] The vases seem to have been sufficiently labor-intensive as to be unprofitable. According to Natale Rossi, "That was another thing of Bob's. He was in the airplane someplace and the guy next to him said that he sells vases to greenhouses. Millions of them. If you can make a vase cheap enough I can sell a million of them," continuing, "I don't care what you make but we make it our way you gotta weld the bottom and you gotta clean it . . . It didn't work out . . . they were made really cheap but they still cost money to make."[80]

Although the postwar spending frenzy had not yet run its course, new difficulties for domestic manufacturers had begun to emerge within two years of the end of the war. One leading trade magazine noted that "last year . . . with merchants still starved for merchandise and willing to buy practically everything in sight, cannot . . . be used as a gauge for 1947 business," and continued with a portent of the future with "of great interest of course, was the influx of imported ware."[81] Just as domestic producers were completing their conversions from war-time production, so were those of Europe and Asia, many of which had faced the additional task of reconstruction.

By 1948, the post-war paper shortages and the Forge's backlog of orders had eased sufficiently to yield an opportunity to produce a formal catalog and brochure featuring a portion of its post-war product line.[82] [83]

Wendell August, now nearing the age of 65 years, began to make arrangements for his retirement, planning to leave the thriving Wendell August Forge in the care of his sons.

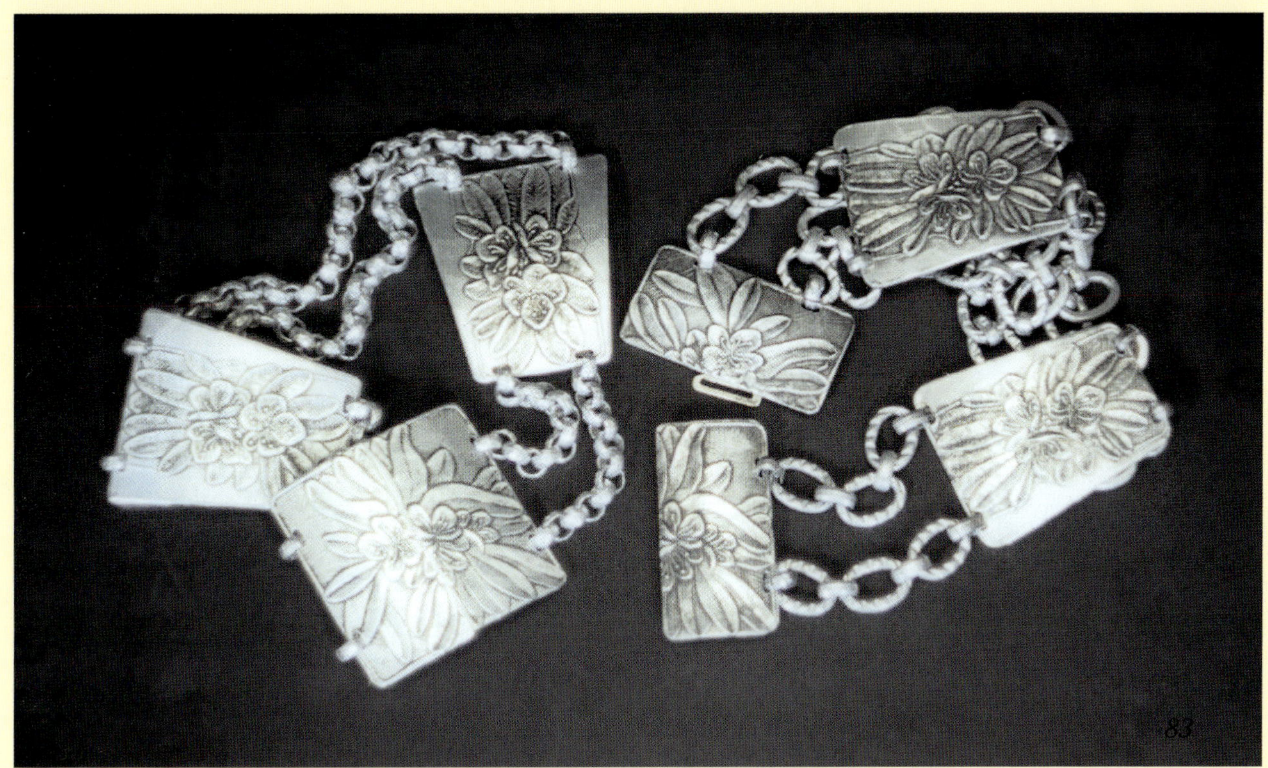

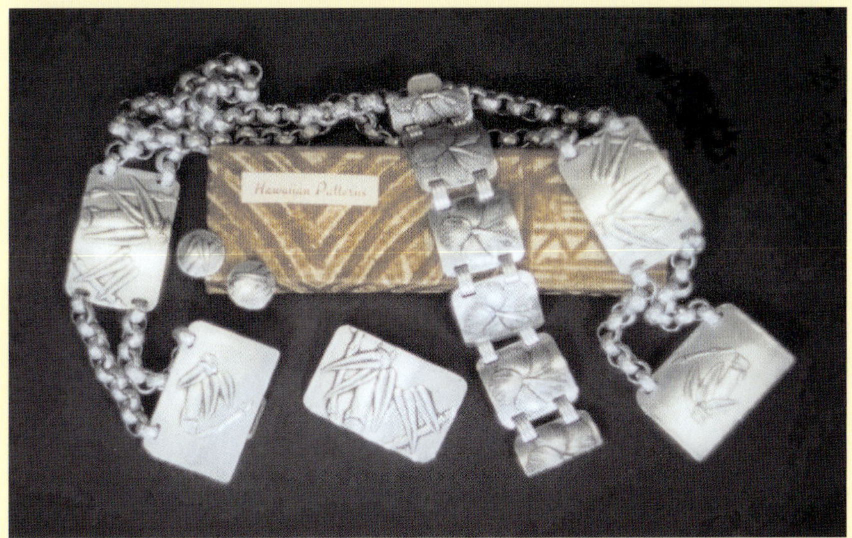

Top, H. Pomerantz, Inc. aluminum belts
(note 73)

Center, Hawaiian Patterns aluminum jewelry and gift box
(note 75)

Bottom, Aluminum planter by REA Metal Products, Inc.
(note 78)

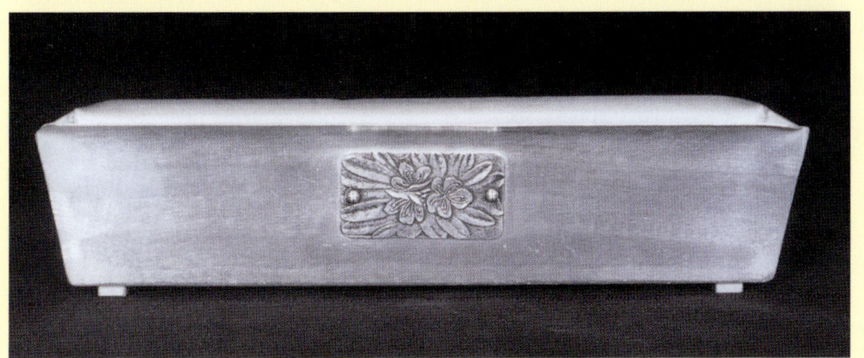

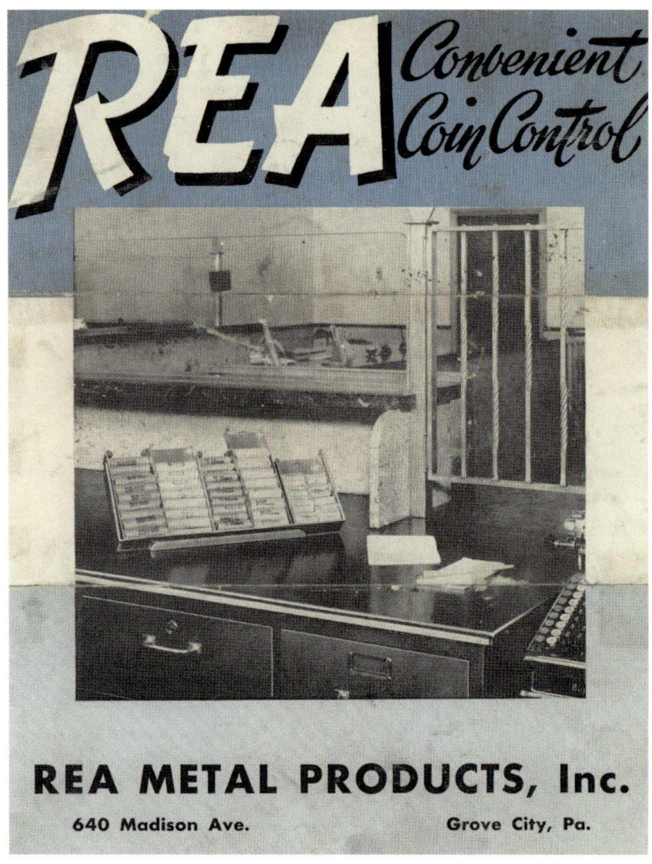

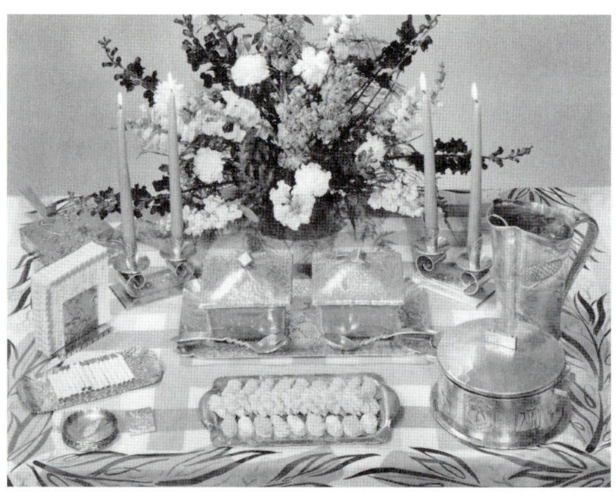

Left, 1948 brochure for REA Metal Products, Inc. coin trays (note 77)

Above, Table setting from 1948 Wendell August Forge catalog (note 82)

Bottom, Sample of 1948 Wendell August Forge aluminum giftwares (note 83)

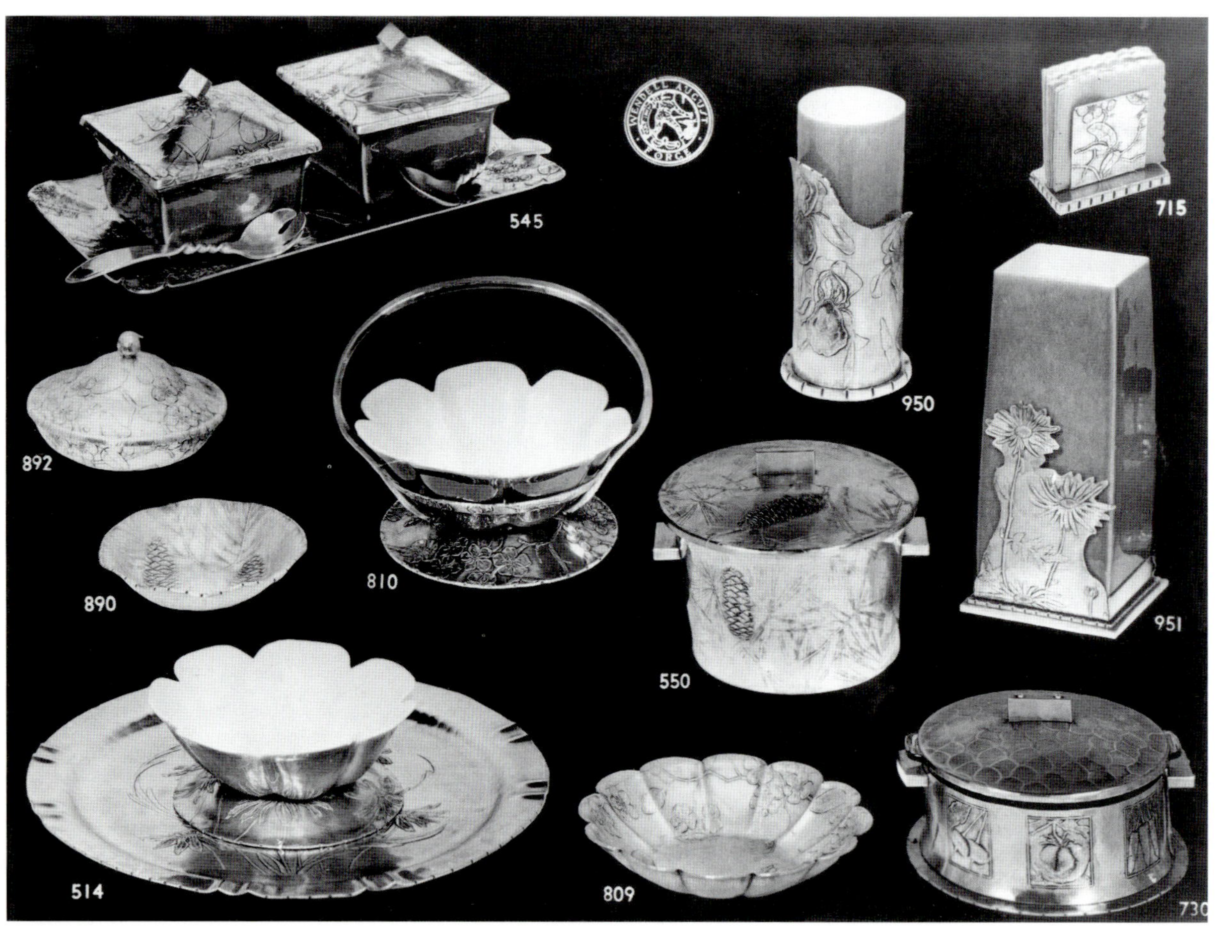

1 Bonita J. Campbell, unpublished aluminum giftware products database.

2 See, e.g., *Official Guidebook of the New York World's Fair* (New York: Exposition Publications, Inc., 1939) and Cheney Archer, "The New York World's Fair - A Major Influence on Design," *The Gift and Art Buyer*, April 1939.

3 Illustrated are: #231 covered bon bon, 7-1/4"x9-1/2", with *Bittersweet* motif; #550 18" diameter *Shell* hors d'oeuvres server; #807 13" diameter star-shape bowl with *Grape* motif. Collection of the Wendell August Forge. Photography by Richard Smaltz, Smaltz Studio, West Middlesex, PA; photograph courtesy of the Wendell August Forge.

4 Illustrated are: #200 14"x25" (including handles) Service Tray with *Pussywillow* motif on #903 Folding Stand, Ball Knobs, "X" hinge, 19" high, used with 200 series service trays; #220 17"x30" (including handles) Service Tray with *Ducks* motif on #905 Folding Stand, Tool Marked, "X" hinge, 19" high, used with 220 series service trays; #205 19" round Service Tray with rope-twist rail handle with *Sailfish* motif and #400 18"x24" oval Service Tray with fish hook handles with *Yacht* motif, both on #913 Folding Stand, Capstan Knobs, Twisted Legs, "X" hinge, 19" high, used with 205 and 400 series service trays. Reverse side of original photograph marked "Wendell August Forge, Grove City, Pa., Sold exclusively by Janis-Tarter, Greeman, Inc., 225 Fifth Avenue, New York City." Photograph courtesy of the Wendell August Forge.

5 Illustrated are: #79 cigarette box, glass box, aluminum lid and 5-1/2"x6" tray with *Bittersweet* motif; #85 5-1/2" *Shell* bon bon; #535 5-1/2"x12-1/2" celery tray with *Bittersweet* motif; #536 5-1/2"x12" celery tray with a *Wheat* motif; #580 mayonnaise and marmalade set, divided glass bowl and glass spoons, aluminum cover and 8" plate with *Bittersweet* motif; #581 mayonnaise set, glass bowl and glass spoon, 7" aluminum plate with *Bittersweet* motif; #591 marmalade set, 7-oz glass and glass spoon, aluminum cover and 6" plate with *Bittersweet* motif; #865 7" diameter covered candy box with *Bittersweet* motif; #866 7"x7-1/2" covered candy box, Heisey *Crystolite* glass, "heart leaf" aluminum cover with *Bittersweet* motif. Reverse side of original photograph marked "Wendell August Forge, Grove City, Pa., Sold exclusively by Janis-Tarter, Greeman, Inc., 225 Fifth Avenue, New York City." Photograph courtesy of the Wendell August Forge.

6 The pottery pieces marked La Mirada were produced by American Ceramic Products in Los Angeles. The firm, which had recently been acquired by Thomas Hamilton, had been known as La Mirada Potteries under its previous owner, Cecil Jones. The La Mirada name was continued on the firm's artwares, using the crackle glaze that Jones had developed. In about 1947, the glaze was changed, eliminating the crackle effect. The earlier of the La Mirada pottery items incorporated in the Wendell August Forge line have a blue-green crackle glaze on both the exterior and the interior. The crackle effect is absent from the later pieces, the exterior color has a slightly more greenish cast, and the interior is a cream color. Although several of the aluminum forms and motifs remained in use for many years, the use of La Mirada pottery in the Wendell August Forge product line was being phased out by late 1950. Some insight into La Mirada Potteries and American Ceramic Products can be obtained from Jack Chipman's *Collector's Encyclopedia of California Pottery* (Paducah, KY: Collector Books, 1992).

7 Illustrated are: #415 14"x18" Buffet Service tray with *Bittersweet* motif and two "Blue Pottery Dishes;" #547 8"x12" Hors D'oeuvre Tray with *Apple Blossom* motif and "Blue Pottery Dish;" #845 Flower Set, tray with *Apple Blossom* motif and "Blue Pottery Frog;" #846 Flower Set, 13"x17" base with *Bittersweet* motif and 9" square "Blue Pottery Dish" and 10" "Pottery Chinese Figure;" #876 15-1/2" diameter, two-gallon punch bowl and ladle with a *Grape* motif. Photograph courtesy of the Wendell August Forge.

8 Illustrated are: #810 8" diameter aluminum basket with *Apple Blossom* motif and #43 9-1/4" diameter scalloped La Mirada pottery bowl (4-1/2" tall from base of holder to top of bowl, 8-1/2" tall overall including handle); #841 aluminum vase holder, 3"x5" base, with "Peacock" motif and 7" tall La Mirada pottery vase; #951 aluminum vase holder with *Apple Blossom* motif and 11-1/4" tall La Mirada pottery vase (5"x3-1/2" at base and 4-1/2"x3-1/8" at top); #955 wall pockets, 10"x12" aluminum holders with *Pine* and *Bittersweet* motifs and 4-3/8"x6-1/4" shell-shape La Mirada pottery; #950 aluminum vase holder with *Iris* motif and La Mirada pottery vase (overall 4-1/2" diameter base and 9" tall). The "Peacock" vase is of pre-World War II design and manufacture. The remaining items are of late 1940s manufacture; however, the wall pockets appear to be of earlier design. Collection of Dennis and Rita Wildnauer, photographs courtesy of Dennis Wildnauer.

9 #840 aluminum vase holder with *Ducks* motif and La Mirada pottery vase, overall measurements 2-1/2"x5" and 7-1/2" tall; Collection of the Wendell August Forge. Photography by Richard Smaltz; photograph courtesy of the Wendell August Forge.

10 Frank Rossi, personal interview, 12 June 1996.

11 Frank Rossi, personal interview, 18 Dec. 1994.

12 Frank Rossi, personal interviews, 18 Dec. 1994, 12 June 1996 and 9 July 1997.

13 Joseph Leone, personal interview, 24 July 1997.

14 Janis-Tarter, Greeman, Inc. advertisement, *The Gift and Art Buyer*, Dec. 1939.

15 James A. McCausland, untitled and handwritten manuscript, 1939; courtesy of the Wendell August Forge.

16 F. Rossi, 12 June 1996.

17 Leone, 24 July 1997.

18 "European War Developes [sic] New Market for Pennsylvania Coal," *Grove City Reporter-Herald*, 17 Nov. 1939.

19 "Cooper-Bessemer Directors Meet, Declare Dividends," *Grove City Reporter-Herald*, 29 Dec. 1939.

20 "Worried About Giftwares' Future?," *The Jewelers' Circular-Keystone*, Jan. 1940.

21 Andrew Duane, "Chicago Features Merchandise Sources for a New American Era," *The Gift and Art Buyer*, July 1940.

22 J. Richard Iander, "Art in Metal," *The Jewelers' Circular-Keystone*, June 1940.

23 Campbell, giftware products database.

24 Wendell August Forge workforce outside of the Grove City facility; *circa* 1940 photograph courtesy of the Wendell August Forge.

25 A.D. LeMonte, "Business Built on Aluminum Art Work," *The Youngstown Vindicator*, June 9, 1940.

26 Anthony J. Pompa, personal interview, 25 July 1997.

27 F. Rossi, 9 July 1997.

28 Frank Rossi, personal correspondence, 6 Feb. 1998.

29 Pompa, 25 July 1997.

30 Robert E. August, *circa* 1944, and Mary (Boquine) August, 1937; photographs courtesy of M.C. (August) Taylor.

31 The name, Pinegrove Company, and the start date, appear in Robert Edward August's *Bureau of Naval Personnel Officer Qualifications Questionnaire* dated 29 Mar. 1945; copy courtesy of M.C. (August) Taylor. The name, Pine Grove Craftsman, appears in the transcription of an interview of John S. "Jack" Carruthers, conducted by Catherine Youngo of the Wendell August Forge on 23 Feb. 1994, and it also appears in *Polk's Grove City Classified Business Directory*, where the company is described as "Metal Ware Manufacturers," located at 401 Erie. The directory identifies (in alphabetical order) Wendell August, Wendell August, Jr., Robert August, John Carruthers, and James McCausland, as being affiliated with the company. After World War II, the property at 401 Erie was conveyed by Wendell August to Howard N. and Eleanor B. Kelly on 11 Apr. 1946.

32 Carruthers, 23 Feb 1994.

33 Catherine Youngo, notes from unrecorded interview with Natale Rossi, 1 Sept. 1992; courtesy of the Wendell August Forge.

34 Natale Rossi, as quoted in Catherine Youngo, transcription of recorded interview, 19 Nov. 1992, courtesy of the Wendell August Forge.

35 Carruthers, 23 Feb. 1994.

36 Carruthers, personal interview, 4 Aug. 1997; Carruthers, 23 Feb. 1994.

37 Illustrated are: 8"x11-1/2" rectangular tray with "pine cone" motif; 4-1/4" diameter coaster depicting "Edison Institute Museum Entrance;" 12" diameter tray with "sail boat" motif. All items marked WENDELL on the reverse side. Collection of the Wendell August Forge. Photography by Richard Smaltz; photograph courtesy of the Wendell August Forge.

38 Illustrated are: 8-1/4"x11-1/2" rectangular tray with "bittersweet" motif; and, all having variations of the "hummingbird(s) with flowers" motif, #175 4-1/4"x6-1/8" rectangular ash tray, #350 11-1/2" diameter tray with fluted edge, #155 6" diameter bowl with scallop-cut edge, #450 7-3/4" diameter tray with scallop-cut edge. All items marked PINEGROVE on the reverse side. Collection of the Wendell August Forge. Photography by Richard Smaltz; photograph courtesy of the Wendell August Forge.

39 Based on: transcriptions of Catherine Youngo's recorded interviews with Lewis "Doc" Rossi (22 March 1994), N. Rossi (19 Nov. 1992), and J. Carruthers (23 Feb. 1994); "Wendell" and "Pinegrove" items in private collections and the Collection of the Wendell August Forge.

40 Carruthers, 23 Feb. 1994.

41 Louis Stark, "Lack of Aluminum Cuts Plane Output," *New York Times*, 22 Sept. 1940.

42 The military service date is based on Carruthers, 4 Aug. 1997; the end date for Pine Grove Craftsman is based on R. August, *Officer Qualifications Questionnaire*, 29 Mar. 1945.

43 A Buenilum line in pewter appeared in *The Gift and Art Buyer*, Aug. 1941, and Everlast Metal Products Corporation advertised an unspecified "Crafted Everlast Metal" in *The Gift and Art Buyer*, Sept. 1941.

44 Tony Barlow, "Dump Aluminum Here," *The Saturday Evening Post*, 27 Sept. 1941, p. 47; Ed Nofziger, "Nix on that one — it's aluminum," *The Saturday Evening Post*, 18 Oct. 1941, p. 37.

45 Information regarding pewter items is sketchy. A Wendell August Forge announcement about the availability of pewter items is dated January 1942. The motif found on the few pewter pieces that have surfaced appears on an aluminum card tray as a *Hobnail* motif in one of the 1950 Wendell August Forge catalogs. The *Hobnail* motif also appears on a *circa* 1936 aluminum cigarette box. Two undated photographs depict pieces, identified as being pewter and bronze, bearing the *Hobnail* motif. The forms of these pieces are consistent with those of the aluminum items produced in the 1939-1941 time period.

46 N. Rossi, 19 Nov. 1992.

47 Letter of 25 Aug. 1945, written by Wendell M. August, requesting the discharge of Robert E. August from active duty in the United States Navy, and submitted to The Chief of Naval Personnel; courtesy of M. C. (August) Taylor.

48 According to Robert August's *Officer Qualifications Questionnaire*, he held these positions with Johnson Housing, Inc., of Sharon, PA, doing "plant layout, scheduling, cost and material estimating and direct supervision of manufacturing processes, and supervision of erection of houses" from July 1941 to Dec. 1943.

49 The property was actually leased directly to the Cooper-Bessemer Corporation. The first lease commenced on 19 Aug. 1942, and the lease was renewed several times during the war. The last lease renewal period was originally for 1 July 1945 to 1 July 1946; however, it was terminated 28 Feb. 1946, thereby enabling the Wendell August Forge to resume production in its original plant. Based on lease documents provided by the Wendell August Forge.

50 "Wendell August Gets Army Contract," *Brockway Record*, 21 Aug. 1942, microfilm transcription courtesy of Adam Sabatose, Brockway Area Historical Society.

51 From West Bend Aluminum Co. advertisements, *The Gift and Art Buyer*, July and Oct. 1944.

52 From Everlast Metal Products Corp. and Kromex Corp. advertisements, *The Gift and Art Buyer*, Nov. 1944.

53 Campbell, giftware products database.

54 Edward Deniker, *circa* 1938 photograph courtesy of Bonnie J. (Deniker) Adams. Information sources include Bonita J. Campbell, *Edward B. Deniker: Hand Crafted Aluminum*, research monograph, 1998, 20 pp.

55 James V. DePonceau, 1994 photograph courtesy of James V. DePonceau. Information sources include James V. DePonceau, personal interviews, 17 June 1997, 27 July 1997, 19 Jan. 1998, 21 Feb. 1998; personal correspondence, 20 and 25 June 1997, 10 July 1997, 2, 14 and 18 Sept. 1997, 10 Feb. 1998.

56 Frank and Lenora (Isacco) Rossi, *circa* 1947, and Eugene and Josephine (Rossi) Town, *circa* 1950; photographs courtesy of Lenora (Isacco) Rossi. Information sources include Bonita J. Campbell, *Eugene Town and Frank Rossi: Hand Made Aluminum Made in California by Town*, research monograph, 1996, 38 pp.

57 Decorative aluminum giftwares from Everlast Metal Products Corp., Lehman Brothers Silverware Corp., and Yorkville Craftsmen were among those already back in production and being advertised in *The Gift and Art Buyer* in July and August of 1945.

58 Letter of certification of condition of Wendell August, by R. H. McDonald, M.D., Cleveland Clinic, Cleveland, Ohio, 22 Aug. 1945; courtesy of M. C. (August) Taylor.

59 Memorandum requesting release from active duty, submitted by Ensign Robert E. August to The Chief of Naval Personnel, dated 29 Aug. 1945, with supporting letters and documents; courtesy of M. C. (August) Taylor.

60 Pompa, 25 July 1997.

61 F. Rossi, 12 June 1996.

62 *Polk's Grove City Classified Business Directory*, 1946; directory search courtesy of Gwen Oberlin.

63 F. Rossi, 12 June 1996.

64 Joseph Ziccardi, personal interview, 5 Aug. 1997.

65 Illustrated are: #75 two-pack smoking set with *Apple Blossom* motif and La Mirada pottery box; #225 8"x13" tray with *Bittersweet* motif (also available in *Apple Blossom, Dogwood* and *Pine*); #235 10" or 12" round tray, available singly or as a two-size set, with *Dogwood* motif (also available in *Bittersweet, Cactus, Apple Blossom* and *Pine*); #240 10" or 12" square tray, available singly or as a two-size set, with *Bittersweet* motif (also available in *Apple Blossom, Cactus, Dogwood* and *Pine*; #350 9"x18" baby tray with *Educational* motif; #514 salad set, 18" diameter tray with *Daisy* motif and La Mirada scalloped bowl; #905 7" wastebasket with *Daisy* motif (also available in *Apple Blossom, Bittersweet, Dogwood* and *Pine*); #950 vase with *Iris* motif and La Mirada insert. 1947 photograph and price list courtesy of the Wendell August Forge.

66 Natale Rossi, "The Conception of a New Business," manuscript, 1988, courtesy of Thomas Armour and Wendell August Forge.

67 Alcoa ran a sequence of advertisements to promote the creation of new firms and products using Alcoa aluminum. One of these advertisements featured the Wendell August Forge and appeared in *American Metal Market*, 5 Feb. 1946; *Daily Metal Reporter*, 12 Feb. 1946; *Wall Street Journal*, 4 March 1946; *Chicago Journal of Commerce*, 11 March 1946; *New York Journal of Commerce*, 18 March 1946; and *Manufacturers Record*, March 1946. Advertisement courtesy of the Wendell August Forge.

68 Brenner Jewelers railing, Youngstown, Ohio; *circa* 1946 photograph courtesy of Henry Swyers and the Brockway Area Historical Society.

69 Wendell August Forge brochure, 1948, courtesy of the Wendell August Forge.

70 Motifs that can be discerned on the jewelry depicted here are *Sail Boat, Pine Cone, Horse and Rider, Oriental Tree, Zinnia, Bittersweet, Apple Blossoms, Wistaria, Italian Floral, Thistle,* and *Shasta Daisy*. Although the photograph is included in a *circa* 1946 sales binder, research data suggest that it was taken before World War II. Photograph courtesy of the Wendell August Forge.

71 Necklace bearing post-World War II Wendell August Forge mark; each *Dogwood* flower is approximately 2-1/2"x3". Collection of the Wendell August Forge. Photography by Richard Smaltz; photograph courtesy of the Wendell August Forge.

72 F. Rossi, 12 June 1996.

73 Two H. Pomerantz, Inc. "Platanelle" belts, with identical hammered aluminum panels and different links. The two panels that hook together at the front of the belt are each 1-1/2"x2-7/8"; the wedge-shaped pieces are each 3" long, 2-1/2" wide at one end and 1-1/4" wide at the other. Collection of William Freeman; photograph courtesy of William Freeman. Three "Platanelle" motifs have thus far been documented as definitely being produced by the Forge: the head of a lion-like animal, a floral similar to a mountain laurel (shown on the belts), and a grape vine with triangular groupings of grapes. H. Pomerantz, Inc. appeared in city directory searches from 1943 to 1947 at 142 West 36th in New York City, the business of the firm being "buckles" and "buttons." Herman Pomerantz also appears in the *Official Gazette* of the U.S. Patent Office, with design patents in 1944, 1946 and 1947 for a buckle and two brooches, none of which are likely to have been Wendell August Forge products. Although it is possible that the "Platanelle" items were produced before World War II, available data suggest a 1946-1947 time frame for their manufacture.

74 As indicated in Note 73, one aspect of the business of the Pomerantz firm involved buttons. In a partially-decipherable portion of Catherine Youngo's 19 Nov. 1992 recorded interview of Natale Rossi, Rossi mentions dies that he engraved for buttons for expensive dresses. Wendell August Forge die photographs depict several items that appear to be buttons.

75 Illustrated are: 7-1/4" long bracelet, 1-1/4" square links, with "hibiscus" motif; 3/4" diameter earrings with "bamboo" motif; pin, 1-1/2"x2-1/4" with "bamboo" motif; belt, 25-1/2" long, four 1-1/2"x2-1/4" panels with "bamboo" motif; original "Hawaiian Patterns" jewelry box, 2-1/4"x 7-1/4". Collection of William Freeman; photograph courtesy of William Freeman. The reasons underlying the "Hawaiian Patterns" mark have not yet been determined. One of these jewelry items bears both a "Hawaiian Patterns" mark and a Wendell August Forge mark that came into use shortly after World War II. Coasters and trays with a "tropical fish" motif have been reported with the notched edges used by the Forge *circa* 1945-1952. The "bamboo" and "hibiscus" motifs seem to have been engraved by Natale Rossi in the immediate post-war period; the "tropical fish" motif appears to be from dies engraved by Louis Donato *circa* 1934.

76 REA Metal Products, Inc., engaged in the business of "Metal Stamping," appears in *Polk's Grove City Classified Business Directory* from 1946 through 1955 with the officers as stated, at which time the formal participation of Wendell M. August, Jr. was discontinued. Wendell M. August then became Secretary/Treasurer until his death in 1963, and then Mary (Mrs. Robert) August assumed the combined position.

77 The REA Metal Products, Inc. coin trays brochure is dated June 1, 1948. Although the brochure states "Patent Applied For," no patent or application for same have yet been found. Brochure courtesy of the Wendell August Forge.

78 Aluminum planter, 4-1/2"x13-1/2"x3-1/4" tall, by REA Metal Products, Inc. Collection of William Freeman; photograph courtesy of William Freeman.

79 Joseph Ziccardi, personal interview, 12 Oct. 1997.

80 N. Rossi, 19 Nov. 1992.

81 Madeline Love, "Gift Chats," *The Jewelers' Circular-Keystone*, March 1947.

82 Illustrated are: #5 4-1/2" coasters with *Pine* motif; #18 match pad cover with *Pine* motif; #55 4"x9" cigarette tray with *Apple Blossoms* motif; #96 6"x7" rectangular silent butler and scraper with *Apple Blossoms* motif; #225 8"x13" tray with *Dogwood* motif; #545 vegetable service with servers, two 6-1/4" square La Mirada dishes, 10"x18" aluminum tray and lids with *Apple Blossoms* motif; #715 napkin holder, 2"x5" base, 4" tall, with *Apple Blossoms* motif; #730 covered casserole and hot mat with *Vegetables* motif and 8" Pyrex dish; #760 9" tall, 2-qt. water pitcher with *Pine* motif; #840 double candlesticks, 3-1/2"x6" base, with *Apple Blossoms* motif. Product data based in part on 1948 Wendell August Forge brochure, courtesy of the Wendell August Forge. Photograph and 1948 Wendell August Forge catalog, dated 20 Apr. 1948, consisting of spiral-bound 8"x10" photographs, courtesy of M.C. (August) Taylor.

83 Illustrated are: #514 salad set, 18" diameter tray with *Shasta Daisy* motif and La Mirada scalloped bowl; #545 vegetable service with servers, two 6-1/4" square La Mirada dishes, 10"x18" aluminum tray and lids with *Bittersweet* motif; #550 covered ice bucket, "holds two trays of cubes," with *Pine* motif; #715 napkin holder, 2"x5" base, 4" tall, with *Apple Blossoms* motif; #730 covered casserole and hot mat with *Vegetables* motif and 8" Pyrex dish; #809 11" diameter scalloped and fluted bowl with *Apple Blossoms* motif; #810 aluminum footed basket holder with bail handle and *Apple Blossoms* motif, and 9" diameter La Mirada scalloped bowl; #890 7-1/2" diameter candy dish with *Pine* motif; #892 7" diameter covered candy dish with *Apple Blossoms* motif; #950 aluminum vase holder with *Iris* motif and La Mirada vase, overall measurements approximately 4-1/2" diameter and 9" tall; #951 aluminum vase holder with *Shasta Daisy* motif and La Mirada vase, overall measurements approximately 4-1/2"x6" and 12" tall. Product data based in part on 1948 Wendell August Forge brochure, courtesy of the Wendell August Forge. Photograph and 1948 Wendell August Forge catalog, dated 20 Apr. 1948, consisting of spiral-bound 8"x10" photographs, courtesy of M.C. (August) Taylor.

Chapter Four

A Struggle to Survive: 1949-1977

In 1949, the Wendell August Forge was abruptly faced with a sharp decline in business. The moderation of post-war consumer spending, the increased competition from other aluminum giftwares manufacturers, and the revival of the imported goods market, resulted in retail merchants with overstocked inventories of Wendell August Forge product. In addition, the emerging conflict in Korea was increasing the demand for strategic war materials, causing the price of aluminum to rise. Orders slowed significantly and, instead of the increased hiring of the preceding three years, the Forge found it necessary to reduce its workforce by more than 25 percent.[1] Joseph Ziccardi, a Forge craftsman at the time, recalled that "Business was so good, I got a mortgage in 1948 to buy a house, and then the next year — boom! I had to find another job over in Mercer for awhile to make the mortgage payments."[2]

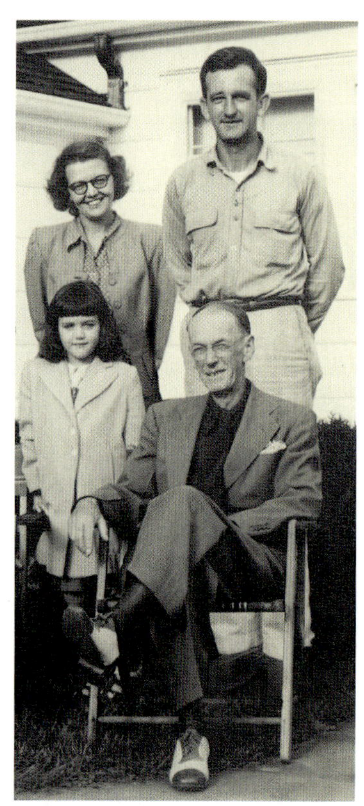

Rear, Mary B. and Robert August; Front, Mary Charlotte and Wendell August (note 4)

Wendell August had apparently begun his retirement when the customary market for Wendell August Forge products vanished. "The shock came when the customers refused [the products]," wrote Natale Rossi, "Mr. August came back and called his best customers, and asked them what's the big idea . . . they told him, if we can't sell them, how are we going to pay you?"[3] Although Robert August would continue to have most of the day-to-day operating responsibilities for the Forge, Wendell August abandoned any further plans of his anticipated Florida retirement, remaining in Grove City to oversee the business.[4]

Believing that more extensive advertising and marketing could improve sales, the Forge did its first national advertising since 1940. This included its first known direct consumer mail-order advertising, touting its baby tray and porringer and, later, a serving tray and coaster set.[5] In one effort to revive the jewelry sales, "Mac" August (Wendell M. August, Jr.), corporation Secretary and Sales Manager, and his cousin, E. Stillman Groves, who was responsible for the sales from the Grove City site, had the Forge make large 17-1/2"x24" hammered aluminum jewelry display cases with hinged lids and filled each with a generous supply of Wendell August Forge aluminum jewelry. Groves then made calls to jewelry stores to place them. According to Natale Rossi, "They made all this jewelry 'cause he is going to sell so many of them. He didn't sell one."[6]

A bright spot in an otherwise disappointing year was a commission from Alcoa for 30 "Pennsylvania Ambassador" plaques. These plaques were to be presented by local Chambers of Commerce during Pennsylvania Week, to selected people born in Pennsylvania who had "gone out of the state and spread the tradition of Pennsylvania in various fields of interest." Each plaque, 10-1/2" by 15" and 1/4 inch thick, featured a cameo portrait of William Penn, surrounded by a stylized landscape symbolizing the farming, city, industry, and mining areas of Pennsylvania. A gold plate in the shape of the Keystone State, engraved with the recipient's name, was mounted in the center of each plaque. Designed by James McCausland, the die was engraved by Natale Rossi from McCausland's drawing. The pieces were forged by Andrew Thomas.[7]

Pursuing a revival of the market for the Forge's aluminum giftwares, glossy catalogs were produced during 1950 in at least January, August, October,[8] and December,[9] and agents were engaged in New York City, Atlanta, Toronto, and Portland, Oregon, as well as in Grove City. The products featured in the catalogs were little changed from those of the previous decade. The forms of most of the pieces dated back to at least 1939-1940, and some of the "stock designs," such as *Dogwood* and *Pine*, were even older. This was quite a contrast to the years preceding World War II, when at least ten new items or motifs were developed each year.[10] The production of items with La Mirada pottery diminished throughout 1950, and the only unusual new items introduced were pieces featuring "Blenko Green Glass."[11] The effort was to no avail, and by the close of 1950, the workforce had been reduced by yet another 20 percent, to about 20 people.[12]

During the difficult years of 1949 and 1950, the Forge metalworkers, who were not affiliated with a formal union, developed a layoff process that persevered for at least the next two decades. The metalworkers were divided into three groups — finishers (or sanders), craftsmen, and blacksmiths — the blacksmiths having the highest priority and the finishers having the lowest priority. Blacksmiths were those responsible for making railings and similar items; the craftsmen were responsible for cutting dies, repoussé (hand hammering), and cutting and forming items; and the finishers were responsible for sanding, carbon coloring, waxing, and polishing. The metalworkers in each group, who were prioritized by seniority, elected a representative to serve on a committee that would meet with Robert August to review the order of layoff. When the layoff of a worker was required, the person

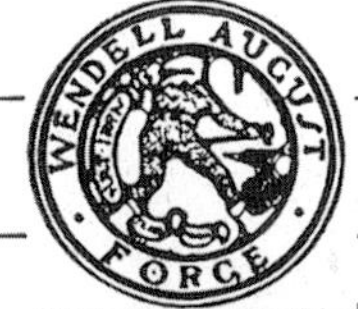

BOWLS AND CANDLESTICKS

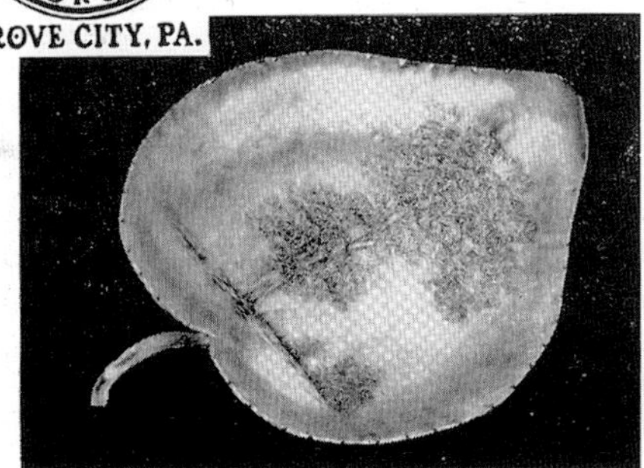

803 Leaf Bowl—12"x18"—Leaf Shape—
List Price $8.75 ea. American Elm

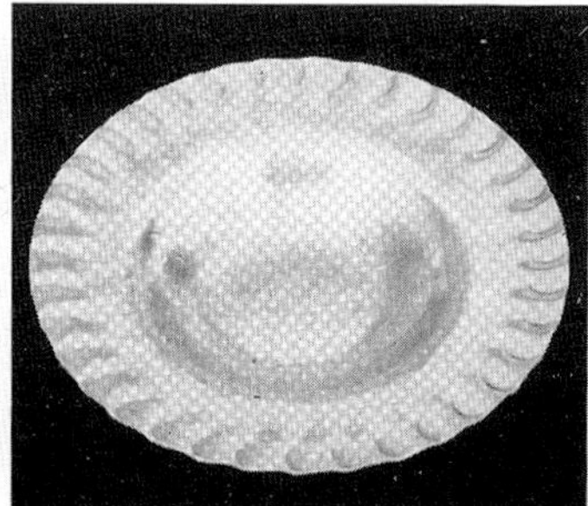

805 Flower Bowl—Round—15"—
Fluted Edge—
List Price $12.50 ea. Plain

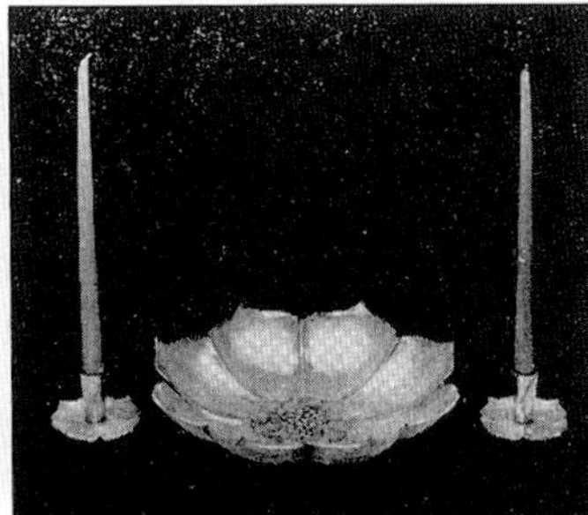

807 Bowl—Star Shape—14"—
List Price $13.75 ea. Grape
807-C Candlesticks—One Candle—
List Price $6.25 pr. Plain

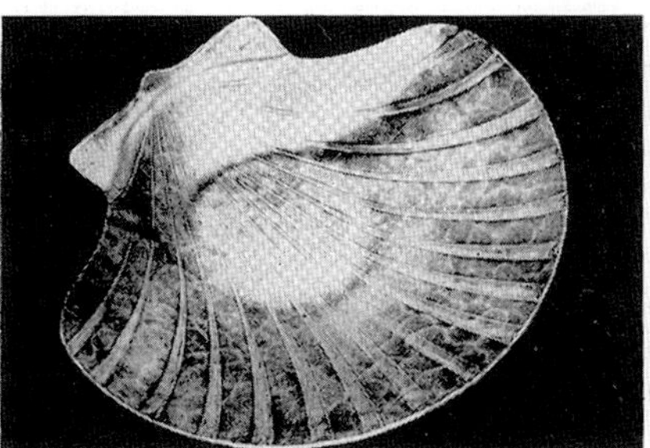

810 Shell Bowl—12"....List Price $7.50 ea. Shell

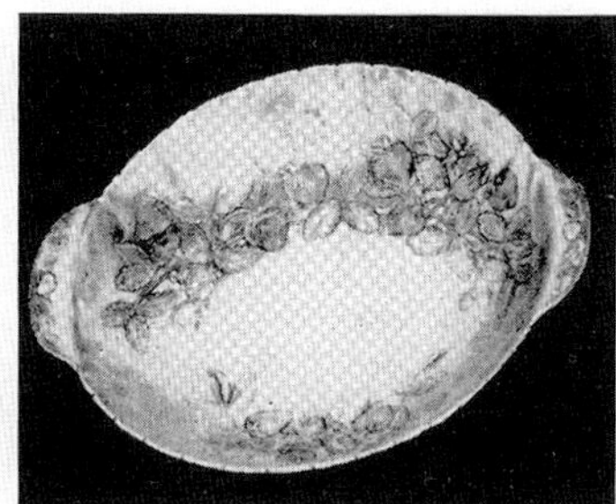

815 Fruit Bowl—Round—11"—
Handles—
List Price $7.50 ea. Strawberry

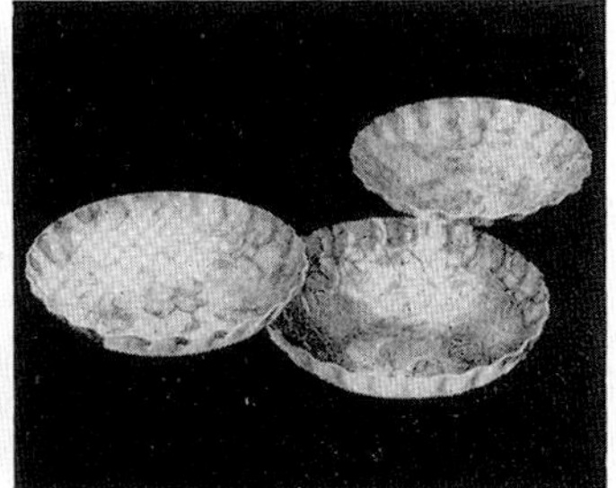

825 Bowls—Fluted Edge—	LIST PRICE
Nest of Three..........	$21.50 set
12" Bowl only..........	8.00 ea.
11" Bowl only..........	7.25 ea.
10" Bowl only..........	6.50 ea.

A, BS, DW, P

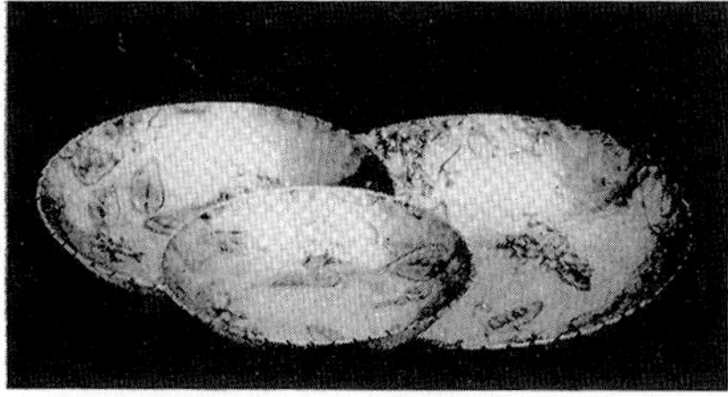

830 Bowls—Round—	LIST PRICE
Nest of Three......	$17.50 set A, BS, DW, P
12" Bowl only.....	7.50 ea.
10" Bowl only.....	6.00 ea.
8" Bowl only.....	4.50 ea.

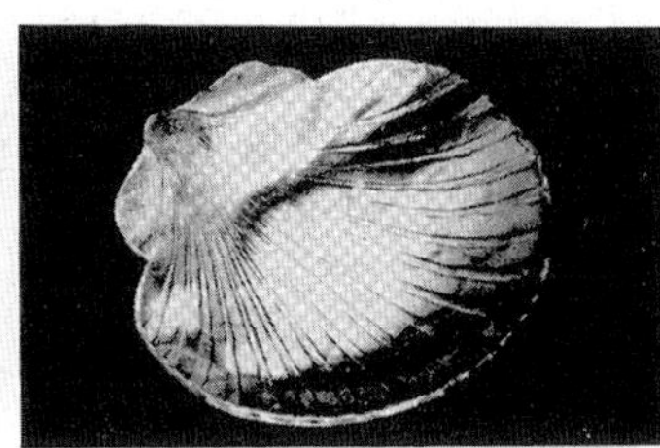

880 Nut Dish—5½"—
List Price $2.75 ea. Shell

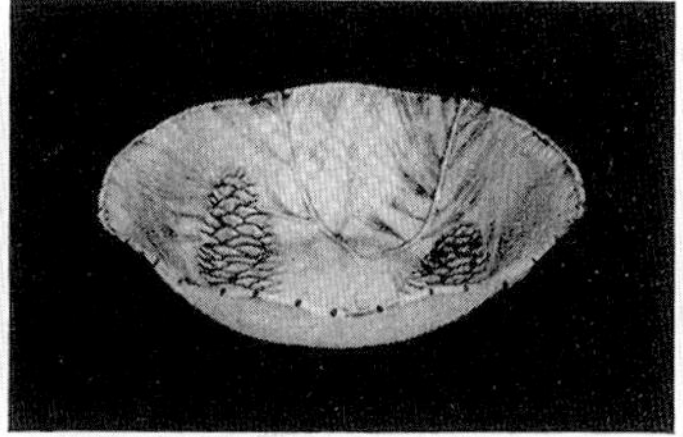

890 Candy Dish—Round—
Fluted Edge—7½"—
List Price $4.00 ea. A, BS, DW, P

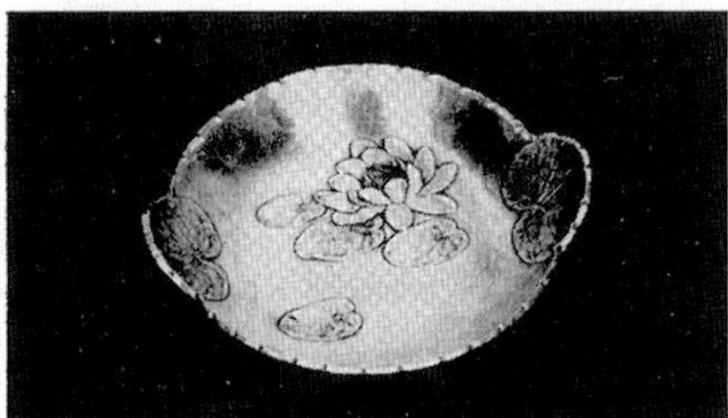

891 Candy Dish—Round—7"—Handles—
List Price $3.25 ea. P, Water Lily

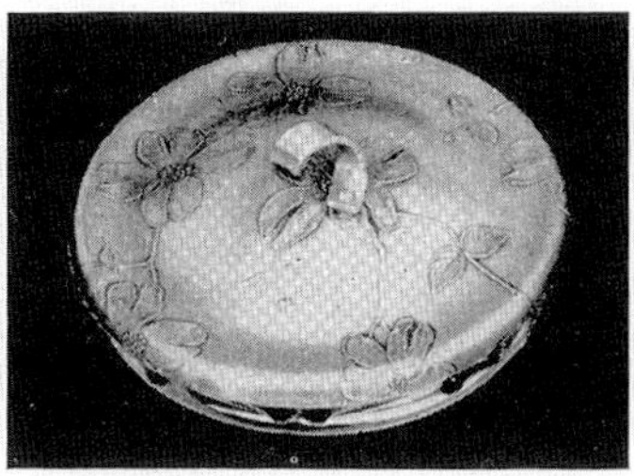

892 Candy Dish—3-compartment
glass with Aluminum Lid—
List Price $5.00 ea. A, BS, DW, P

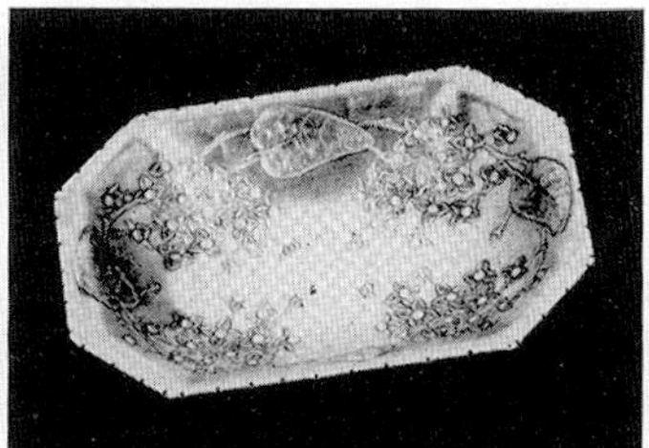

893 Candy Dish—Rect.—
5"x7"—Handles—
List Price $3.75 ea. A, BS, DW, P

Page 6 of the October 1950 Wendell August Forge catalog (note 8)

SMOKING ACCESSORIES & JEWELRY SETS

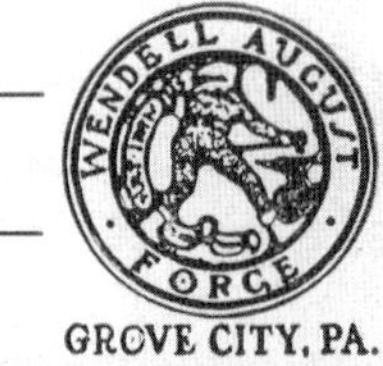

GROVE CITY, PA.

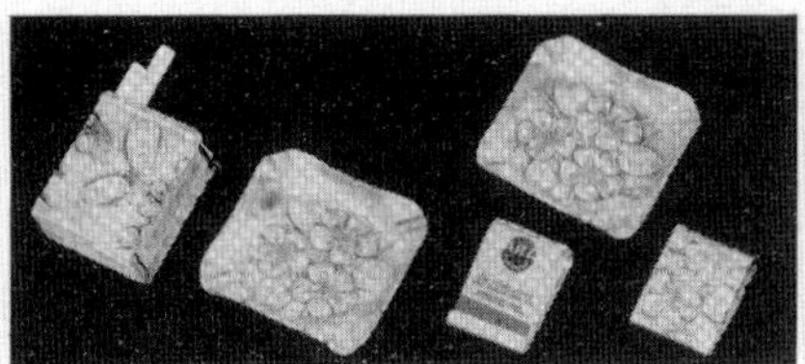

77 Smoking Set........List Price $5.00 set Boxed
A, BS, DW, P

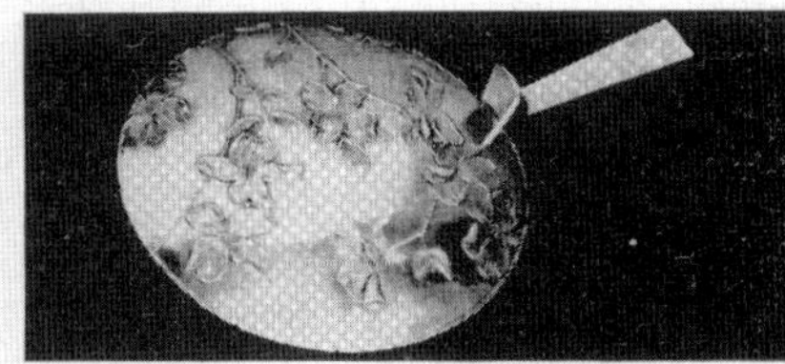

95 Silent Butler—7½"—
List Price $12.50 ea. A, BS, DW, P

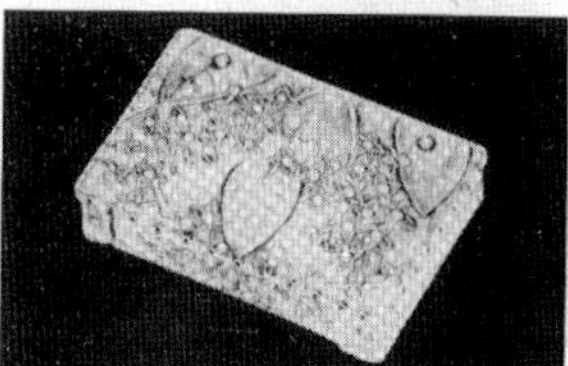

50 Cigt. Box—2 Pack—Hinged Lid.........List Price $10.00 ea.
A, BS, DW, P

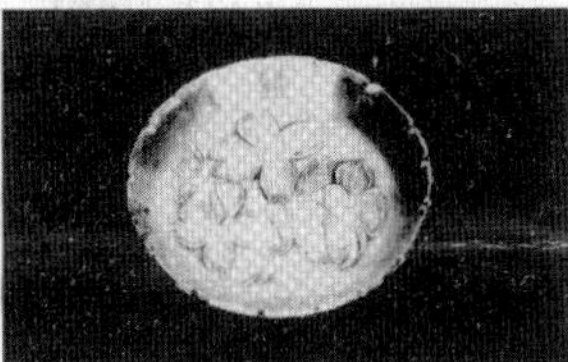

1 Coaster—
3½"........List Price $9.60 dz.
A, BS, DW, D, P, S

5 Coaster—
4½"......List Price $12.00 dz.
A, BD, BS, DW, D, P, S

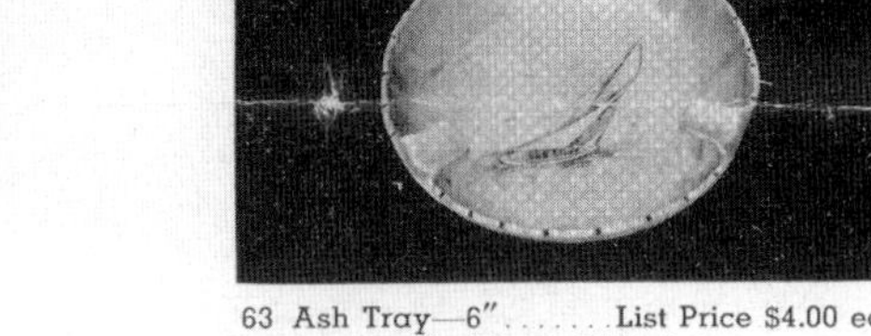

63 Ash Tray—6".......List Price $4.00 ea.
A, BS, D, DW, P, S

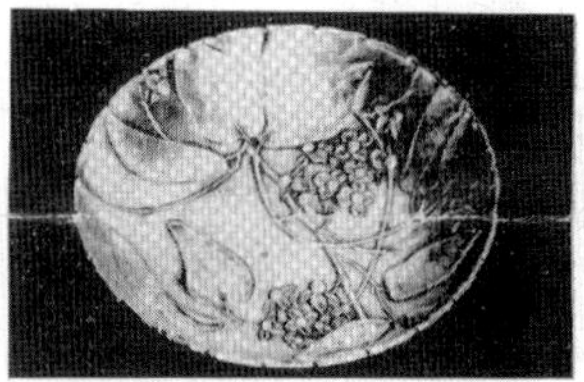

10 Ash Tray—6"—
List Price $2.50 ea. A, BS, DW, P

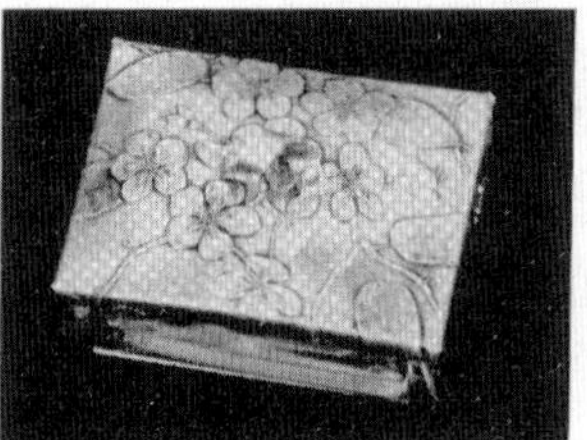

51 Cigt. Box—Glass—Alum. Lid........List Price $5.00 ea.
A, BS, DW, P

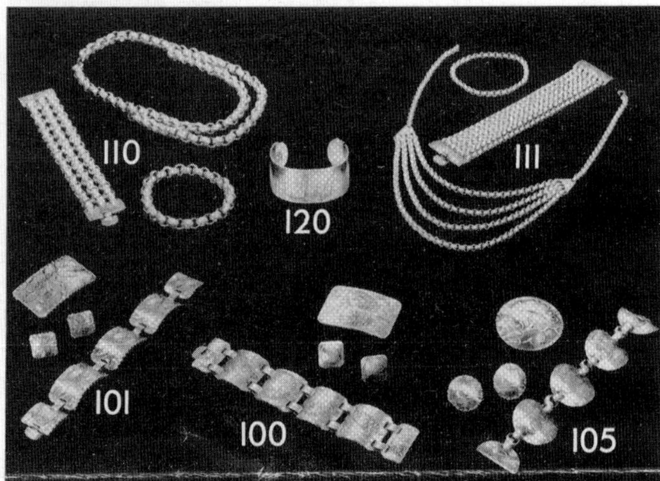

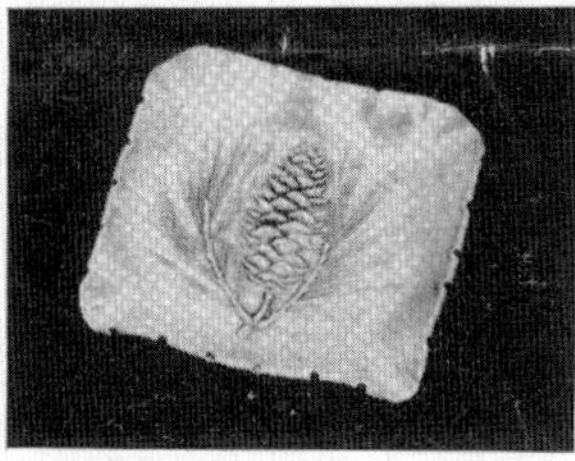

64 Ash Tray—3½"—
4 Rests............$15.00 dz.
A, BS, DW, P

66 Ash Tray—4½"—
4 Rests.........$18.00 dz.
A, BD, BS, DW, D, P, S

JEWELRY

No.	Item	Price	Designs
100	Bracelet—Plain Edge—5 Medallions—1¼" square	$5.00 ea.	Apple, Oriental Tree, Pine, Sail
100	Earrings—Plain Edge—¾" square	2.50 ea.	Apple, Oriental Tree, Pine, Sail
100	Pin—Plain Edge—1½"x2¼"	1.25 ea.	Apple, Oriental Tree, Pine, Sail
101	Bracelet—Reeded Edge—4 Medallions—⅞"x1½"	4.00 ea.	Apple, Oriental Tree, Pine, Sail
101	Earrings—Reeded Edge—¾" square	3.00 ea.	Apple, Oriental Tree, Pine, Sail
101	Pin—Reeded Edge—1½"x2¼"	1.50 ea.	Apple, Oriental Tree, Pine, Sail
105	Bracelet—Plain Edge—4 Round Medallions—1¼"-dia.	3.00 ea.	Apple, Oriental Tree, Pine, Sail
*105	Earrings—Plain Edge—Small Round—¾"-dia.	2.50 ea.	Apple, Oriental Tree, Pine, Sail
105	Earrings—Reeded Edge—Large Round—1"-dia.	3.00 ea.	Apple, Oriental Tree, Pine, Sail
105	Pin—Plain Edge—2" Round	1.25 ea.	Apple, Oriental Tree, Pine, Sail
110	Bracelet—Large Chain—3 Strands	2.50 ea.	Apple, Oriental Tree, Pine, Sail
110	Bracelet—Large Chain—1 Strand	.75 ea.	Apple, Oriental Tree, Pine, Sail
110	Necklace—Large Chain—2 Strands	2.50 ea.	Apple, Oriental Tree, Pine, Sail
*110	Necklace—Large Chain—1 Strand	1.50 ea.	Apple, Oriental Tree, Pine, Sail
111	Bracelet—Small Chain—6 Strands	3.00 ea.	Apple, Oriental Tree, Pine, Sail
111	Bracelet—Small Chain—1 Strand	.75 ea.	Apple, Oriental Tree, Pine, Sail
111	Necklace—Small Chain—4 Strands	3.00 ea.	Apple, Oriental Tree, Pine, Sail
*111	Necklace—Small Chain—2 Strands	2.00 ea.	Apple, Oriental Tree, Pine, Sail
120	Bracelet—One Piece—1¼" Wide	1.50 ea.	Apple, Oriental Tree, Pine, Sail

*Not shown in photograph

Page 7 of the December 1950 Wendell August Forge catalog (note 9)

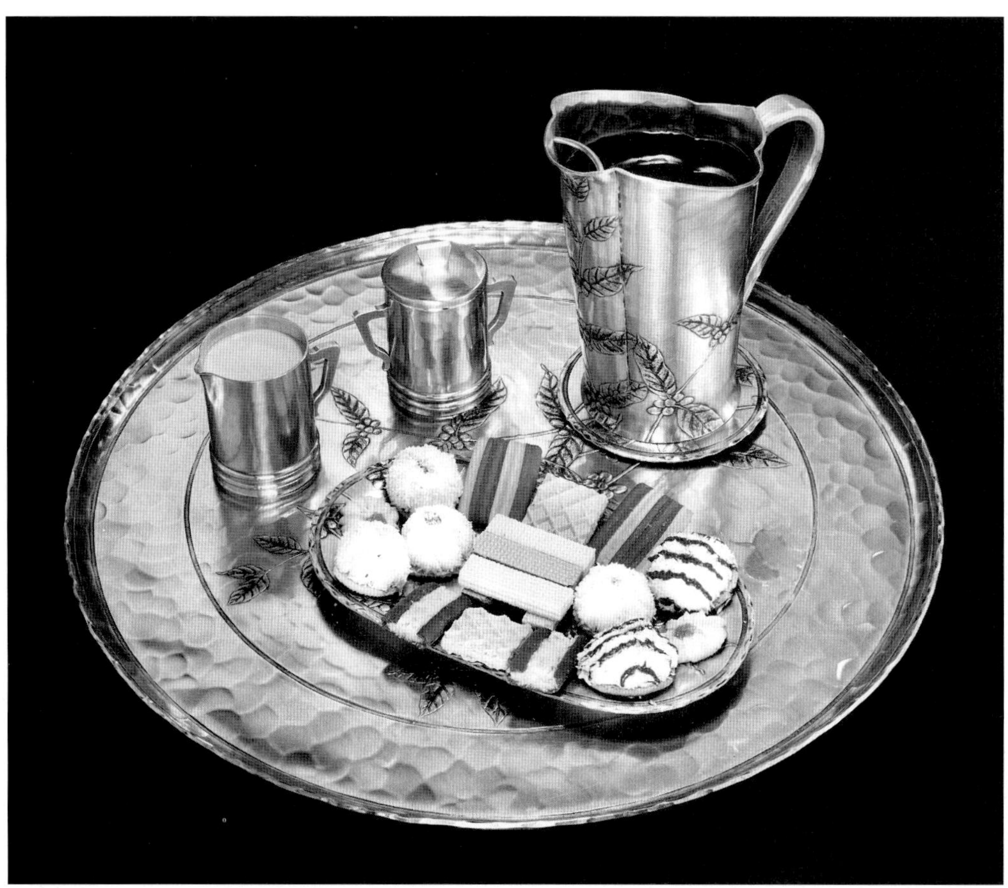

Mid-1950s aluminum wares with Coffee Bean *motif (note 15)*

with the lowest seniority in one group could "bump" down to the next group, thus setting off a domino effect as the layoff rippled through the groups. As might be expected under such strained circumstances, there was dissension amongst some of the metalworkers, which included complaints of unfairness in the layoff process.[13]

Most consumer products of the early 1950s tended toward a "modernist" style, with an emphasis on spare, functional, clean-lined design. Square and rectangular furniture, relieved only by an occasional curved boomerang or amoeba shape, featured limed oak as the preferred wood. Decorative items were used sparingly, and the emphasis in metals was on brass, chrome, stainless steel, and stark black wrought iron. Smooth plastic and sleek color-anodized aluminum forms graced the kitchen and spoke of the casual living styles that were permeating the nation. The country's designers were "singing the praises of functionality, machine technology, and mass production, and they were deriding decoration."[14] Objects that reflected a craft tradition like that of the Wendell August Forge were held in increasingly low regard by the decorating communities. The Forge's repoussé-embellished products were out of step with the trends.

In an attempt to respond to the changed consumer taste, and to meet the need to reduce costs, the Forge's giftware was modestly redesigned. The "tool-marked" (notched) edges, for example, became an embellishment of the past. The forms of the pieces were simplified and streamlined.[15] [16] The delicately made and subdued jewelry, which could not compete with the flashy rhinestones that were coming into vogue, was removed from the product line. Despite these changes, the Forge's products were not competitive. The essence of the Wendell August Forge product was its handcrafted aluminum repoussé work, and the consumer public had turned its attention away from both the metal and the type of decoration.

Although the Wendell August Forge had in the past produced some commissioned commemorative and advertising items, it was not a line of business that had been actively pursued. With its workforce at the lowest level since 1932, and little demand for its regular consumer giftwares or ornate decorative architectural work, new markets had to be explored. Although intricate repoussé work might no longer be in the mainstream of the consumer decorative arts movement, it was ideally suited for the more traditional forms of embellishment considered appropriate for business gifts and commemorative awards.

Natale Rossi volunteered to go on the road to seek out customers, particularly those who might provide a steady and reliable stream of income. Rossi would later write "Mr. August said he supposed I could sell, and I told him that I knew I could. Well he said, 'I tell you what. We will do it. We do not want to lose any money. We'll give you 15% of what you sell, and you pay all your expenses.' I said okay . . . I saw Triangle Springs [Triangle Auto Spring Company, DuBois, Pennsylvania] . . . I told him I was selling hand made aluminum items which could be used for special customers . . . They said their Christmas worries were over, they [would] buy a different item every year with the same design. I asked them how many they needed and they said 750. I jumped! . . . That really started my selling experience . . . From then on, I concentrated on industries, tool and die companies, and machine shops."[17]

One morning in mid-January of 1958, James McCausland, the Forge designer and plant superintendent, told his wife, Mary, that he wasn't feeling very well, and was going to rest in bed until about noon and then go to work in the afternoon. According to Mary McCausland, "Well, I kept going upstairs and checking on him, you know, and he didn't like

Below, left to right, Aluminum lazy susan with Dogwood *motif; aluminum pitcher and bowl with* Coffee Bean *motif (note 16)*

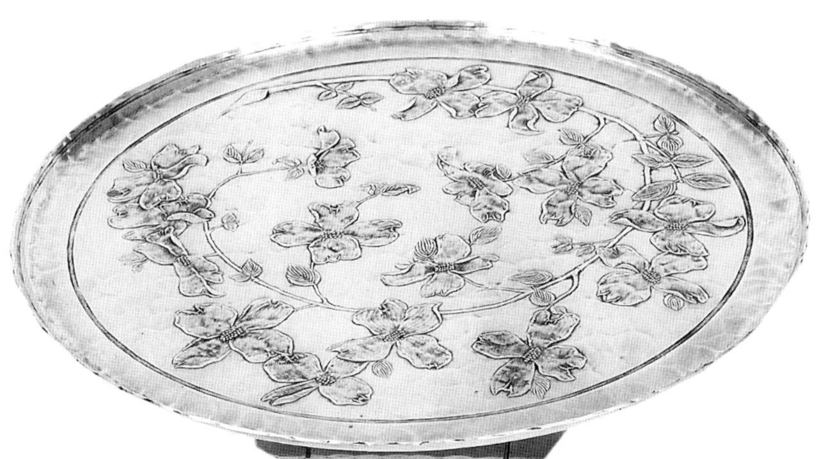

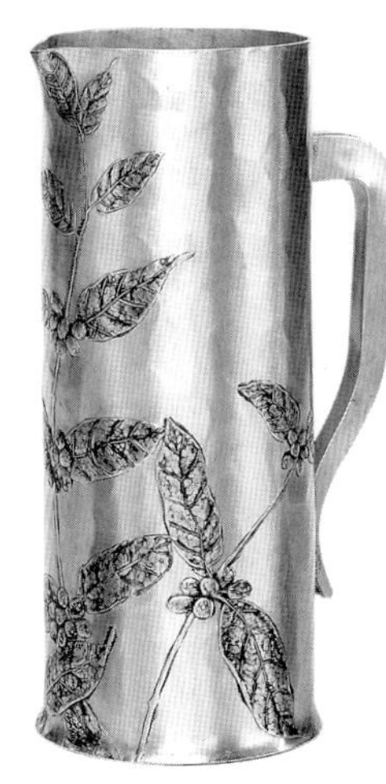

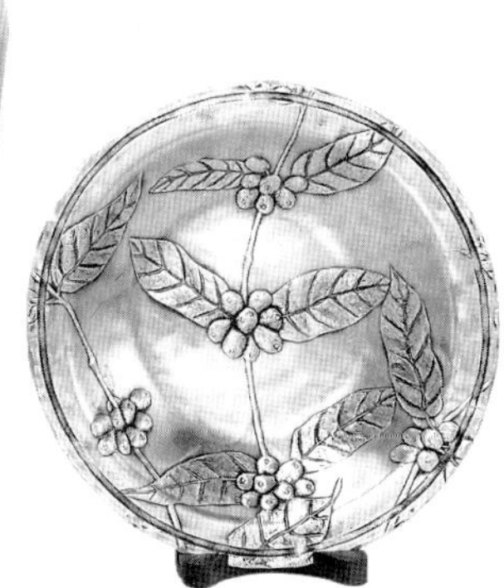

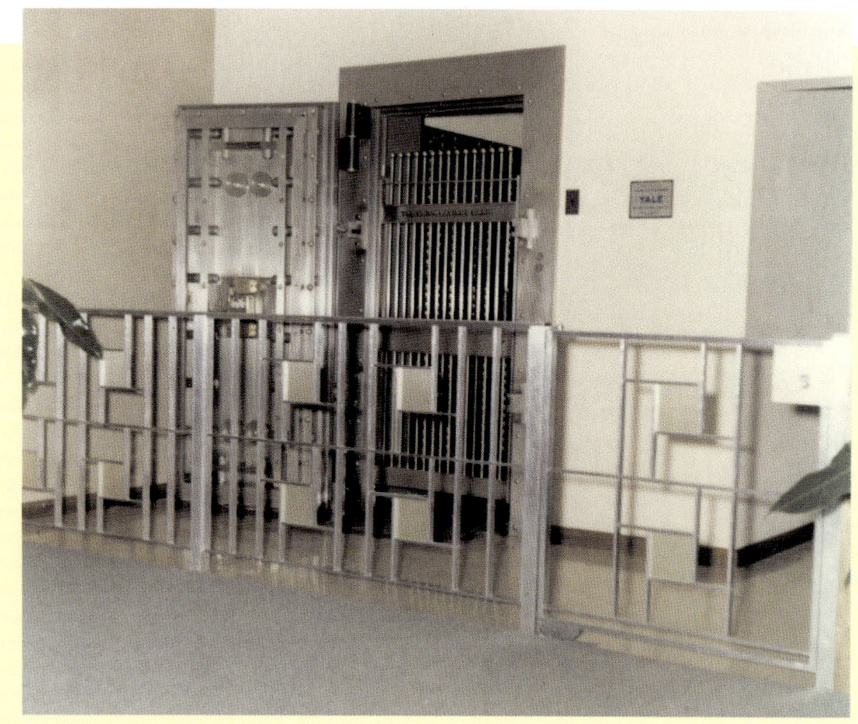

Left, Railing at Union Savings Bank (note 21)

Below, Check desks at Oak Hill Savings Bank (note 20)

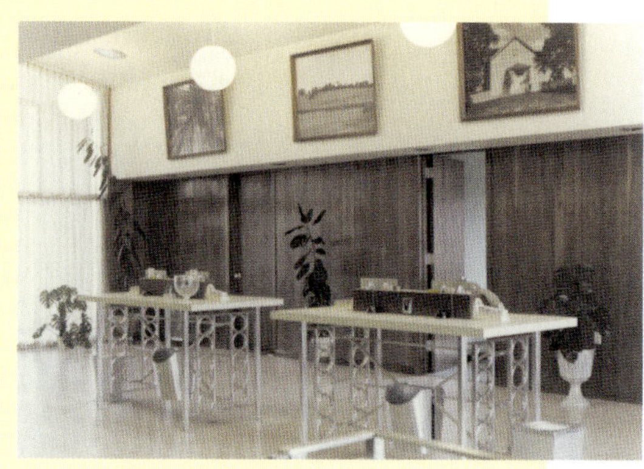

that . . . the last time I went up, he was gone in his sleep."[18]

The unanticipated death of James McCausland deprived the Wendell August Forge of the primary artistic talent that had guided its designs since 1928. Within a few years, virtually all traces of his influence vanished from the product line, with only echoes of his talent remaining in a few of the motifs and forms that he had developed. McCausland's death also put an added strain on Wendell August, who had not been in good health since the 1940s and was by now 73 years old.

When the giftware business had started to decline, Robert August began to capitalize on his previous engineering education and construction experience, by doing general contracting for bank remodeling. Working with an architect from Niles, Ohio, he subcontracted the carpentry, plumbing, and electrical work to Grove City firms. The Wendell August Forge developed and made the aluminum architectural appointments and accessories for the remodeled banks, such as signs, stair railings, gates, balconies, benches, check desks, waste baskets, and lighting fixtures.[19] Although very little precise documentation of this work has been located, two images from the early 1960s provide some insight into the types of remodeling projects in which the Forge was engaged.[20] [21]

By 1960, the Forge's cadre of skilled metalworkers had dwindled to about eight, with the threat of the company's possible closure on the horizon. Its circumstances were not unique; by this time, only a handful of the aluminum giftware firms that had crowded the market in the late 1940s were still in production.[22]

Efforts were made to diversify the giftware product line by expanding into other metals. One such diversification was the result of a late

1959 Natale Rossi sales visit to a tool and die shop in Dunkirk, New York, where he learned about Anthony Serio, the owner of a tool and die company in Elmira, New York, who was seeking someone to make a bronze railing for his home. Serio ended up with not only the elaborate railing, but also a pair of fireplace andirons, a marble-topped coffee table, some lighting fixtures, a back splash and range hood for the kitchen, and some repoussé-embellished items for his customers — all in bronze.[23] [24] According to Rossi, "He told me if I made them in bronze he would buy, so I made them in bronze."[25]

Another area of diversification in metals involved stainless steel. Sharon Steel, of Sharon, Pennsylvania, was among the companies that were approached in the early 1960s by Natale Rossi and Joseph Ziccardi about purchasing hand made aluminum giftware items for their customers. "I went to sell at Sharon Steel, and when the girl asked what I was selling, I said hand made aluminum items," recalled Rossi, continuing, "She said why don't you make them in stainless because they will not buy aluminum."[26] Rossi and Ziccardi decided to give it a try.

Ziccardi said that when they first tried making Wendell August Forge pieces from stainless steel, "the dies cracked in two, you had to hammer so hard . . . we broke a lot of dies." He and Rossi contacted a friend who worked at Cooper-Bessemer Corporation for advice. Suggestions for tempering the dies to withstand the intensity of the hammering that

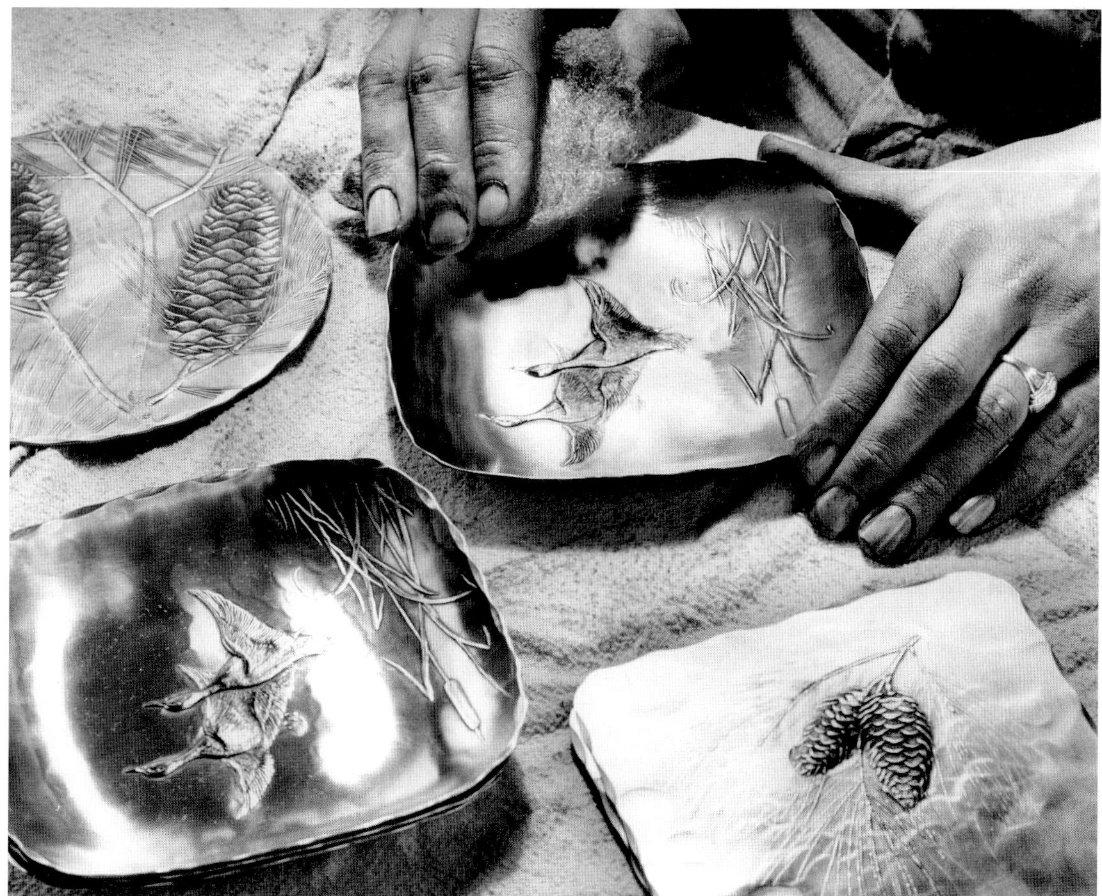

Stainless steel giftwares
(note 28)

was required were soon forthcoming, and the Wendell August Forge embarked on a new venture in the production of stainless steel giftwares.[27] Broken dies were not the only problem associated with the repoussé of stainless steel. Because stainless steel was so much harder than aluminum, the repoussé process yielded less detail in the motif. This meant that the motifs also had to be worked with tools from the front of the piece in order to bring out the detail.

The steel manufacturing industry was quite taken with this application, and in a mid-1962 press release, Allegheny Ludlum Steel Corporation featured the Wendell August Forge, noting that "The hand craftsman is gone. The able artisan and pride of workmanship is passe, and so goes the lament. Not so! In a small shop in Grove City, Pa. is a plant that makes unusual gifts of stainless steel, all by hand."[28] Certainly the shop was now small, with only about eight or nine metalworkers still remaining on a fulltime basis.[29]

With fewer people needing to produce more product in order to survive, the pressures to reduce production costs were intense. One of the more significant changes made was the gradual abandonment of the use of the Forge's large, elaborately engraved dies. "It took a lot of time for the antiquing and sanding with some of those large dies like *Apple Blossom*," Joseph Ziccardi remarked. "Natale started cutting some small dies and we saved time on the sanding."[30] In addition, the practice of numbering each item began to decline, since the process of striking a style number on the back of each item required time. A third significant change was the movement toward piecework, with metalworkers being paid in accord with their individual production. This change led to some conflicts among the remaining metalworkers and reduced quality control for the items produced.

Although the decade of the 1960s had started on a more hopeful commercial note, another significant change was looming. One day in early April of 1963, Anthony Spatara and Joseph Ziccardi were the only people in the shop. Robert August was out of town to bid on a bank remodeling job, Natale Rossi was making calls on prospective industrial clients, and the remaining craftsmen were working at various bank remodeling sites. Ziccardi, recalling that day, said "When lunch time arrived, Wendell chatted with me and Tony before heading downtown to Dad's Restaurant for his usual ham sandwich lunch. He never came back."[31] Wendell August's obituary noted that he "died suddenly at 2 p.m. Wednesday of a heart attack suffered while eating lunch," and went on to summarize his founding of the hand-wrought aluminum industry and the Forge's recognition for its unique decorative work in banks and churches.[32]

With Wendell August's death, the shares in the family-owned corporation were now held by Robert August (President), Wendell M. "Mac" August, Jr. (Vice-President), and Mary August (Secretary-Treasurer).[33] Because neither Mac August, Bob's brother, nor Mary August, Bob's wife, were actively engaged in running the Wendell August Forge, the burden of its operation fell solely to Bob August.

Shortly after Wendell August's death,

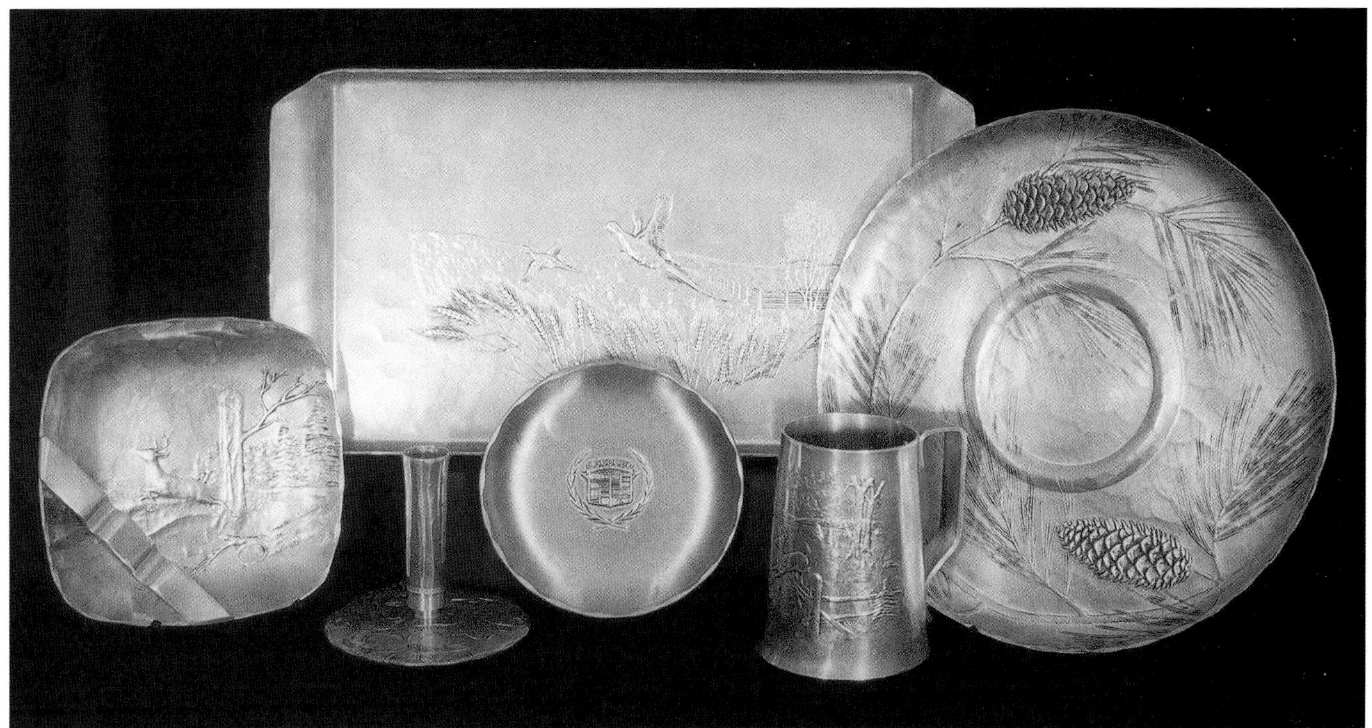

Items typical of those produced in the 1960s (note 34)

Bob August opened a tiny gift shop in the Grove City plant, where there had previously existed only a small display area in which potential customers could view samples of the types of items that were available.[34] Now potential customers visiting the site could actually shop. This represented the first step taken by the Forge into the retail business.

With the addition of the gift shop, the Wendell August Forge became a popular place for local people to visit and shop with out-of-town guests. Watching the metalworkers engrave dies, hammer out pieces, and fire them at the forge was an experience enjoyed by both young and old. Catherine Youngo, who would later work for the Forge, remembered when, after moving to Grove City in the late 1960s, "I searched out interesting things to do with the boys and discovered this neat little business that made a lot of noise, had a lot of nice people working in it and was located on a dead end street, in a red brick, ivy covered building. Wendell August Forge was my once or twice a week stop. My son Chris insisted we come to see the men who hammer on metal."[35]

But how, on a dead-end street in the heart of a small town in northwestern Pennsylvania, did one draw a retail crowd beyond the local citizens and their visitors? One solution to this problem emerged in the mid-1960s, shortly after Mary Lou McNaughton joined the Forge as Robert August's secretary and office manager. McNaughton related "My husband and I liked to go on automobile trips. I would go to the Mobil gas station and get their tour guide books that told you about places to visit and things to see on the way. I thought we should have the Forge included in these books, and so I suggested it to Bob. He decided that it was a good idea, so I wrote it all up, and we paid the fee to be included in the tour book, and sent it in. We began to get visitors and, because we were in the book, we started getting calls from bus companies and travel agents. By 1967 or so, tour buses were stopping at the Forge."[36]

During the balance of the 1960s and into the 1970s, the basic giftware product line continued to feature the highly simplified forms with small motifs that had come into use after McCausland's death. Relying primarily on business-related gifts, awards, and commemoratives, there was no direct advertising. Natale Rossi continued to work as a salesman, eventually creating and using his own unique "calling card,"[37] traveling throughout the eastern United States.[38] For catalogs, McNaughton recalled, "We had photographers come in, and then the glossies were put in books to carry around."[39] One piece of 1967 literature prepared by an Alcoa distributor depicts a few items identical to those made in 1962.[40]

Bob August's general contracting services for the remodeling of banks continued throughout the decade. It was wearing work. M.C. Taylor, Bob August's daughter, remembered that he "was on the road half the time to every town within a radius of about 200 miles, bidding on jobs or working on them."[41] "Those bank jobs," said Joseph Ziccardi, "We'd go to one small town and stay there and work for about a week, and then we'd go to another small town and stay another week. We did lots of aluminum work for the Mellon Banks; the panels were gold anodized, but we couldn't install it, 'cause they had a contract with the union and so the union workers installed it. We must have done hundreds of little banks."[42] According to McNaughton, "We must have done the work for at least 150 different banks, mainly in West Virginia, Ohio, New York, and, of course, Pennsylvania."[43] Although Forge records of the bank remodeling projects no longer exist, among

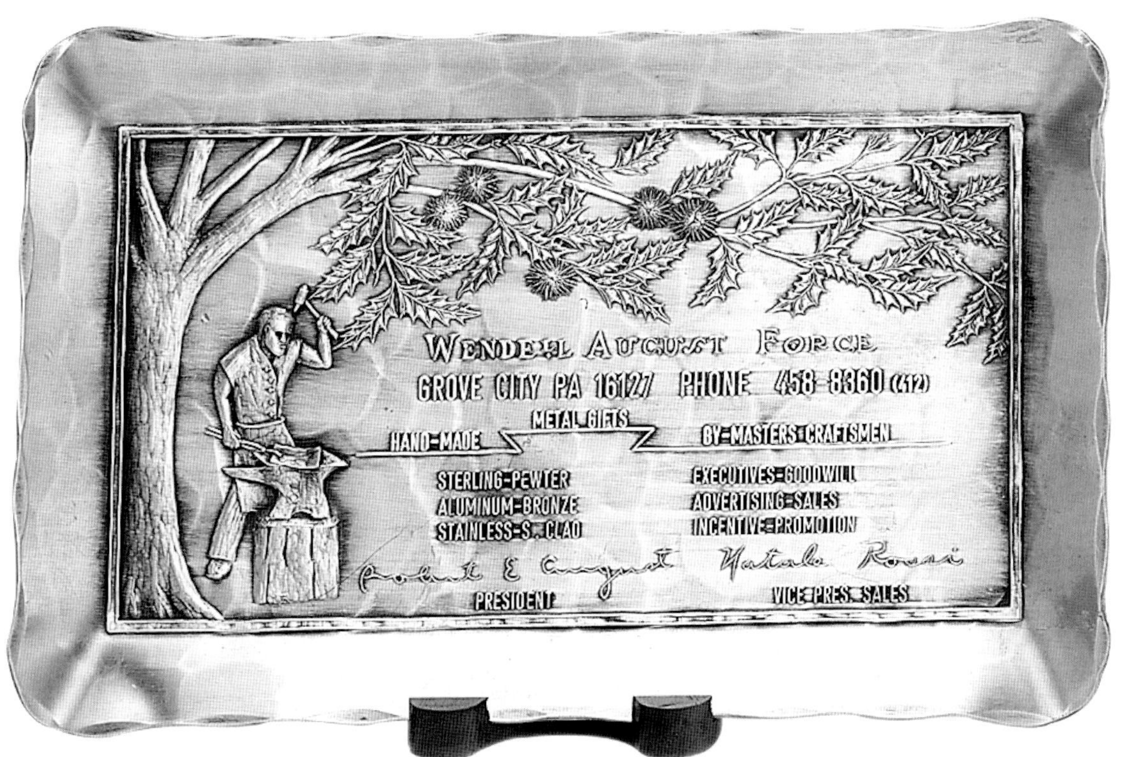

Robert August - Natale Rossi "calling card" *(note 37)*

the towns in which remodeling projects are thus far believed to have been executed are Moundsville, West Virginia; Bowling Green, Findlay, Lowellville, Nelsonville, and Reynoldsburg, Ohio; Butler, Franklin, Galeton, Greenville, Grove City, Indiana, Oil City, Sandy Lake, and State College, Pennsylvania; and several locations in the Pittsburgh metropolitan area.[44]

Unfortunately, despite all of the efforts expended, the decade of the 1960s was not much kinder to the Wendell August Forge than had been its predecessor, and sales continued to decline. The affluence of the United States did not find its way to the Forge, as consumer taste and designer opinion continued to move inexorably away from the type of product that was the embodiment of the Wendell August Forge. The spare, stark Modernist lines popularized in the 1950s dominated home decor, and manufacturers turned increasingly to synthetic materials, with new forms of plastic appearing everywhere and for everything. Bright, bold colors were accented by gleaming stainless steel and mirrored chrome. The line between the popular and fine arts continued to blur with the Pop Art works of Andy Warhol and Roy Lichtenstein, and the Op Art works of Victor Vasarely and Bridget Riley.[45] In comparison, hand crafted metalware decorated in repoussé with intricate flora and fauna was perceived as a quaint curiosity.

As the 1960s drew to a close, the continued existence of the Wendell August Forge was not at all assured. The bank remodeling work that had largely sustained the Forge was dwindling. The REA Metal Products aluminum coin boxes and trays that had provided extra after-hours work for Wendell August Forge metalworkers were being replaced by plastic ones. The sales from the gift shop were small. The market for business gifts and commemorative items, although stable, was not sufficiently large to sustain the company. The Wendell August Forge needed to find new and different markets, or close down.

In the late 1960s, Limited Edition collectibles began to become widely popular and attractive in the United States. Mary Lou McNaughton, the office manager, was a serious collector of plates, including Limited Editions. For some time since she had joined the Wendell August Forge, she had wondered why it was not producing collector plates. Robert August was apparently not in favor of the idea, considering many of the plates on the market to be lacking in the dignity that he considered appropriate for the Forge's product. However, the Bicentennial of the United States was approaching. McNaughton wondered why the Forge could not issue a commemorative series of collector plates in honor of the Bicentennial. Such an issue would be in keeping with other special commemorative items that the Wendell August Forge had produced, and certainly the Bicentennial of the United States was a sufficiently substantial event to be worthy of the work of the Forge. Natale Rossi was enthusiastic about the idea, and between McNaughton and Rossi a list of potential events to depict on a Limited Edition series was developed. After a few revisions, Robert August finally approved the creation of a series of Limited Edition plates, called *Great Moments in History*. One plate would be

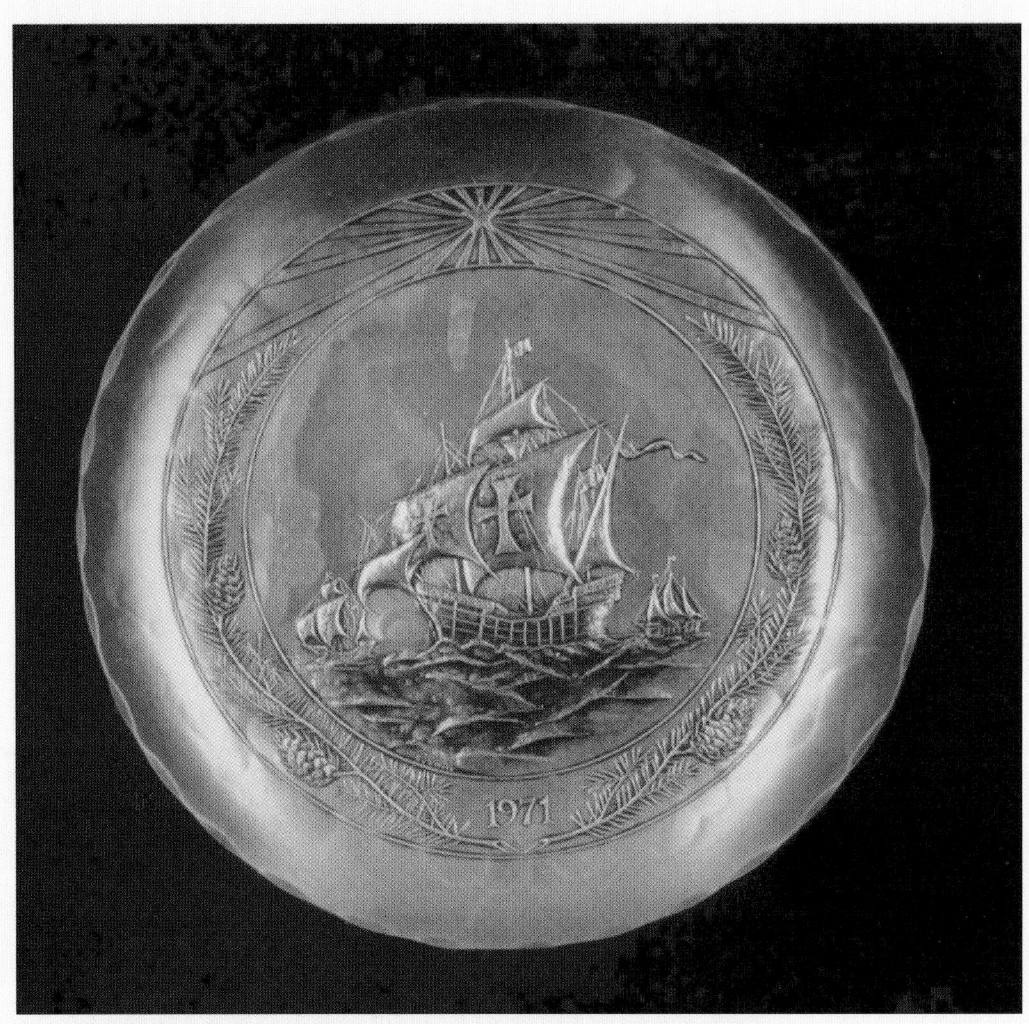

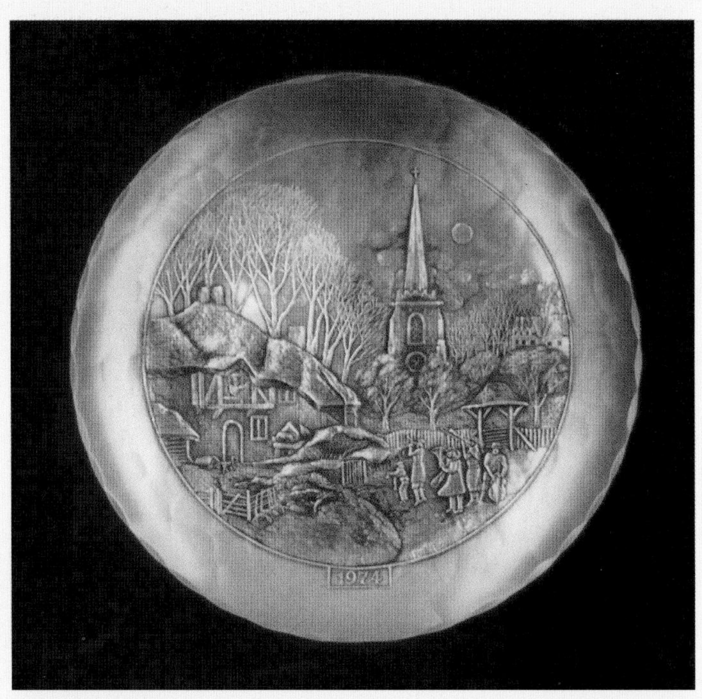

Limited Edition Plates
(note 49)

Above, Columbus *in pewter*
Left, The Carolers *in pewter*

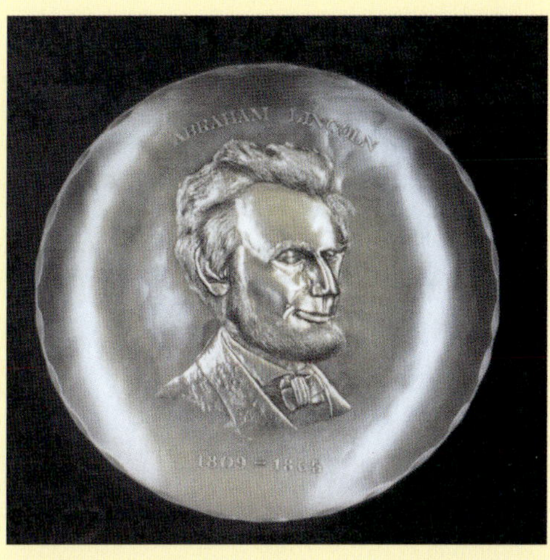

Left, Abraham Lincoln *in pewter*
Below, Signers of the Declaration *in bronze*

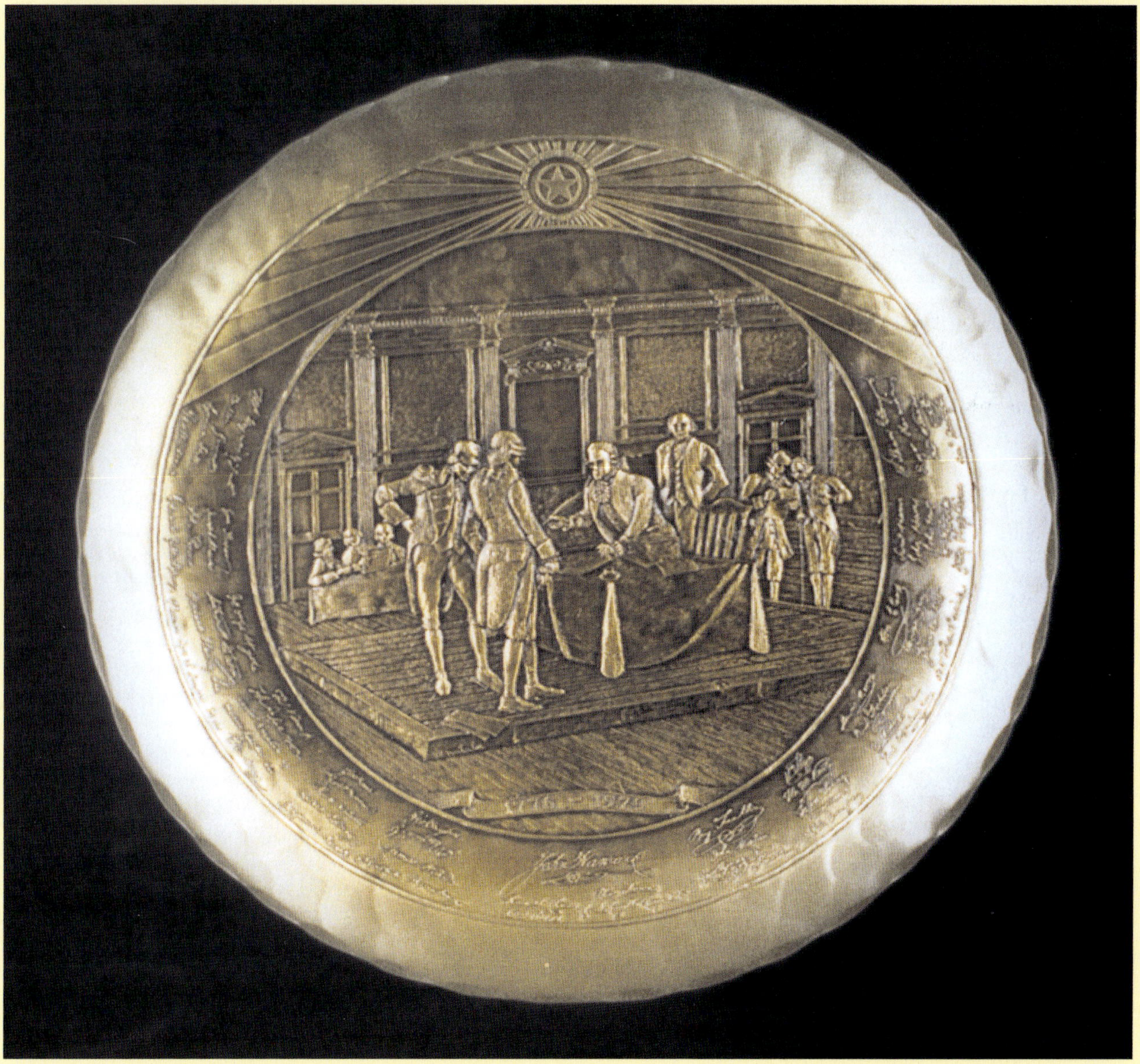

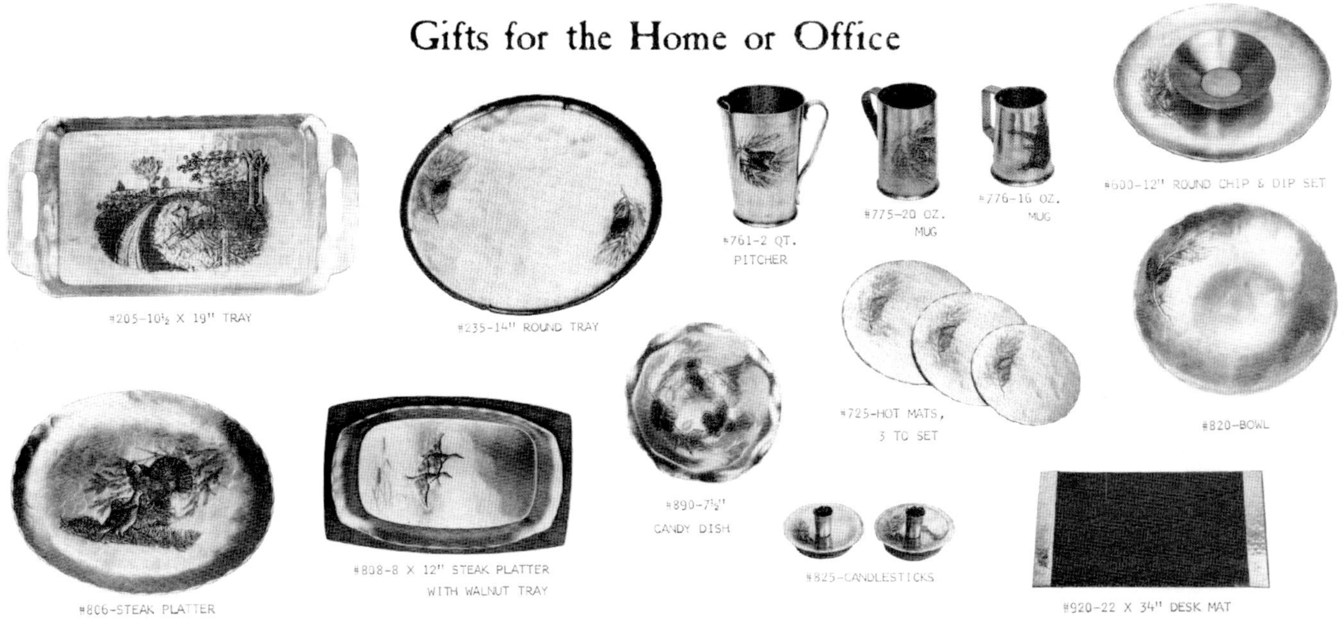

Portion of the 1971 Wendell August Forge brochure (note 51)

issued each year from 1971 through 1976 in aluminum, bronze, pewter, and sterling silver.[46]

The *Great Moments in History* series led the Wendell August Forge into a market in which its strengths were highly valued. The permanence of the metal, the hand crafting of the items, and the intricate motifs reminiscent of an earlier era converged with a Limited Edition market that was expanding and a special event that encouraged consumers to acquire mementos in its honor. The series was carried by the America House division of the Franklin Mint in Philadelphia, thereby providing the marketing that was needed.

The introduction of the series met with success, and the Wendell August Forge issued two President plates, *John Fitzgerald Kennedy* in 1971 and *Abraham Lincoln* in 1972.[47] The demand for the Forge's Limited Editions was sufficient to require the hiring of additional metalworkers, leading to the first increase in the workforce since the 1940s. "I started in March, 1972, when business was picking up again," said Ron Winder, a master craftsman with the Forge, "and I was the twelfth person in the shop."[48] As a result of the positive reception given to the new product line, the Forge continued to add new Limited Editions,[49] such as *Peace on Earth* in 1973; *Model T Ford, Best Friend,* and the first of the *Christmas* series in 1974; and the first of the *Wildlife* series in 1977.[50]

The revitalization associated with the development of the Limited Editions appears to have also emerged in the regular giftwares line, as it is the years of 1971 and 1974 for which printed brochures of the Forge's product have surfaced. These two brochures are very similar, and illustrate 40 to 50 items having about 12 different motifs. Most of the pieces were available in aluminum, bronze, stainless steel, and stainless clad aluminum.[51] The corporate giftwares clientele had continued to grow, and by this time the Forge had produced pieces for such firms as Allegheny-Ludlum Steel, Anheuser-Busch, Brockway Glass, Brockway Pressed Metals, Coors Brewery, Ford Motors, General Motors Cadillac Division, Lockhart Iron and Steel, Stackpole Carbon Company, Triangle Auto Spring, United States Steel, and many insurance companies such as John Hancock, Liberty Mutual, Metropolitan Life, and New York Life.[52]

By 1977, the Wendell August Forge was the only firm known to be producing the repoussé metalwares that it had introduced in the early 1930s. It could look back on its serious struggles since 1949, breathe a sigh of relief, and celebrate its survival. It provided an opportune time for Robert August, now in his sixties, to begin contemplating retirement.

Chapter Notes

1 Based on *Polk's Grove City Classified Business Directory*, searches of 1946, 1949 and 1954 by Gwen Oberlin, searches of 1950 and 1952 by William Freeman; Frank and Lenora (Isacco) Rossi, personal interviews, 12 June 1996; Frank Rossi and Josephine (Rossi) Town, personal interviews, 9 July 1997; Anthony Pompa, personal interview, 25 July 1997; Joseph Ziccardi, personal interview, 5 Aug. 1997; Natale Rossi, "The Conception of a New Business," manuscript, 1988, courtesy of Thomas Armour and the Wendell August Forge.

2 Ziccardi, 5 Aug. 1997.

3 Rossi, "Conception."

4 Wendell M. August, Robert E. August, Mary (Boquine) August and Mary Charlotte "M.C." August (daughter of Robert and Mary August); *circa* 1950 photograph courtesy of M.C. (August) Taylor.

5 "Tough Trencherman Baby Set," a baby tray and porringer for $9.95 postpaid, appeared in *House Beautiful*, Dec. 1949, p. 51. "Cocktails for Four," a rectangular handled tray with four coasters in the *Pine* motif for $9.95 postpaid, appeared in *House Beautiful*, Feb. 1950, p. 27. These advertisments indicated that the items could be ordered from Mail-A-Way Supply Co., Grove City 1, Pa., which was the Wendell August Forge.

6 Natale Rossi, as quoted in Catherine Youngo, transcription of recorded interview, 19 Nov. 1992; courtesy of the Wendell August Forge.

7 "Forge Makes Ambassador Plaques," *Grove City Reporter-Herald*, 30 Sept. 1949.

8 Page 6 of the eight-page October, 1950 Wendell August Forge Catalog, courtesy of Henry Swyers and the Brockway Area Historical Society. The "A, BS, DW, P" appearing with the #825 bowls refers to *Apple Blossom, Bittersweet, Dogwood* and *Pine* motifs.

9 Page 7 of the eight-page December, 1950 Wendell August Forge Catalog, courtesy of the Wendell August Forge. For jewelry, "Apple" refers to the *Apple Blossom* motif, and "Sail" refers to the *Sailboat(s)* motif. For other items, "A, BD, BS, DW, D, P, S" corresponds to *Apple Blossom, Bird Dog, Bittersweet, Dogwood, Ducks, Pine* and *Sailboat(s)* motifs.

10 Rossi, interview, 19 Nov. 1992.

11 Two items known to feature pale green Blenko glass were a 7-inch aluminum underplate holding a small bowl with an aluminum lid, called a "Salad Dressing, Jelly," and a one-quart glass "Water Bottle" with an aluminum top and undertray. The highly regarded Blenko Glass Company was established in 1893 and is located in Milton, West Virginia.

12 See Note 1.

13 Pompa, 25 July 1997; and Ziccardi, 5 Aug. 1997, personal correspondence, 15 Aug. 1997.

14 Richard Horn, *Fifties Style*, (New York: Friedman/Fairfax Publishers, 1993).

15 Grouping of items bearing the motif referred to by collectors as "Coffee Bean." The 1/8"-wide grooves on some of the items decorated many of the Wendell August Forge pieces from the mid-1950s through the early 1960s. July 1954 photograph courtesy of the Wendell August Forge.

16 Illustrated are: #550 17" diameter lazy susan with a square wood pedestal and square aluminum foot, *Dogwood* motif; cylindrical pitcher, 10-3/4" tall and 4-1/2" diameter at base, "Coffee Bean" motif; #895 7" diameter bowl, "Coffee Bean" motif. Collection of the Wendell August Forge. Photography by Richard Smaltz, Smaltz Studio, West Middlesex, PA; photograph courtesy of the Wendell August Forge.

17 Natale Rossi, "Conception."

18 Mary (Mrs. James) McCausland, as quoted in Catherine Youngo, transcription of recorded interview, 19 Nov. 1992; courtesy of the Wendell August Forge.

19 Mary Charlotte "M.C." August Taylor, personal interview, 24 June 1997; Mary Lou McConnell McNaughton, personal interview, 27 Sept. 1997; Ziccardi, 5 Aug. 1997.

20 Portion of interior of Oak Hill Savings Bank, Moundsville, West Virginia; April 1962 photograph courtesy of the Wendell August Forge.

21 Portion of interior of Union Savings Bank, location currently unidentified; April 1961 photograph courtesy of the Wendell August Forge.

22 By 1960, decorative art aluminum pieces were still being produced by Wendell August Forge, Arthur Armour, Buehner-Wanner, Kensington, Krischer Metal Products, and possibly Cellini-Craft and Wilson Specialties. Everlast Metal Products Corporation was in the process of closing its doors.

23 James W. Serio, son of Anthony Serio, personal interview by William Freeman, 20 Feb. 1998. The bronze items still remain in the Serio home, now occupied by James W. Serio.

24 Teresa Spatara, " 'Guess I'm just lucky,' Natale Rossi says," *The Herald*, Sharon, Pa., 3 May 1980.

25 Rossi, "Conception."

26 Rossi, "Conception;" Joseph Ziccardi, personal interviews, 5 Aug. 1997 and 12 Oct. 1997.

27 Ziccardi, 5 Aug. 1997 and 12 Oct. 1997.

28 Small stainless steel pieces; photograph from Allegheny Ludlum Steel Corporation press release packet dated 17 May 1962; courtesy of the Wendell August Forge.

29 *Polk's Grove City Classified Business Directory*, searches of 1962 and 1965 by Gwen Oberlin; Ziccardi, 5 Aug. 1997.

30 Ziccardi, 12 Oct. 1997.

31 Ziccardi, 5 Aug. 1997.

32 "W. M. August, Grove City, Dies at 78," *The Butler Eagle*, 12 Apr. 1963.

33 *Polk's Grove City Classified Business Directory*, search of 1964 by William Freeman.

34 Illustrated are: #820C aluminum candleholder with *Dogwood* motif, 4-1/2" diameter base and 4" tall, with 1/8" groove characteristic of 1950s design, *circa* early 1960s; #70 aluminum ashtray with a deer motif, 6-3/4" square, *circa* early 1960s; 5-3/4" diameter stainless clad dish with Cadillac emblem, *circa* mid-1960s; 8-3/4"x15-5/8" aluminum tray with a pheasant motif, *circa* mid-1960s; mug, 3-3/4" diameter base and 4-3/4" tall, with a deer motif (different from the one on the ashtray), *circa* late 1960s; #545 11-3/4" diameter underliner tray with *Pine* motif, *circa* early 1960s. Collection of William Freeman; photograph courtesy of William Freeman.

35 Catherine Youngo, personal correspondence, 27 July 1997.

36 Mary Lou McConnell McNaughton, personal interview, 10 Oct. 1997.

37 Robert August - Natale Rossi "calling card," a 4-7/8"x8" tray. Collection of the Wendell August Forge. Photography by Richard Smaltz; photograph courtesy of the Wendell August Forge.

38 "Grove City Man Named Officer of State K of C," *Grove City Reporter-Herald*, 7 May 1964.

39 McNaughton, 10 Oct. 1997.

40 "Craftsman's Touch," *Aluminum Views*, advertising pamphlet, Williams and Company, Inc., Pittsburgh, PA, Second Quarter, 1967; courtesy of Thomas Armour.

41 Taylor, 24 June 1997.

42 Ziccardi, 5 Aug. 1997.

43 McNaughton, 27 Sept. 1997 and 10 Oct 1997.

44 Taylor, 24 June 1997; Joseph Leone, personal interview, 24 July 1997; McNaughton, 27 Sept. 1997; Ziccardi, 5 Aug. 1997, 15 Aug. 1997 and 12 Oct. 1997.

45 Polly Powell and Lucy Peel, *50s & 60s Style* (London: Quintet Publishing Limited, 1988), p. 82.

46 McNaughton, 10 Oct. 1997.

47 An advertisement for Reizenstein's, in the 11 Dec. 1972 *Pittsburgh Post Gazette*, includes *Columbus* and *Landing of the Pilgrims* from *Great Moments in History*, and the two President plates, *John Fitzgerald Kennedy* and *Abraham Lincoln*. Courtesy of Thomas Armour.

48 Ronald E. Winder, personal interview, 7 Aug. 1997.

49 Illustrated are: 9" diameter pewter plate, Serial No. 90, *Abraham Lincoln* (1972 initial issue); 8-3/4" diameter pewter plate, Serial No. 1935, *Columbus* from the *Great Moments in History* (First Edition, 1971 initial issue); 8-1/2" diameter bronze plate, *Signers of the Declaration* from the *Great Moments in History* (Last Edition, 1976 initial issue); 8" diameter pewter plate, Serial No. 169, *The Carolers* from the *Christmas* series (First Edition, 1974 initial issue). All dies engraved by Natale Rossi. All items from the Collection of Joseph A. McCandless and Sarah Jane McCandless; photographs courtesy of William Freeman.

50 Based on Wendell August Forge Plate Collectors Division Catalog, 1 July 1984, courtesy of Mary Ann Felegy; and Ronald E. Winder, personal correspondence, 2 Oct. 1997.

51 Portion of the 1971 Wendell August Forge brochure. The 1971 and 1974 brochures consist of single 11"x17" folded sheets. Each brochure depicts the *Schooner* motif and the Wendell August Forge trademark representation of 1932 on the front; an adaptation of the 1946 Alcoa advertising series that featured the Wendell August Forge on the back; and small images of various items on the inside, with a price list in the center. Courtesy of the Wendell August Forge.

52 Sources of information pertaining to commercial clients include Marion Dick, personal interview, 4 Aug. 1997; McNaughton, 10 Oct. 1997; Ziccardi, 12 Oct. 1997; Rossi, "Conception;" and the Collection of William Freeman.

Chapter Five

The Knecht Transformation: 1978-1994

In 1977, F. W. "Bill" Knecht III, a native of Warren, Ohio, was completing his eighteenth year in marketing and sales with the IBM Corporation, having joined the company after his graduation from Youngstown State University.[1] Although he enjoyed his work at IBM and was successful at it, he had long dreamed of having his own firm.

At the same time, in Grove City, Robert August was considering potential options for his retirement. Since there were no August family members interested in running the Wendell August Forge, his options were limited to either closing down the business completely when he retired, or selling it to someone outside of the family. The success of the Limited Editions that the Forge had begun offering in 1971 ensured that the Forge, although small, was an enterprise with financial value. A reasonable course of action was to seek a buyer.

In late 1977, Robert August discussed with his accountant, Jerry DeNicholas, the possibility of selling the Wendell August Forge. Around the same time, Bill Knecht was discussing his potential interest in acquiring a small company with his accountant, the same Jerry DeNicholas. As a result of the common DeNicholas connection, August and Knecht were brought together.

Knecht had first visited the Wendell August Forge on his sales route some 15 years before when a new typewriter was needed, and "fell in love with the place."[2] In February, 1978, within 90 days of learning about the opportunity to purchase the Wendell August Forge, Knecht acquired the firm, with a three-year employment contract while he learned the business from Robert August. During this period, August would remain as President, with Knecht as Executive Vice-President.

Knecht left a large corporation with thousands of employees for a small family-owned firm of about 20 people, a considerable change for him. While scientists and engineers at IBM were designing computers, the craftsmen at Wendell August Forge were "still cutting oval dishes with tin snips."[3] Knecht's daughter, Deborah, later recalled "I can still remember my first impression when I came over to visit when I was still in high school. I looked around and thought, oh my gosh, we're going to starve."[4]

Had Knecht realized how challenging his

F. W. "Bill" Knecht III and F. W. "Will" Knecht IV

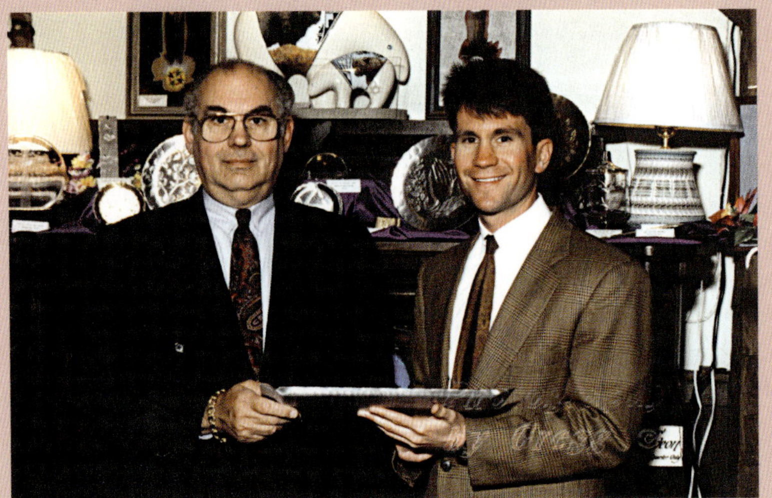

Frank Wilson "Bill" Knecht III is the Chairman and Chief Executive Officer of the Wendell August Forge. A native of Warren, Ohio, he graduated from the Mercersburg Academy and earned his B.S. degree at Youngstown State University in Youngstown, Ohio in 1959. He is married to Connie (Ralston) Knecht, and they have two children, Deborah (Knecht) Fetter and Frank Wilson "Will" Knecht IV.

Prior to purchasing the Wendell August Forge in 1978, Bill Knecht was in marketing and sales with IBM for 18 years, specializing in educational sales. During his tenure with IBM, he became a member of IBM's Golden Circle, recognizing him as one of the top 100 salespersons in the country.

He has long been actively involved in community affairs in both Mercer County, Pennsylvania and Mahoning County, Ohio, including service on the boards of the Penn Northwest Development Corporation, Grove City College, The Youngstown Club, and the Boardman (Ohio) Board of Education. Among the several awards that he has recently received as a result of his community service are the 1992 Man of the Year Award from the Mahoning County American Cancer Society, 1995 Boy Scouts of America French Creek Council Distinguished Citizen Award, 1995 Community Service Award from the Boardman Civic Association, and 1996 Louis and Barbara Thiel Distinguished Service Award from Thiel College of Greenville, Pennsylvania.

Frank Wilson "Will" Knecht IV (1966-present) is the President of the Wendell August Forge. He earned his B.A. degree in Business Administration from Wake Forest University in Winston-Salem, North Carolina in 1988, where he served as President of the Student Body. Will Knecht has been with the Forge since 1989. In addition to fulfilling his responsibilities as President of the Wendell August Forge, he is active in the Boardman (Ohio) Civic Association, Downtown Youngstown (Ohio) Revitalization and Greater Youngstown (Ohio) Coalition of Christians, the Fellowship of Companies for Christ, and the Old North Baptist Church.[1]

new career would rapidly become, he might have reconsidered the change. Forty-five days after Knecht reported for work at the Wendell August Forge, Robert August was diagnosed with a terminal illness, and died in December of 1978. Knecht recalled "Bob became very ill very quickly, and couldn't come to the shop. I would go over to his house every day with the mail so that we could keep the business going."[5] Cathy Youngo remembered that "Bob was to stay on for three years or so to help Bill learn the company, but the fates said no and Bob died within that first year."[6] In spite of the tragic situation, the firm survived. Mary Lou

McNaughton, Robert August's office manager in the 1960s and 1970s, contended that the Wendell August Forge would have ceased to exist had Knecht not taken it over when he did and had the dedication to keep it going.[7]

Not only did the Forge survive, but Knecht was also able to make some progress with his conception of what the Wendell August Forge could become. To begin, a new trademark handstamp was created to commemorate the change in ownership, with the added virtue that it incorporated the "Handmade" and "Grove City, PA" references into the single handstamp. A marketing and publicity plan was developed and implemented, and the Limited Editions series were sustained. Additional employees were hired, which required the establishment of a formal apprenticeship training program, since few craftsmen existed who were skilled in the handworking of metals. To ensure that the Forge's product would consistently be of the highest quality, a stringent quality control system was implemented. Knecht established a process whereby each craftsman or finisher was authorized to reject an item in which they found some flaw, and the lacquering[8] and packing people were given the same authority.[9] In his fourth month with the Forge, Knecht began grappling with the essentially nonexistent employee fringe benefits; there was only an outdated health plan with just two employees enrolled.[10] By the time that his first anniversary with the Wendell August Forge arrived, Knecht was able to report that the Forge had had an excellent year in 1978. The forecast for 1979 was even better, with plans to have a small museum by 1980.[11]

In 1979, the Forge received an important commission from the U.S. Arms Control and Disarmament Agency to produce 12 solid bronze plates commemorating the Strategic Arms Limitation Talks II treaty between the United States and the Soviet Union.[12] The plates were presented to diplomats at the signing of the treaty in Geneva, Switzerland. The die, engraved by Natale Rossi, was destroyed after the plates were created. Knecht involved as many of the craftsmen as possible in the honor of producing the plates by having twelve different people work on the historically-significant commission.[13]

The year, 1979, was also one of construction and remodeling. A portion of the space between the Wendell August Forge building and the old REA Metal Products building was enclosed, joining the two buildings together and expanding the facility in the process. The former REA Metal Products building was then remodeled to create a combined showroom and museum. Exceeding his own timetable, Knecht was able to open the showroom/museum facility to the public by November.[14] Catherine "Cathy" Youngo, remembering the former need "to wear longjohns under slacks to keep warm; we had no central heat, but we did have a kerosene stove,"[15] did not regret leaving the old showroom for a new one that included not only heat but also fine antique furnishings. The Forge production area was also revamped to more readily accommodate visitors, who could watch the metal pieces being made by hand.

Knecht also undertook the effort to develop the nomination of the Forge's master die

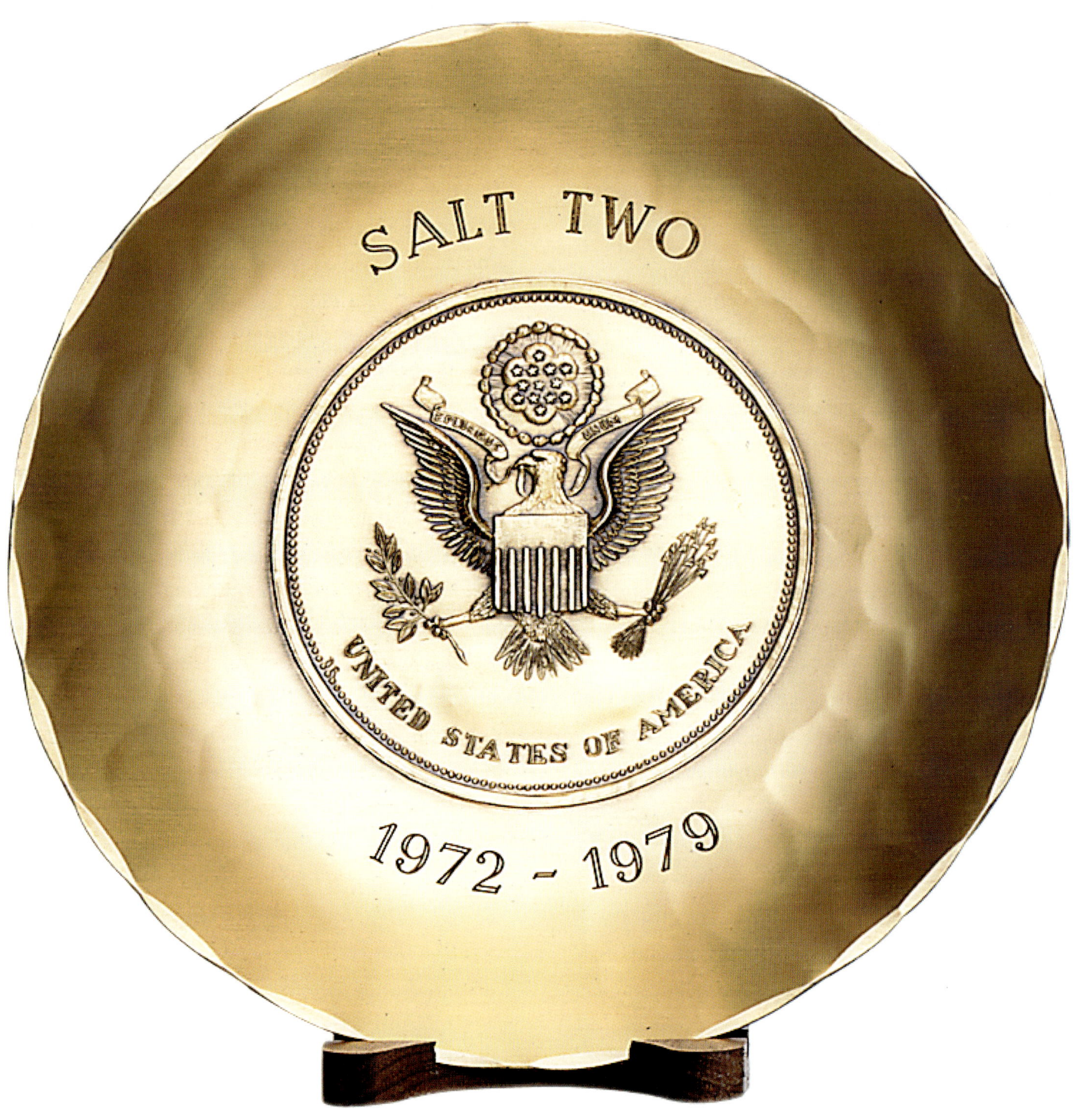

Bronze plate commissioned to commemorate the Strategic Arms Limitation Talks II treaty (note 12)

engraver, Natale Rossi, for a prestigious Hazlett Memorial Award for Excellence in the Arts in Pennsylvania. Rossi had engraved hundreds of dies since the late 1940s, and been the primary salesperson for corporate accounts since the 1960s. The Pennsylvania Council on Arts soon selected him as the recipient of one of the ten awards that were granted early in 1980.

By the end of 1979, the Forge had produced several newly-engraved dies, such as a view of the Three Rivers area of Pittsburgh, and issued a new Limited Edition of *Neuschwanstein Castle* plates and tankards featuring the famous Bavarian royal castle of Ludwig II. An increased number of special corporate gift and commemorative items had been completed, and a growing number of visitors had been accommodated, including an average of three tour buses full of passengers each week during the Summer months.

However, December of 1979 also presented Knecht with an unexpected challenge when Natale Rossi, by now 71 years old, decided to retire from the Forge, shortly thereafter starting his own business, By Natale.[16] He had been with the Forge almost continuously since the late 1920s, and now knew more about the Forge and its work than did anyone else. The impact of his departure was heightened when it became apparent that his son, die engraver Christopher Rossi, would also be leaving, thereby leaving the Forge with no die engravers. Thus 1980 would commence with a crisis, as new die engravers were needed immediately. Steven Adams and David Bruck,[17] both of whom had been finishers for the Forge in the late 1970s and had recently completed commercial art degrees, were quickly recruited to begin honing their die engraving skills.

Much of the early 1980s swept past in an enormous blur of work. Most of the "old-timers" had retired by this time, taking their expertise with them. At the same time, advertising and marketing were increasing sales and, of course, the amount of work to be done. Cathy Youngo, reminiscing about the first few years, recalled that "Bill and I would spend each morning writing job cards for the men to tell them what to produce for the day. We had many commercial orders as well as customer and retail orders to make, and as we hobbled along making each of those *long* days count, we were able to accomplish miraculous things. We were never late with an order and rarely did we have to send something back to the shop to be *fixed*. Customers came and went in the showroom, never realizing that we were operating with a novice boss, a single internal sales person, a plant manager, and a lot of good men."[18]

By 1985, Forge products[19] could be embellished with at least 100 different motifs from numerous newly-engraved dies. Among the many motifs that made their appearance during this period were *Antique Bicycle, Butterfly Daisy, Cardinal, Clown Face, Irish Setter, Lily of the Valley, Owl, Pineapple, Santa Mouse, Unicorn* and *Double Wedding Bells*. Items were carried in a wide range of metals, including aluminum, bronze, copper, pewter, stainless clad aluminum, stainless steel, and sterling silver. The creation of Limited Edition and collector series had been continued and

By Natale

By Natale, Inc. was formed in May of 1980 by Natale Rossi and two of his eight children, John Mark and Christopher. The corporate officers were Natale's wife, Irene (Panada) Rossi as President, Christopher Rossi as Vice President, and John Mark Rossi as Secretary, with Natale Rossi serving as advisor and salesman. The organization of the firm changed several times during the 1990s, with the deaths of both Irene and Natale Rossi, and the departure of John Mark Rossi from the business, leaving Christopher Rossi as the sole proprietor.

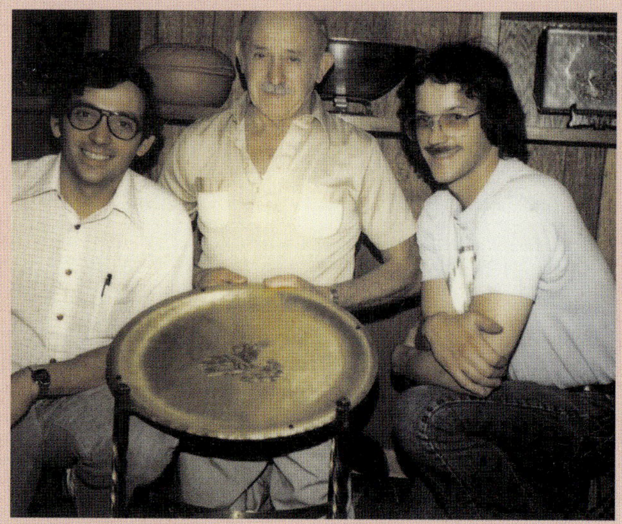

John Mark, Natale and Christopher Rossi

When By Natale was started, Natale Rossi had recently retired from the Wendell August Forge after his lengthy career there as a blacksmith, die engraver, and salesman. Christopher Rossi had been a craftsman and die engraver with the Forge for six years. The die engraving for By Natale was shared between them. They were soon joined by Natale's brother, master craftsman Lewis "Doc" Rossi, when he retired from the Wendell August Forge.

The initial location of the By Natale forge and showroom was at the rear of 405 Rainey Avenue in Grove City. In about 1986, the firm relocated to the Schoolhouse Shops in Leesburg, and in 1995 it relocated again to the Old Company Store on Old Ash Road outside of Grove City. By Natale, Inc. discontinued operations in early 1997.

The By Natale product line included a wide range of hand crafted metalwares, such as coasters, trays, clocks, desk sets, vases, bowls, hot mats, tankards, vases, magazine racks, ornaments, jewelry, candleholders and sconces, produced primarily in aluminum, bronze and pewter. Although documentation of formal motif names is incomplete, the repoussé motifs that were developed included a variety of animals, such as deer and eagles, florals, such as dogwood and wild rose, and scenes, such as golfing, a covered bridge, and some cityscapes. Motifs with moving religious themes were also created, the most notable of which was Natale Rossi's 10-1/4"x15-3/4" intricate engraving of The Last Supper. By Natale also produced a few series of Limited Edition plates, numerous custom-engraved dies, and several special commissions such as a five-foot, three-dimensional bronze dogwood tree.[16]

expanded.[20] *Raccoons*, the first of the *Animal Kingdom* plate series, was introduced in 1981. The *Christmas Heirloom* ornament series, an idea of Knecht's wife, Connie,[21] was introduced in 1980.[22] In 1982, Steven Adams[23] had left to join the Silver Spring Forge, creating the need for a new die engraver, and finisher Leonard Youngo[24] was selected for the position. Collectors were becoming entranced with the pieces that the Forge had produced in

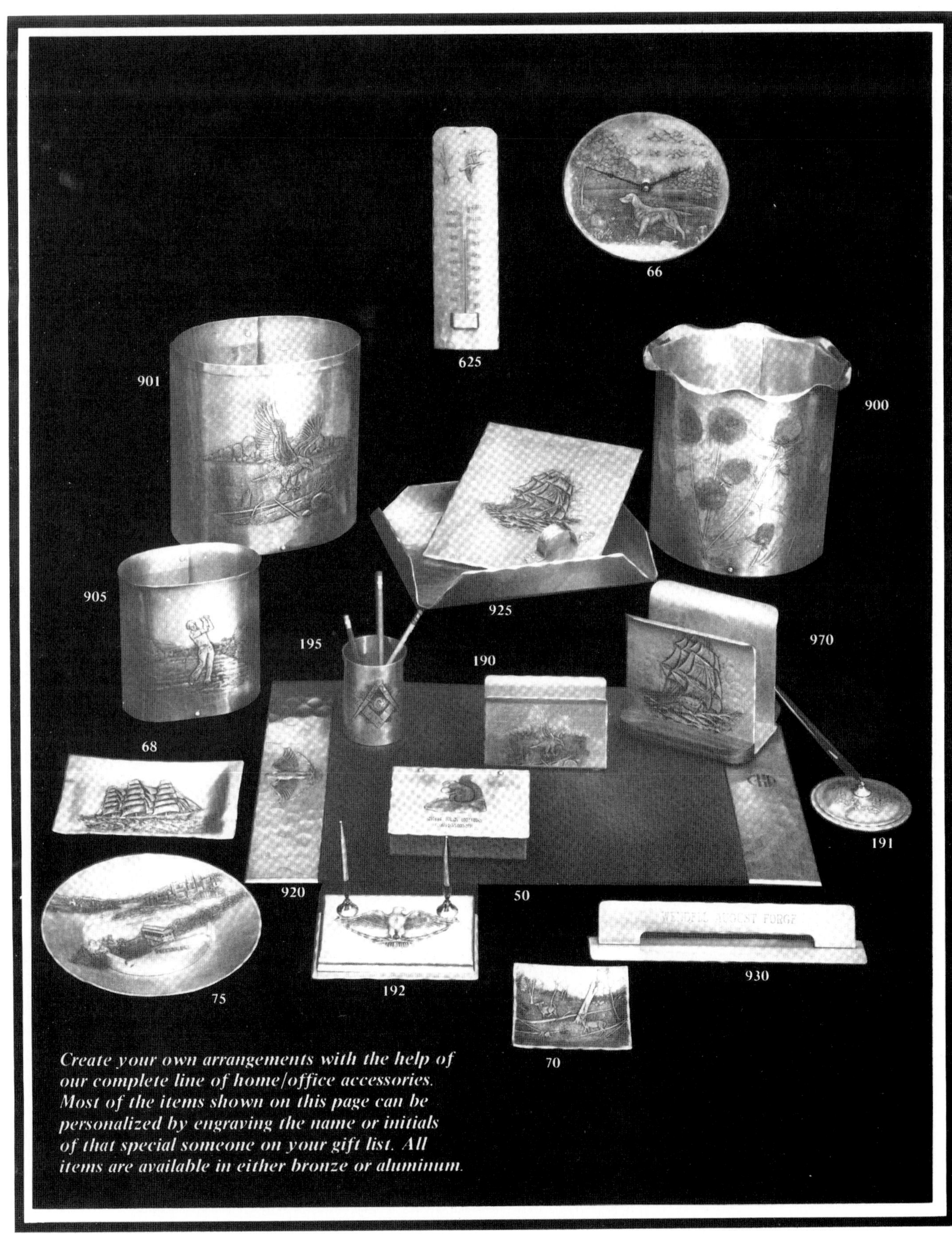

Page 3 of the 1985 Wendell August Forge catalog *(note 19)*

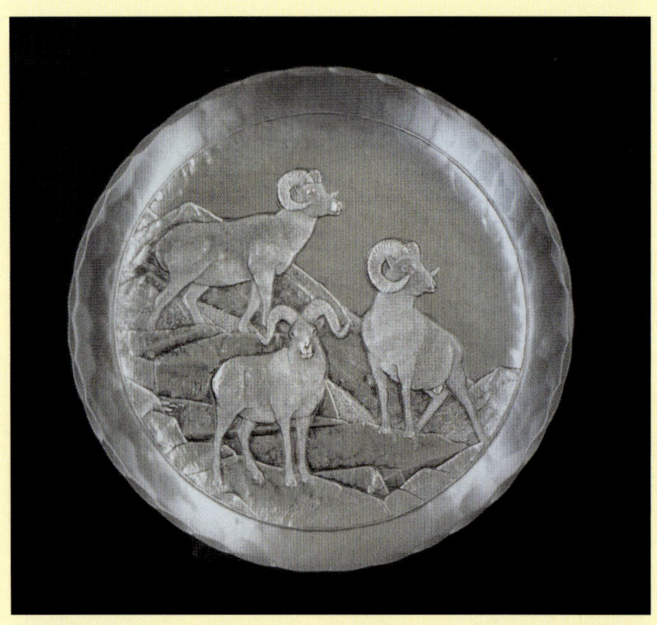

Limited Edition Plates (note 20)

Left, Big Horn Sheep *in aluminum*
Below, The Skating Pond *in bronze*

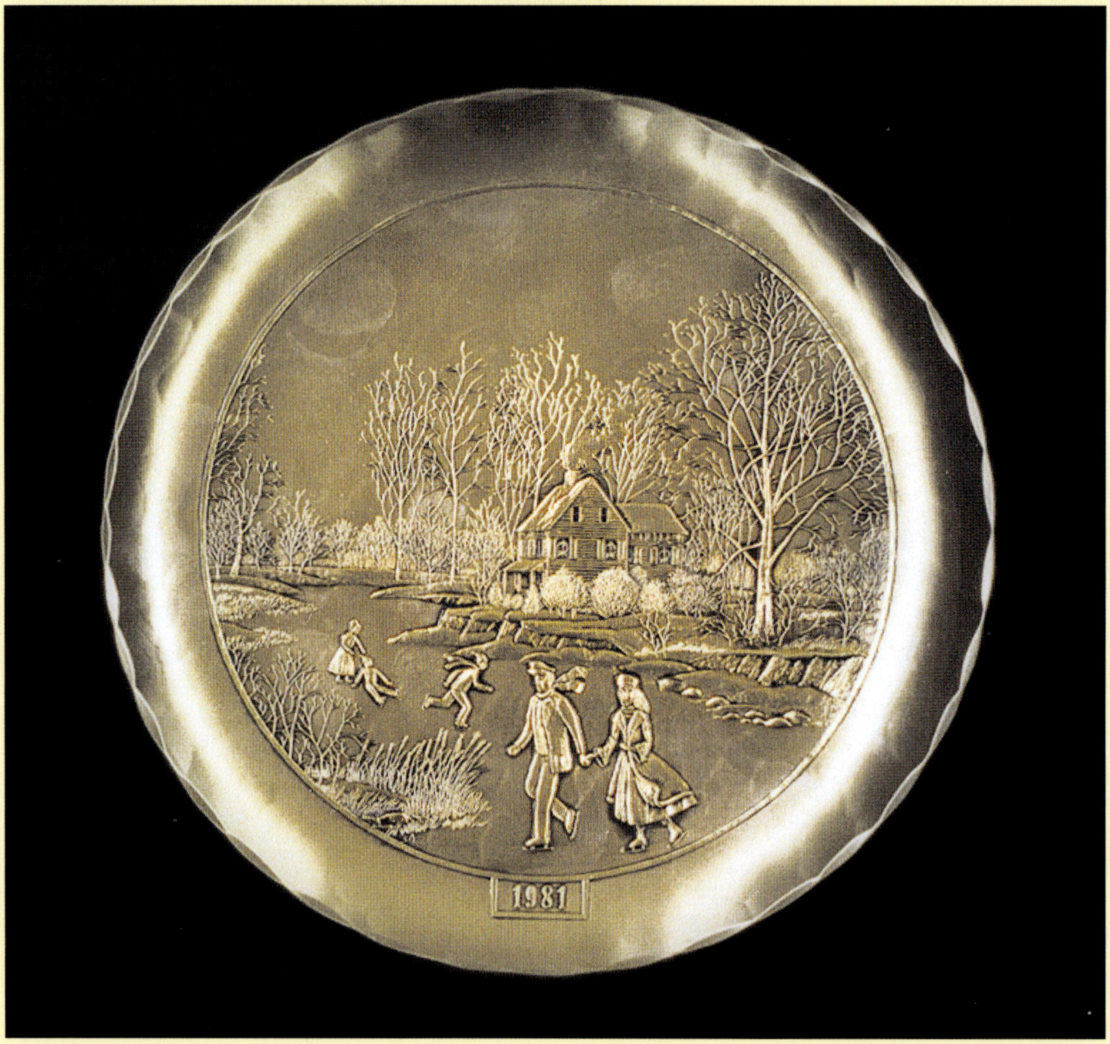

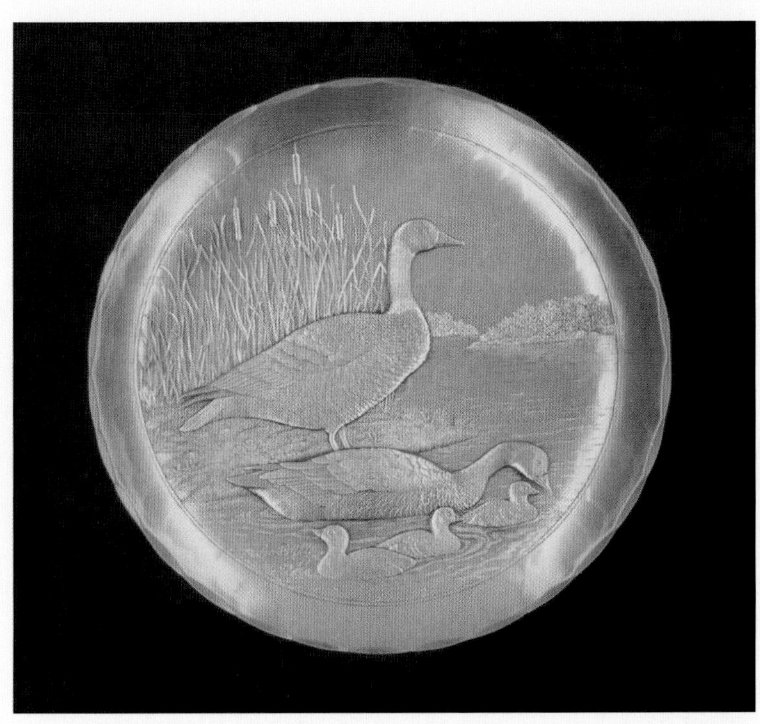

Left, Canadian Geese *in aluminum*
Below, Journey to Bethlehem *in bronze*
Bottom, Honey Bear *in pewter*

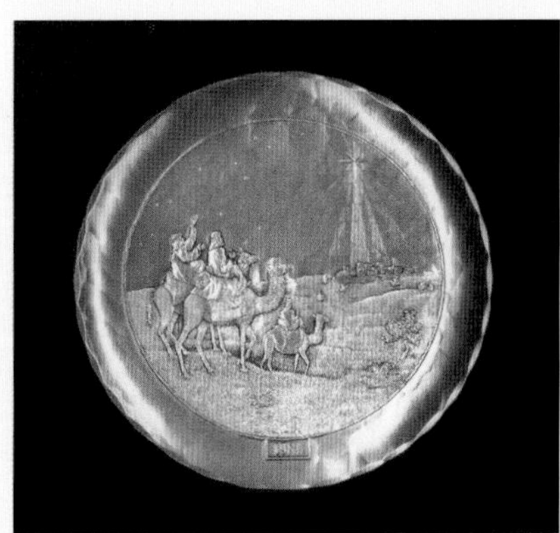

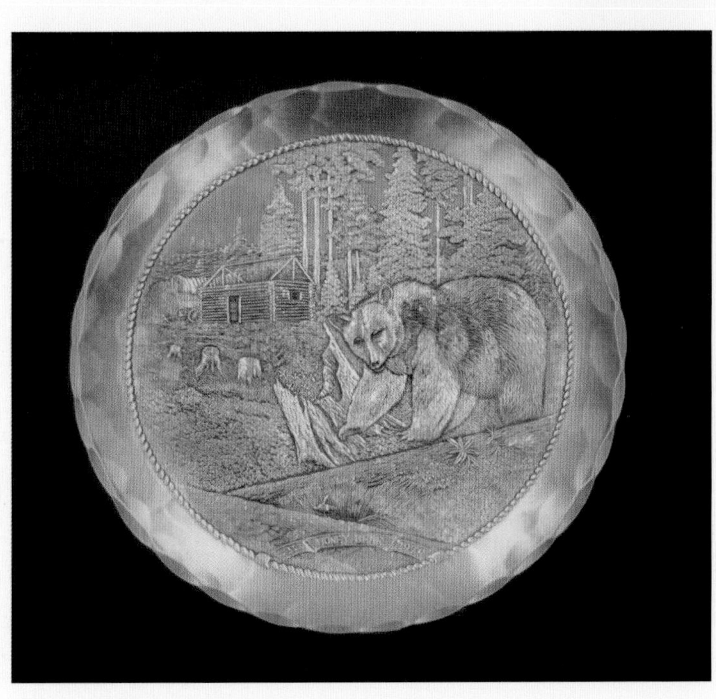

decades past, and both Bill Knecht and Cathy Youngo frequently found themselves attempting to respond to inquiries about the undocumented history of the Forge in addition to their many other responsibilities.

During the early 1980s, Knecht was able to implement another facet of his idea to reposition the Wendell August Forge from being a manufacturer providing quality hand crafted metalwares at wholesale to being a manufacturer providing those products at retail. As one result of his strong personal commitment to sustain and perpetuate American artisanship, he hired the renowned stone wheel glass engraver, Gene Scala. The Grove City facility, now bursting at the seams, was once again expanded and remodeled, in order to create work space for Scala and for the Crystal Gallery display area. The relationship would continue until 1993, when Scala would leave to start his own business. In the intervening years, thousands of visitors would watch the engraving of stemware, beverage sets, vases, trays, rose bowls, pendants, boxes, prisms, and countless other forms of leaded glass, using styles ranging from the ornate Bohemian to the clean-cut contemporary.[25]

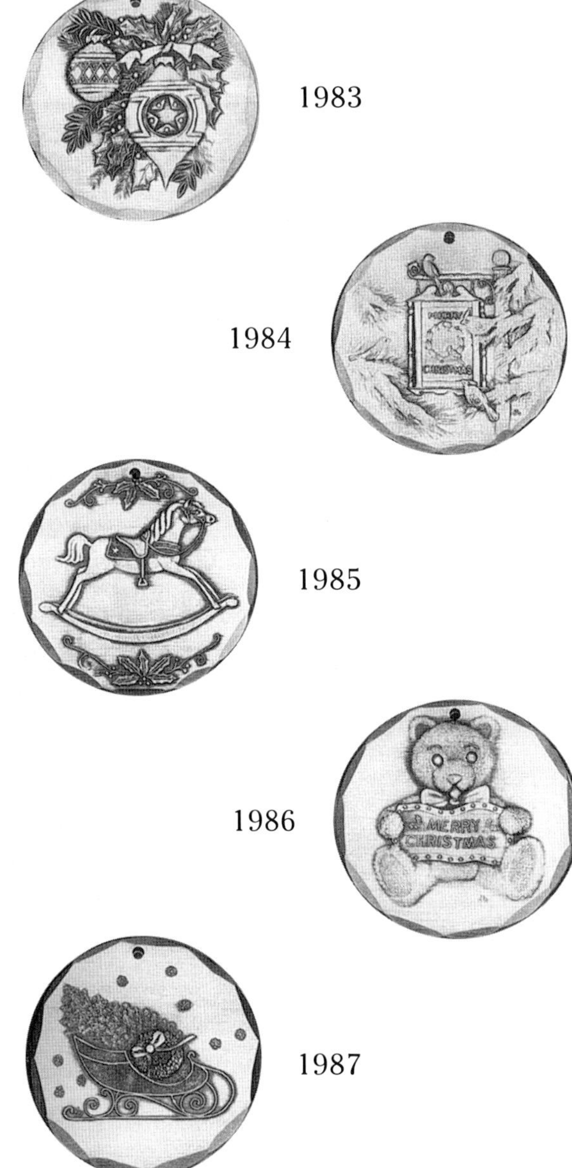

Christmas ornaments (note 22)

Knecht's concept of moving the Forge toward a more retail-oriented mode of operation clearly demonstrated its value when, in the mid-1980s, "the economy went down, corporate accounts went down, and retail carried it."[26] Throughout the last half of the 1980s and into the beginning of the 1990s, the Forge not only remained healthy, but also maintained a modest level of growth.

The product line continued to develop during the late 1980s,[27] although some items, such as boot jacks, youth chairs, and ornamental basket stands, were discontinued. The range of metals offered to the public was also limited somewhat, as copper, stainless clad aluminum, and stainless steel were removed from the regular retail product line, although special orders continued to be accommodated. New motifs continued to be developed, with nearly

David Bruck

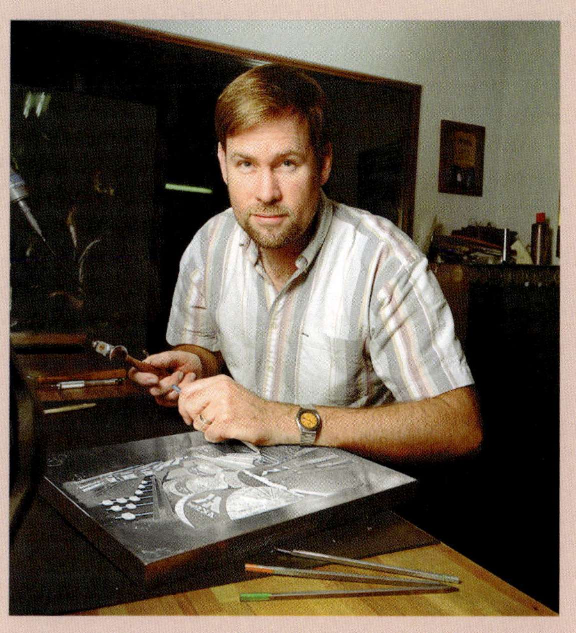

Master die engraver David Bruck (1956-present) grew up on a farm near Leesburg, Pennsylvania, which his parents had purchased when they were married in the early 1950s. His interests in art emerged while he was attending local schools, and he takes pride in his high school art teacher's favorable comments about some of his work for the Wendell August Forge.

Bruck first worked for the Forge in 1976 and 1977 as a finisher, and then continued to work some weekend and summer hours while studying commercial art at the Community College of Allegheny County in Pittsburgh. After completing his degree in 1979, he designed and lettered signs for a shop in the Pittsburgh area, until the opportunity unexpectedly arose in early 1980 to become a die engraver for the Wendell August Forge.

Since joining the Forge as a die engraver, Bruck has engraved about 600 dies, most of which have been based on his own original art work. The first motif that he engraved for a showroom piece was Wedding Bell. Among the motifs that he has developed and with which he is the most pleased are Farm Scene, Ruffed Grouse, Satellite Drive-In, Love Birds, Timber Wolves *for the* Wildlife *Limited Edition Series, and* A Savior is Born *for the Limited Edition Christmas Plate Series.*

Although other responsibilities leave him with little time for artistic pursuits, he retains an interest in watercolors, sculpture, and sign design and lettering. Bruck and his wife, Eva (Edwards) Bruck, reside in the Grove City area with their three sons, Aaron, Tristan, and Clinton.[17]

200 of them identified in the catalog by 1990. *American Quarter Horse, Applebasket, Capricorn, Geese and Teddy Bear, Hummingbird, Kitten in a Basket, Lion, Mountain Laurel, Skating Partners, Tulips* and *Windy Day* are among the motifs introduced in this time period. Continuing the quest to perpetuate fine American craftsmanship, coin necklaces, intricately cut by hand, were added to the product line.

Throughout most of the 1980s, it was still feasible for Dave Bruck and Len Youngo to do all of the die engraving and virtually all of the art work for the dies, not to mention an occasional unique commission. Dave Bruck, for example, created a special piece to commemorate a Super Bowl victory for the San Francisco 49ers, which entailed meticulous overlays of sterling silver on bronze, right down to the laces on a football. Len Youngo is partial to his engraving of an expansive private residence, produced in bronze, which was embellished by using a series of sterling silver overlays for the stone facade, roof shingles hand-carved from slate, and inset pearls for lamp globes. He is also pleased with his engraving of a Native American on horseback, themed to a painting, "Awakening of a Higher Spirit," which was rendered in bronze with sterling silver inlays and set with turquoise.[28]

Much of the original art work that Bruck and Youngo created for their die engravings was drawn from their own personal lives and experiences. For example, Youngo's models for his art and engraving for the 1989 Limited Edition Christmas plate, *Building a Snowman*, were his own children. He remembered that "getting the youngest, a two-year-old at the time, to pose" was quite an experience. Bruck draws on scenes and images from his own past for most of his original art. He credits his natural inclination for being very observant for his ability to recall scenes and images, "seeing them in great detail in my mind's eye," and then create his art work from those memories. Examples of such work include his *Farm Scene* motif and 1986 Limited Edition Christmas plate, *Sounds of Christmas*.[29]

Improvements in production methods continued, with the acquisition of additional equipment to assist in cutting and forming some types of pieces. The hearing losses that Forge craftsmen had suffered in earlier decades from the constant clang of metal on metal had been dealt with by protective ear devices and enclosure of the repoussé area. In 1988, the potential for injuries such as carpal tunnel syndrome were addressed with the acquisition of air-powered hammers.[30]

After celebrating its 65th Anniversary in 1988, the Wendell August Forge became a family affair for the Knechts the following year. Bill Knecht's daughter, Deborah, and son, Will, joined the company. Deborah had been working in Scottsdale, Arizona for three years following her graduation from Youngstown State University, and returned to assume marketing responsibilities. Will, who had graduated from Wake Forest University in 1988, began working in operations. Shortly thereafter, the entire Wendell August Forge workforce posed for an impromptu group photograph.[31]

In 1990, despite the languishing economy, Bill Knecht continued his vision of repositioning the Wendell August Forge and providing it with opportunities to prosper. The catalog style was changed markedly. The intermittent catalogs of the 1980s had been augmented by separate and lengthy annually-issued lists of motifs and prices, with the Limited Edition pieces featured in a separate publication. The 1990 catalog assumed a more consumer-oriented look, with product images, available motifs, prices, and ordering information all integrated into a single catalog,[32] although the Limited Edition items continued to appear in a separate brochure.

Bill Knecht had successfully addressed the needs of corporations and other organizations for awards, commemoratives, and special gift items by developing an extended network of about 200 distributors to handle Wendell August Forge products for such clients in their territories. By this time, the Forge had crafted presentation pieces for virtually all of the Fortune 500 firms and countless smaller

Silver Spring Forge

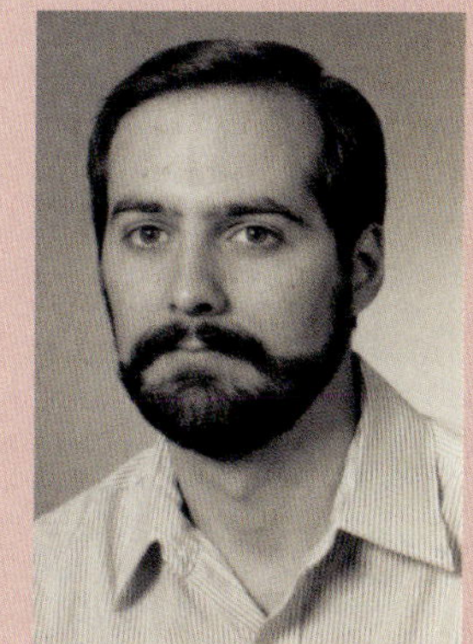

Steven Adams

The Silver Spring Forge was established in late 1982 by Fred P. Todarello and his brother-in-law, James McCutcheon. Todarello (1923-1990) was born and raised in Grove City, graduating from Grove City High School in 1941. He served in the military during World War II, and then was employed with the Wendell August Forge for about 32 years after the war. While at the Wendell August Forge, he specialized in welding, and also supervised the shop for several years in the 1970s.

The die engraver for the Silver Spring Forge was Steven G. Adams (1957-present), who was apprenticed as a craftsman at the Wendell August Forge in 1975, leaving in 1977 to study commercial art at the Community College of Allegheny County in Pittsburgh. He returned to the Wendell August Forge in late 1979, after Todarello's retirement, and was rapidly thrust into die engraving in early 1980.

Adams continued as an engraver with the Wendell August Forge until the opportunity arose in 1982 to join with the Silver Spring Forge as the craftsman and die engraver for the new concern when it started. The Silver Spring Forge was located in New Kingstown, Pennsylvania, near Harrisburg. After about 1-1/2 years, Todarello and McCutcheon decided to leave the business, and Adams continued to run it on his own for an additional six months before closing it down.

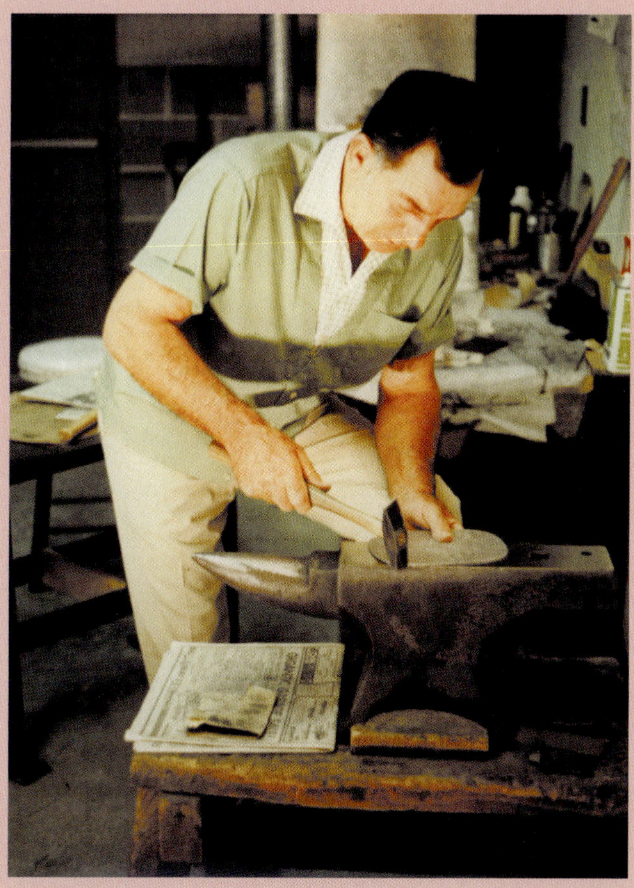

Fred Todarello

The Silver Spring Forge produced hand crafted metal items of aluminum bearing repoussé motifs designed and engraved by Adams, and occasional special commissions in bronze, copper and pewter. The product line included coasters, small trays, an oval dish, napkin holders and rings, clocks, cuff bracelets, plates, hot mats, an oval waste basket, a candleholder, bookmarks, and a letter file. Motifs used included Appleblossom, Dogwood, Strawberry, Hummingbird with Rose, Raspberry with Butterfly, Butterfly with Wild Rose, Butterfly with Daisy, Pine, Eagle, Deer, Ducks, Trout Scene, Clipper Ship and Wedding Bells.

Adams currently resides in Wisconsin where he is a sculptor/engraver for the Medalcraft Mint. He has also done some free-lance engraving for the Wendell August Forge, and produces his own line of aluminum wares bearing the SGA Sculpture/Engraving logo.[23]

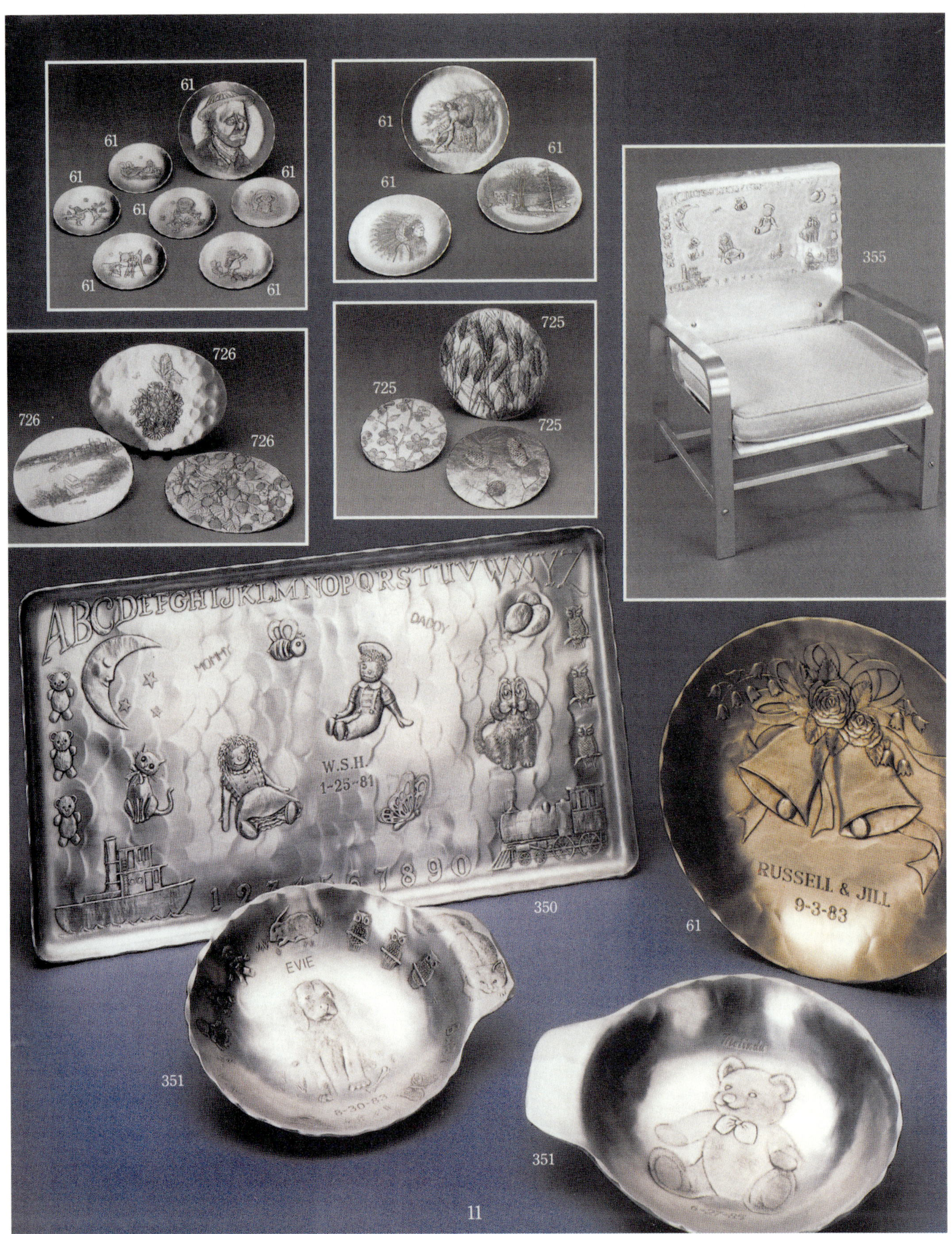

Page 11 of 1987 Wendell August Forge catalog (*note 27*)

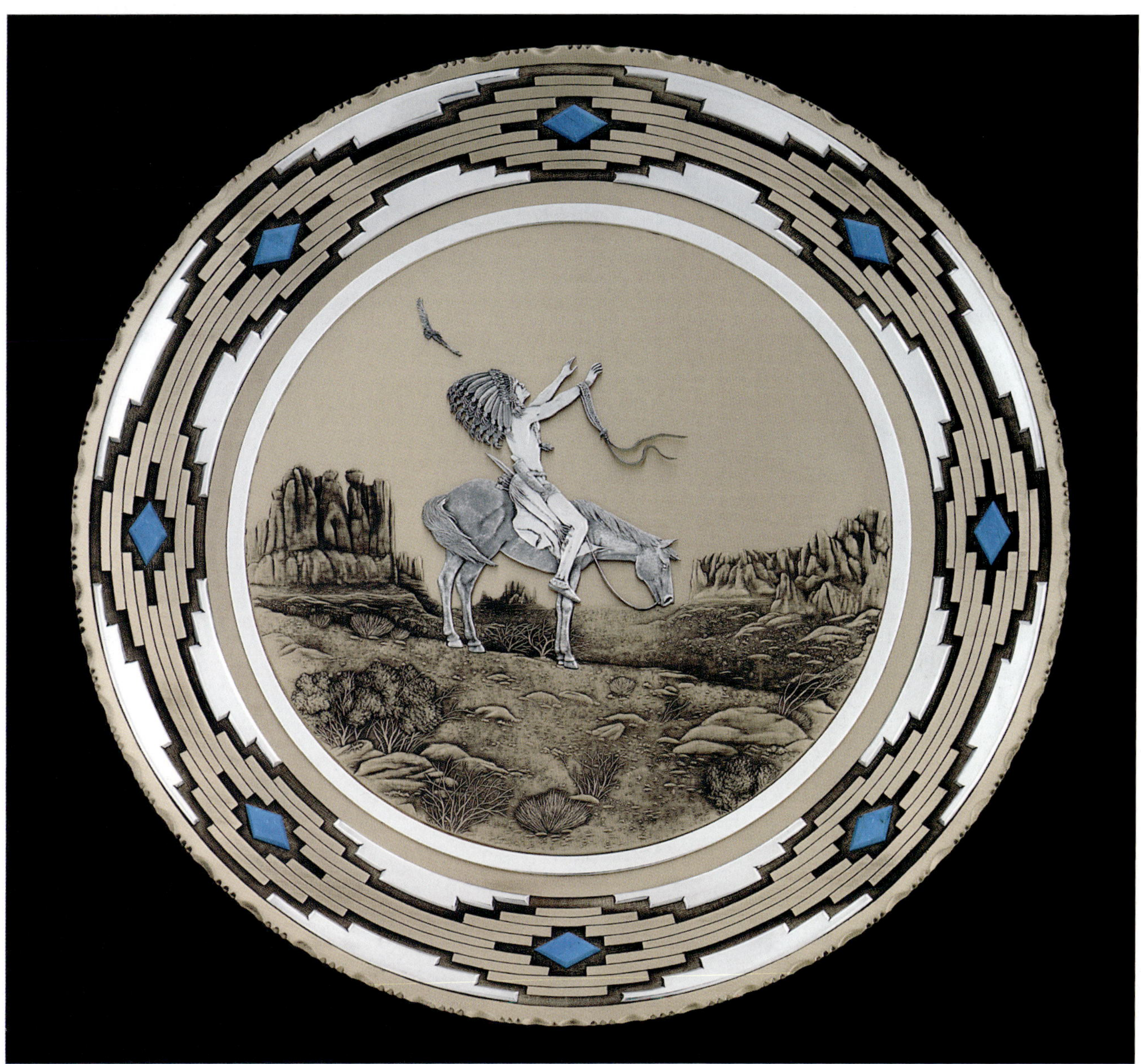

Custom plate in bronze with sterling silver and turquoise (note 28)

businesses, religious organizations, social groups, and others, with customers in every state in the United States and more than 25 foreign countries.[33] For more than a decade, however, Knecht had been continually pondering how to address two fundamental retail issues. First, the Wendell August Forge of Grove City, Pennsylvania was situated off the beaten track on a dead-end street in a residential area of a small borough of about 8,000 people. The site and the borough were steeped with the history of the Forge and the industry that it had started. Expansion opportunities were highly limited, yet it was unthinkable that the firm be relocated in order to improve accessibility. Second, in a time period in which hand crafted American products had almost completely vanished

Leonard Youngo

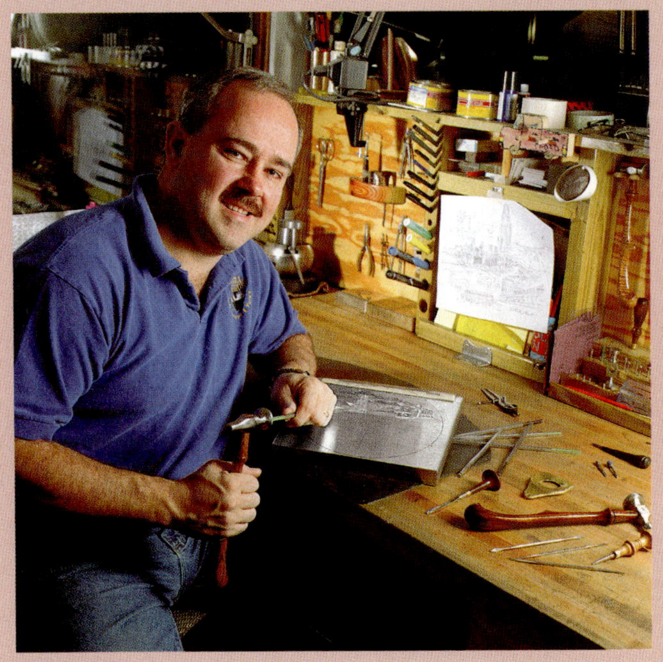

Master die engraver Leonard Youngo (1959-present) was born in Richmond, Kentucky where his father was a student at Eastern University. In 1967, after living in Greenville, Pennsylvania and Warren, Ohio, the Youngo family settled in Grove City, Pennsylvania.

Youngo attended Westminster College in New Wilmington, Pennsylvania, majoring first in biology and then in mathematics, earning his degree in 1981. Somewhat undecided as to his future career direction, he obtained a position as a finisher at the Wendell August Forge while he explored various career options. While at the Forge, his earlier talents and interests in art and hand crafts were stimulated, and when a die engraving opportunity became available in 1982, Youngo applied and was selected as an apprentice for the position.

Youngo's Renaissance background and talents have served him well in the intervening years, during which time he has executed almost 800 die engravings at the Wendell August Forge, for some of which he has developed the original art work. The first motif that he engraved for a showroom piece was the face of Emmett Kelly for Clown Face. *Favored among the motifs that he has engraved are* Indian on Horseback, McConnell's Mill, Pittsburgh Collage, Oriental Village, Building a Snowman *for the Limited Edition Christmas Plate Series, and, most recently,* Creekside Memories.

Youngo's personal interests include athletics, dinosaurs and archaeology, and woodworking. He and his wife, Beth (Michels) Youngo, reside in the Grove City area with their two children, Andrea and Michael.[24]

from the marketplace, the proper presentation of the Forge's work was essential. Traditional retail outlets, such as department stores, were unable to position the product appropriately and, with the typical markup, priced the Forge's product beyond the access of the average consumer. American-made hand crafted products, requiring high investments of individual skill and labor, simply could not be successfully marketed in the same manner as mass-produced products. How could the Forge's high-quality, American-made, hand-crafted objects best be brought to the public?

Knecht decided to experiment by opening a satellite retail outlet in a shopping mall. Named The Signature Collection, the shop opened in 1990. It was located in the Great Lakes Mall in Mentor, Ohio, situated on Interstate 90 along the south shore of Lake Erie, at the northeastern edge of the Cleveland metropolitan area. He was soon dissatisfied with both the concept and the store's performance. Mulling over the problem in a conversation with Natale Rossi in 1992, Knecht commented "we have a store in a mall around Cleveland . . . they don't know that it [the product] is handmade; it is just another store in the mall," continuing, "the reason I don't think it is ever going to work in the store well is that we can't get the message across that it is handmade. That's the problem."[34]

While Bill Knecht continued to seek the best way to bring American hand-crafted metalwares to the public, the Forge continued to make improvements in its service for its visitors and international direct-mail clientele. The Forge's Grove City showroom was remodeled again in 1993. The renovated area was described in the *Allied News* as radiating "warmth, spaciousness, beauty, and an awareness that every customer is treated as a VIP."[35] Attractive catalogs began to be issued annually. Rather than focusing almost exclusively on general types of items, such as coasters, trays, or plates, which could then be requested in several motifs, the Forge's catalogs also began to feature collections consisting of a variety of pieces available in a single motif. In 1993, for example, the customer could readily select from the catalog one or more pieces from the *Amish, Appleblossom, Dogwood, The Gathering, McConnell's Mill, Skyline,* and *Strawberry* collections.[36] The Limited Edition offerings were also incorporated into the same catalog, including the first of the *Olde World Santa* ornament series, *Father Christmas*. By this time, all of the older dies had been retired, and Knecht could take pride in the fact that the dies for the more than 120 different motifs identified in the catalog had been created after his acquisition of the Forge.

The 1993 catalog was also a bellwether of a sort, as the art work for some of the new motifs, such as *The Gathering*, was done by area artist Linda Hoover, rather than die engravers Dave Bruck and Len Youngo. The demand for Wendell August Forge creations had reached the point at which it was no longer possible for Bruck and Youngo to do both the initial art work and the die engraving. In fact, to help ease their workload, the Forge had in recent years obtained a Panagraph-type milling machine to facilitate the initial transfer of the art work outlines to the die surface. Although they both miss the creative outlet provided by doing their own art work for most of the Forge's pieces, their die engraving talents and expertise are sufficiently rare that alternatives are scarce. The relentless nationwide decline in hand craftsmanship has greatly diminished the availability of expert die engravers, particularly for the types of work done at the Wendell August Forge. As Dave Bruck commented, "Working as a finisher back in the late 1970s provided me with valuable insight into how and where the coloring adheres, and how best to engrave a die so that the finished piece looks like I intended it to look."[37]

By 1993, Bill Knecht had realized that the key to most effectively retailing Wendell August

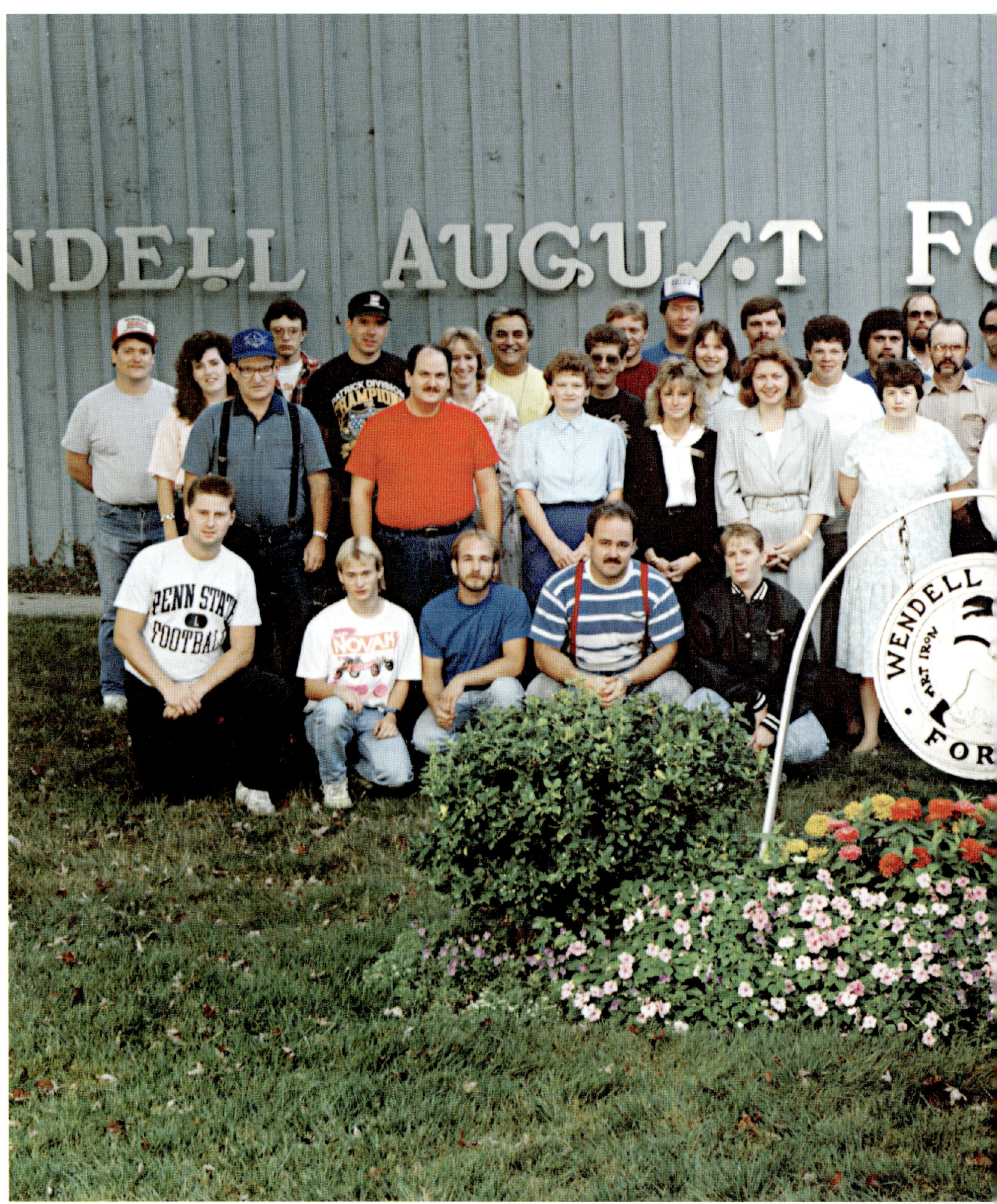

Wendell August Forge workforce circa *early 1990s* *(note 31)*

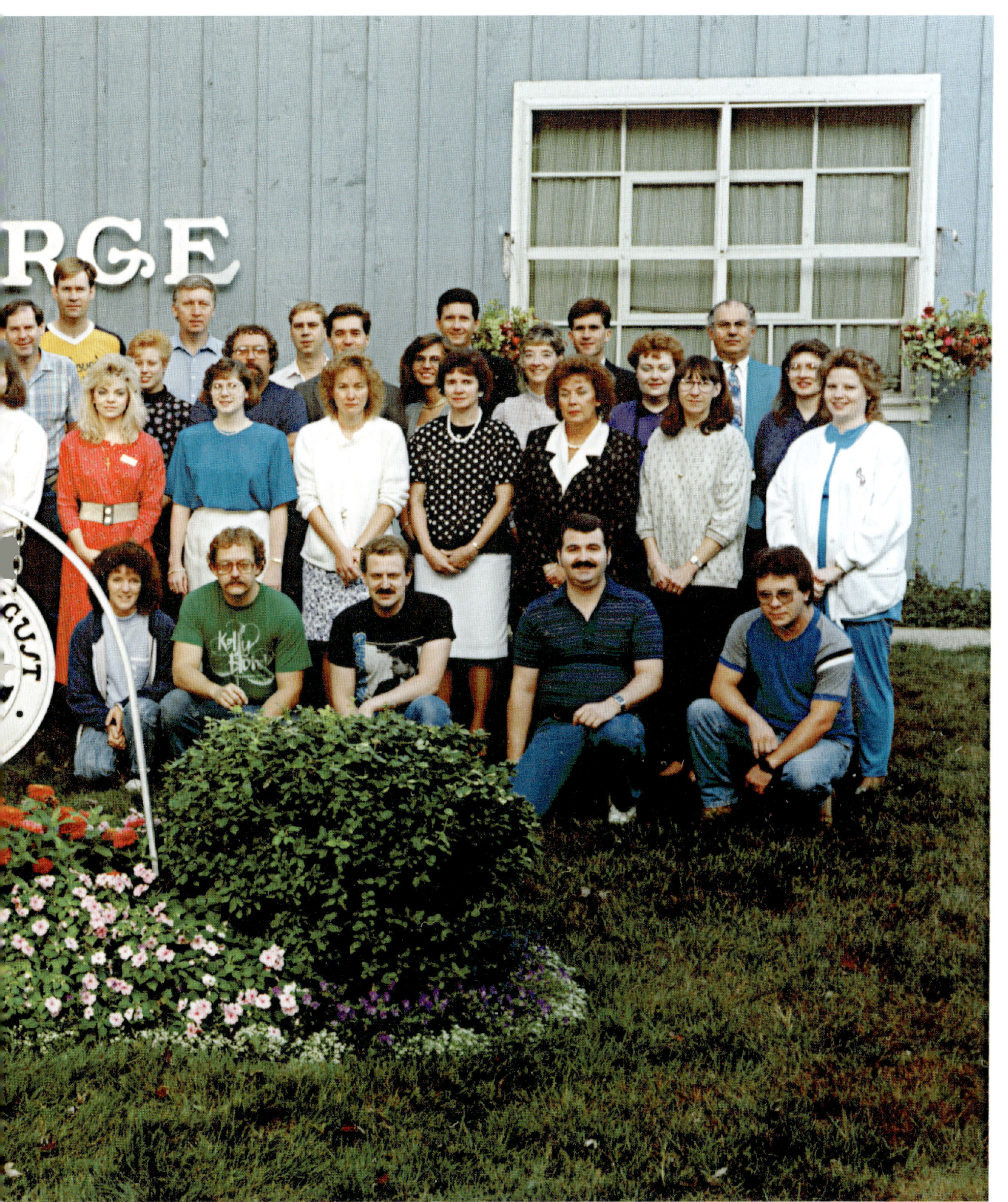
RGE

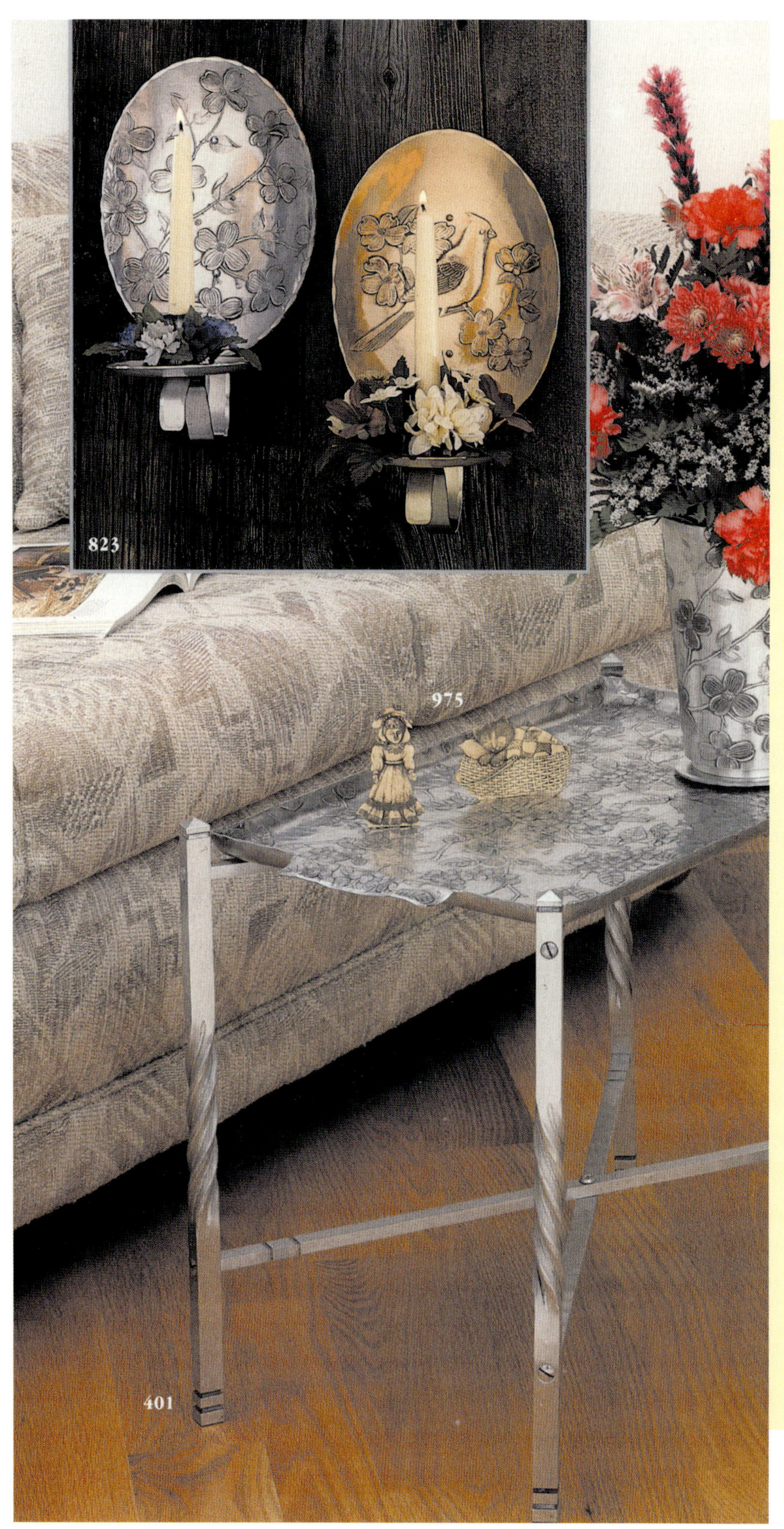

Portion of page 16 of 1990 Wendell August Forge catalog (note 32)

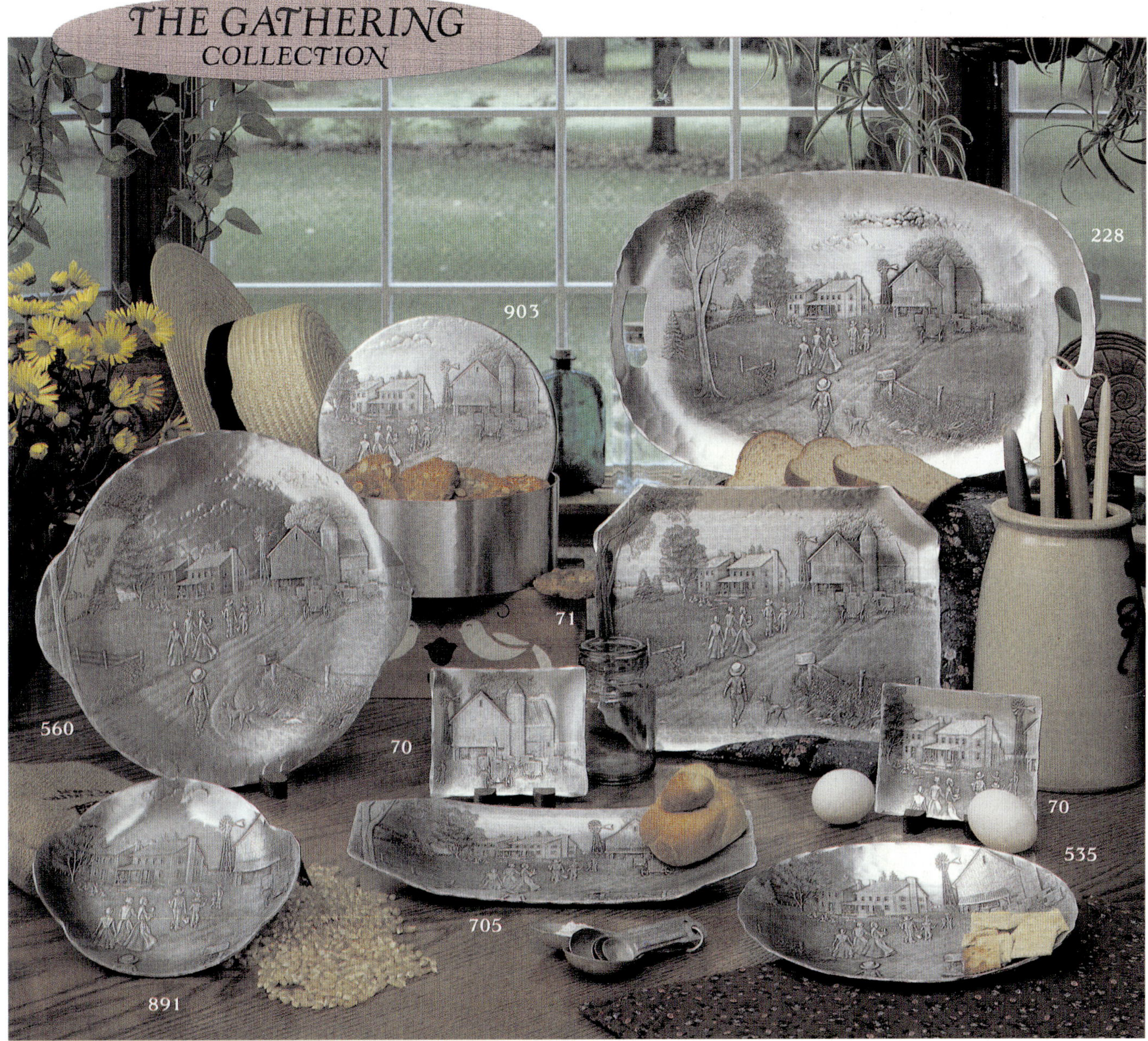

Portion of page 18 of 1993 Wendell August Forge catalog (note 36)

Forge products to the public was to be found at the thriving Grove City facility. A major reason that the Grove City site had become so successful was because visitors could actually see the skilled craftsmen at work at the same processes of decades long since past. The continually increasing number of tour buses brought passengers to the Forge to not only acquire their own piece of history, but also see history being created. Visitors *knew* that they were acquiring something very special when they could watch the pieces being made, one by one. They could see on display the heirloom pieces that the Forge had made 50 and 60 years ago and, in most cases, consider with some sense of awe the objects that had been created when they were children or even before they were born. The value of the experience itself played directly into the yearning for simpler times and the resurgence of interest in American craftsmanship that had slowly begun to emerge across the country in the late 1960s.[38] With this concept in mind, Knecht decided to create a second facility where he could take the Wendell August Forge product *and* experience directly to the people.

Seeking a site with which the unique handcrafted metalwares of the Wendell August Forge would be compatible, Knecht soon focused on the town of Berlin, in Holmes County, Ohio, in the center of the world's largest Amish community. In the early 1980s, the Forge had created a motif that commemorated Amish ways, which had been featured in the product line ever since. The perpetuation of old world craftsmanship in America, one of Knecht's personal objectives, was an integral part of the Amish way of life. He charged his son Will, recently promoted to Vice President, with the project management responsibilities of planning and opening a second Wendell August Forge facility in Holmes County, including a production area and gift shop.

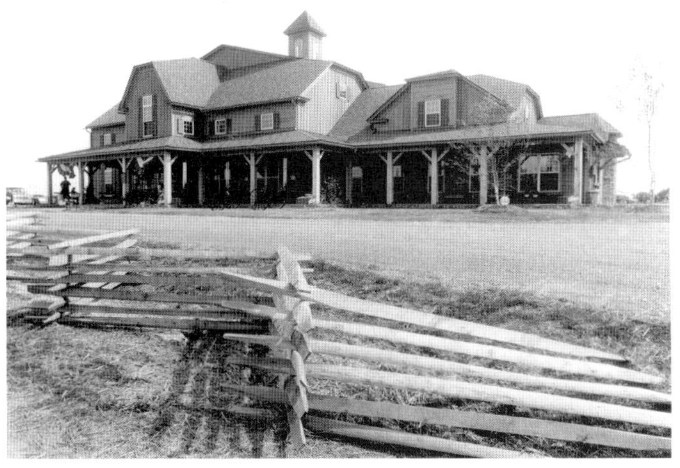

Exterior of Wendell August Forge facility in Berlin, Holmes County, Ohio (note 39)

1994 was a banner year. In Grove City, the space that had been vacated by the glass works was remodeled to provide improved metalware production facilities and additional space for the rapidly-increasing number of employees. The satellite retail operation in Mentor was closed. *Immanuel,* the first edition of the new *Spirit of the Holiday* series, was created, and the new *Barn Raising* collection appeared.

The Wendell August Forge facility in Berlin, Ohio opened on July 1, 1994.[39] The 13,000 square foot open beam timber structure, built especially for the Forge, houses a gift shop, a production work shop, a mini-theater, a snack area and lounge, and a museum area. The gift shop features a tremendous variety of Wendell August Forge metalwares and other unique American-made items, and visitors can observe Forge items being made in the production area. The museum area significantly furthers the dreams of both Bill Knecht and Cathy Youngo of preserving "the history of Wendell August Forge and handcrafted aluminum products."[40] Nearly 100 Wendell August Forge creations, dating from the 1920s through the 1950s, together with a selection of Forge photographs from the era, are on display.

As 1994 drew to a close, and the success of the new Berlin facility and the record-breaking performance of the Grove City facility became evident, the more than 100 people of the Wendell August Forge knew that the pieces of the puzzle were finally falling into place.

1 F.W. "Bill" Knecht III and F.W. "Will" Knecht IV; 1995 photograph courtesy of the Wendell August Forge. Information sources include biographical sketches provided by Catherine Youngo; "Forging Friendships," newsletters, Issue 1, Fourth Quarter 1995, and Issue 6, First Quarter 1997, Wendell August Collectors Guild, Wendell August Forge.

2 Gary Brown, "Wendell August Forge: Creators of artistic hand-wrought metals," *Pennsylvania Power Company Profile*, Winter Issue, 78/79.

3 Ronald E. Winder, personal interview, 7 Aug. 1997.

4 Deborah (Knecht) Fetter, as quoted in David Smith, "Forge keeps fire of the past burning," *Gateway Publications*, 10 Feb. 1993.

5 F.W. Knecht III, personal interview, 28 July 1997.

6 Catherine Youngo, personal correspondence, 27 July 1997.

7 Mary Lou McConnell McNaughton, personal interview, 04 Aug. 1997.

8 Bronze and sterling silver items are lacquered to deter oxidation.

9 Brown, "Wendell August Forge."

10 F.W. Knecht III, personal interviews, 28 July 1997 and 14 Jan. 1998.

11 "Forecast excellent for Forge," *The Sharon Herald*, 2 Feb. 1979.

12 SALT commemorative plate from the Collection of the Wendell August Forge. Photography by Richard Smaltz, Smaltz Studio, West Middlesex, PA; photograph courtesy of the Wendell August Forge.

13 "Forge Helps Commemorate SALT II," *Greenville Record-Argus*, 15 May 1979.

14 "Museum Opens at the Wendell August Forge," *The Sharon Herald*, 27 Nov. 1979.

15 C. Youngo, 27 July 1997.

16 Christopher, John Mark, and Natale Rossi; 1981 photograph courtesy of Adeline (Rossi) Reese. Information sources include Adeline (Rossi) Reese, personal interview, 19 Sept. 1997; Christopher Rossi, personal interviews, March 1994 and 23 July 1996; Frank Rossi, personal interviews, 26 June 1995 and 12 June 1996; Irene Rossi, personal interview, 28 April 1996; Bruce Ballentine, "By Natale Inc. to Open This Weekend," *Allied News*, 19 May 1980; Teresa Spatara, " 'Guess I'm just lucky,' Natale Rossi says," *The Sharon Herald*, 3 May 1980; 1980 By Natale product brochure, courtesy of Frank and Lenora (Isacco) Rossi; 1994 and 1996 By Natale product brochures.

17 David A. Bruck, 1997. Photography by Richard Smaltz; photograph courtesy of the Wendell August Forge. Information sources include biographical sketch provided by Catherine Youngo; "Forging Friendships," newsletter, Issue 7, Third Quarter 1997, Wendell August Collectors Guild, Wendell August Forge; and David Bruck, personal interviews, 16 Jan. 1998 and 3 Feb 1998.

18 C. Youngo, 27 July 1997.

19 Illustrated are: #50 4-1/2"x6" novelty box with hinged lid, *Squirrel* motif; #66 clock, available in 8" or 9" diameter, *Irish Setter* motif; #68 5"x8" ashtray, *Schooner* motif; #70 4-1/2" square tray, *Deer* motif; #75 9" diameter ashtray, *Pittsburgh Renaissance II* motif; #190 letter/napkin holder, *Flying Eagle* motif; #191 single pen holder, *Holly* motif; #192 double pen holder, *Colonial Eagle* motif; #195 pencil holder, *Masonic* motif; #625 thermometer, *Ducks* motif; #900 round fluted wastebasket, *Thistle* motif; #901 oval wastebasket, *Flying Eagle* motif; #905 letter basket, *Golfer* motif; #920 17"x24" desk mat, *Colonial Eagle* motif (and monogram); #925 correspondence box (both letter and legal sizes), *U.S.S. Constitution* motif; #930 name plate; #970 bookends (pair), *U.S.S. Constitution* motif. From page 3, 1985 Wendell August Forge Catalog, loaned by Mary Ann Felegy.

20 Illustrated are: 9" diameter pewter plate, Serial No. 221, *Honey Bear* from the *Wildlife* series (Third Edition, 1979 initial issue); 8" diameter bronze plate, Serial No. 756, *The Skating Pond* from the *Christmas* series (Eighth Edition, 1981 initial issue); 9" diameter aluminum plate, Serial No. 1081, *Canadian Geese* from the *Animal Kingdom* series (Second Edition, 1982 initial issue); 9" diameter aluminum plate, Serial No. 796, *Big Horn Sheep* from the *Animal Kingdom* series (Fourth Edition, 1984 initial issue); 8" diameter bronze plate, Serial No. 102, *Journey to Bethlehem* from the *Christmas* series (Eleventh Edition, 1984 initial issue). All items from the Collection of Joseph A. McCandless and Sarah Jane McCandless. Photographs courtesy of William Freeman.

21 Smith, "Forge keeps fire of the past."

22 Illustrated are: 1983 *Christmas Ornaments*, 1984 *Season's Greetings*, 1985 *Rocking Horse*, 1986 *Christmas Teddy Bear*, 1987 *The Sleigh*. All ornaments shown are 2-1/8" diameter. From page 15, 1993 Wendell August Forge Catalog, loaned by William Freeman.

23 Steven Adams, *circa* 1990, and Fred Todarello, *circa* 1982; photographs courtesy of Steven and Heidi Adams. Information sources include Steven Adams, personal interview, 24 Aug. 1997; Heidi Adams, personal correspondence, Sept. and Nov. 1997; "Helen Louise Todarello," *Allied News*, 15 Nov. 1995; Fred Todarello obituary records, Mercer County Genealogical Society.

24 Leonard Youngo, 1997. Photography by Richard Smaltz; photograph courtesy of the Wendell August Forge. Information sources include biographical sketch provided by Catherine Youngo; "Forging Friendships," newsletter, Issue 2, First Quarter 1996, Wendell August Collectors Guild, Wendell August Forge; and Leonard Youngo, personal interviews, 23 Sept. 1997 and 15 Jan. 1998.

25 Catherine Youngo, personal interviews, Dec. 1997 and Jan. 1998; Wendell August Forge: The Crystal Collection catalogs for 1987, 1989, and 1990, loaned by Mary Ann Felegy, and Wendell August Forge catalog for 1992, loaned by Wendell August Forge.

26 Knecht III, 28 July 1997.

27 Illustrated are: #61 6", 8" and 9" round plates; #350 8"x15" baby tray; #351 6"x7"x1-1/2" baby porringer with side handle, *Puppy* and *Teddy Bear* motifs shown; #355 youth chair; #725 7", 8" and 9" round hot mats, *Dogwood*, *Pine* and *Wheat* motifs shown; #726 9" and 10-1/2" oval hot mats, *Pittsburgh Renaissance II*, *Butterfly Daisy* and *Strawberry* motifs shown. From page 11, 1987 Wendell August Forge Catalog, loaned by Mary Ann Felegy.

28 Plate, late 1980s, 14" diameter, themed to "Awakening of a Higher Spirit," designed and engraved by Leonard Youngo. Photography by Richard Smaltz; photograph courtesy of the Wendell August Forge.

29 Information pertaining to the art work and engraving of David Bruck and Leonard Youngo based on Bruck, 16 Jan. 1998 and L. Youngo, 15 Jan. 1998.

30 Winder, 7 Aug. 1997.

31 Wendell August Forge employees outside of the Grove City facility; *circa* early 1990s photograph courtesy of the Wendell August Forge.

32 Illustrated are: #401 aluminum butler table, 18" high, 14"x24" serving tray, *Appleblossom* motif; #823 oval sconces, 6-3/4"x12-1/2"x6", *Cardinal* motif in bronze and *Dogwood* motif in aluminum; #850 aluminum vase, 8" high, *Dogwood* motif; #975 bronze stand-up cutouts, *Antique Doll* and *Kitten in a Basket* motifs. From page 16, 1990 Wendell August Forge Catalog, loaned by Mary Ann Felegy.

33 F.W. Knecht III, personal interview, 29 July 1997.

34 Catherine Youngo, transcription of recorded conversation between Natale Rossi and F.W. Knecht III, 17 Nov. 1992.

35 Lynda Guthrie, "No 2 gifts are alike at Wendell August Forge," *Allied News*, 31 Mar. 1993.

36 Illustrated are: #70 4"x5" mint trays, #71 9"x12" beveled cookie tray, # 228 12"x18" luncheon tray (with slotted handle), #535 9" oval mint/nut dish; #560 11" diameter luncheon/pastry platter (with handles); #705 6"x12" bread tray; #891 6-1/2" diameter bon-bon dish (with handles); #903 9" diameter x 3" deep cookie tin. All items in *The Gathering* motif. From page 18, 1993 Wendell August Forge Catalog, loaned by William Freeman.

37 Bruck, 16 Jan. 1998.

38 Assessments of the renewed interest in skilled handcrafts have appeared in the popular press in recent years; e.g., see John Balzar, "Rebirth of a Yearning for Quality," *Los Angeles Times*, 10 Jan. 1997, and Esther Schrader, "What Time Has Wrought: Blacksmiths Enjoy Revival Thanks to Boom in Interior Design," *Los Angeles Times*, 9 June 1997.

39 Exterior of the Wendell August Forge facility at Berlin, Holmes County, Ohio, 1994. Photography by Doyle Yoder; photograph courtesy of the Wendell August Forge.

40 Kelly Smith, "Wendell August forges ahead with museum," *Allied News*, 2 Mar. 1994.

Chapter Six

Building The Future

Sooner or later, all family firms face the question of succession. The Knecht family intentions for the future were clearly signaled in 1995 when Bill Knecht assumed the position of Chief Executive Officer and his son, F. W. "Will" Knecht IV, became President of Wendell August Forge, Inc. Will Knecht had joined the Forge in operations in 1989, progressing to operations manager and then to Vice President.[1] Bill and Will Knecht were demonstrating their desires to keep the firm within the Knecht family, and to maintain the prized "small company, family-friendly atmosphere"[2] of the Wendell August Forge.

The Forge's endeavors since 1978 have evolved into a focus on four primary markets for its metal giftwares: Retail Sites, Custom Products, Fundraising, and Direct Marketing.[3] The approach to the development of retail sites began with the success of the Grove City facility as a tourist attraction in its own right.[4] In Grove City, visitors can not only acquire Wendell August wares, but also watch them being made and become familiar with the Forge's history of hand craftsmanship. The success of the retail operation in Grove City formed the rationale for establishing a second facility at Berlin in Holmes County, Ohio in 1994.[5] With its emphasis on American hand craftsmanship and its metalwares commemorating the Amish way of life, the Wendell August Forge was well suited to place its second gift shop and forge in the heart of the world's largest Amish community. The successful extension of the Wendell August Forge concept to Holmes County demonstrated the feasibility of establishing additional retail sites in heritage-rich locations where hand craftsmanship and Americana are both recognized and valued. One result of this approach was the 1998 opening of a new Wendell August Forge facility on King Street in Charleston, South Carolina.

The market for custom products is, of course, one for which the Wendell August Forge has been providing meticulously crafted items for decades.[6] The Forge's outstanding reputation in the production of business gifts and awards is the result of such customer accolades as "These beautiful awards have become highly coveted" and "The award designed by Wendell August Forge was breathtaking," together with such tributes to customer service as "Hand carrying the

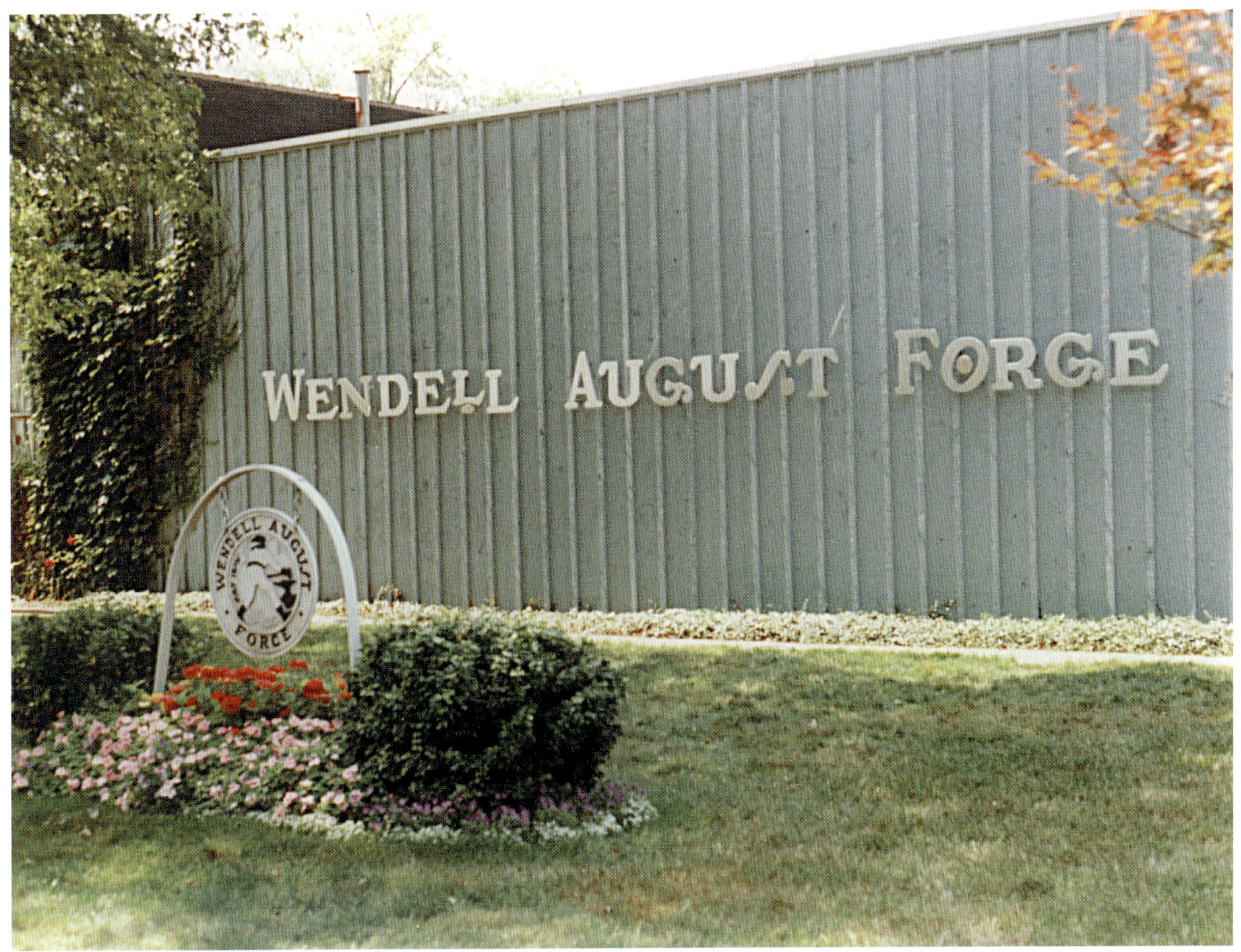

Exterior of Wendell August Forge facility in Grove City, Pennsylvania (note 4)

medallions . . . was a big life saver" and "when our number count grew unexpectedly . . . I had called on a Monday and somebody from your staff had already delivered them to me by Friday."[7] Organizations of all types, ranging from corporations to social clubs to school districts, have commemorated milestones, achievements and people with special hand forged Wendell August mementos. Countless pieces of metalware, in a variety of forms, including plates, trays, coasters, plaques, and other accessories, have been engraved with citations or marked with special handstamps. Hundreds of dies have been custom-engraved with either specified art work designated by the customer or original art created at the Forge, resulting in unique items produced primarily in aluminum, bronze and pewter. Capitalizing on its reputation, the Wendell August Forge has continued to expand its custom products services, first regionally and then nationally.

The creation of a fundraising program, whereby non-profit groups could share in the sales proceeds of special wares, was a completely new venture for the Wendell August Forge. Since 1995, many non-profit organizations have raised funds to support their objectives by selling and sharing in the

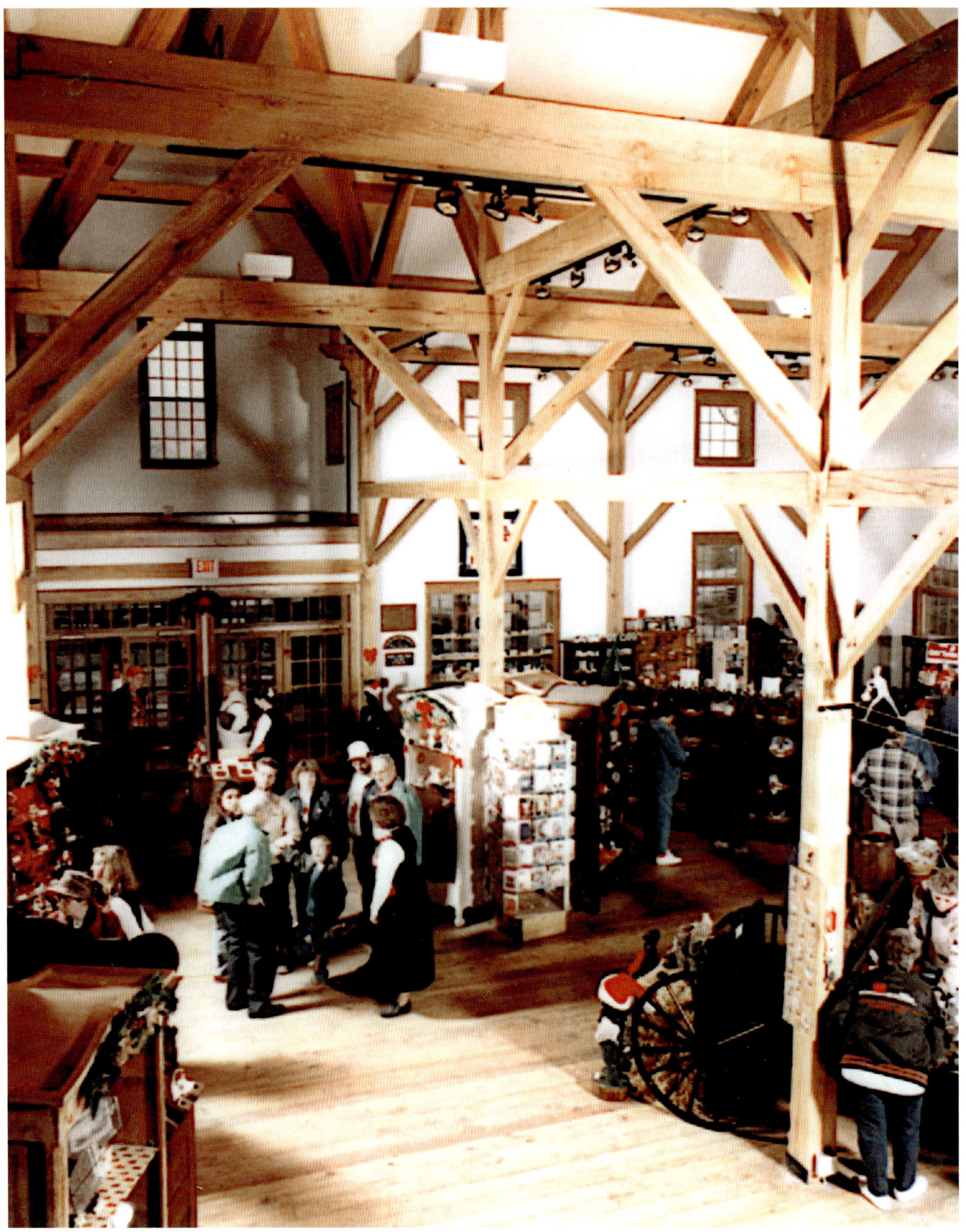

Interior of Wendell August Forge facility in Berlin, Holmes County, Ohio (note 5)

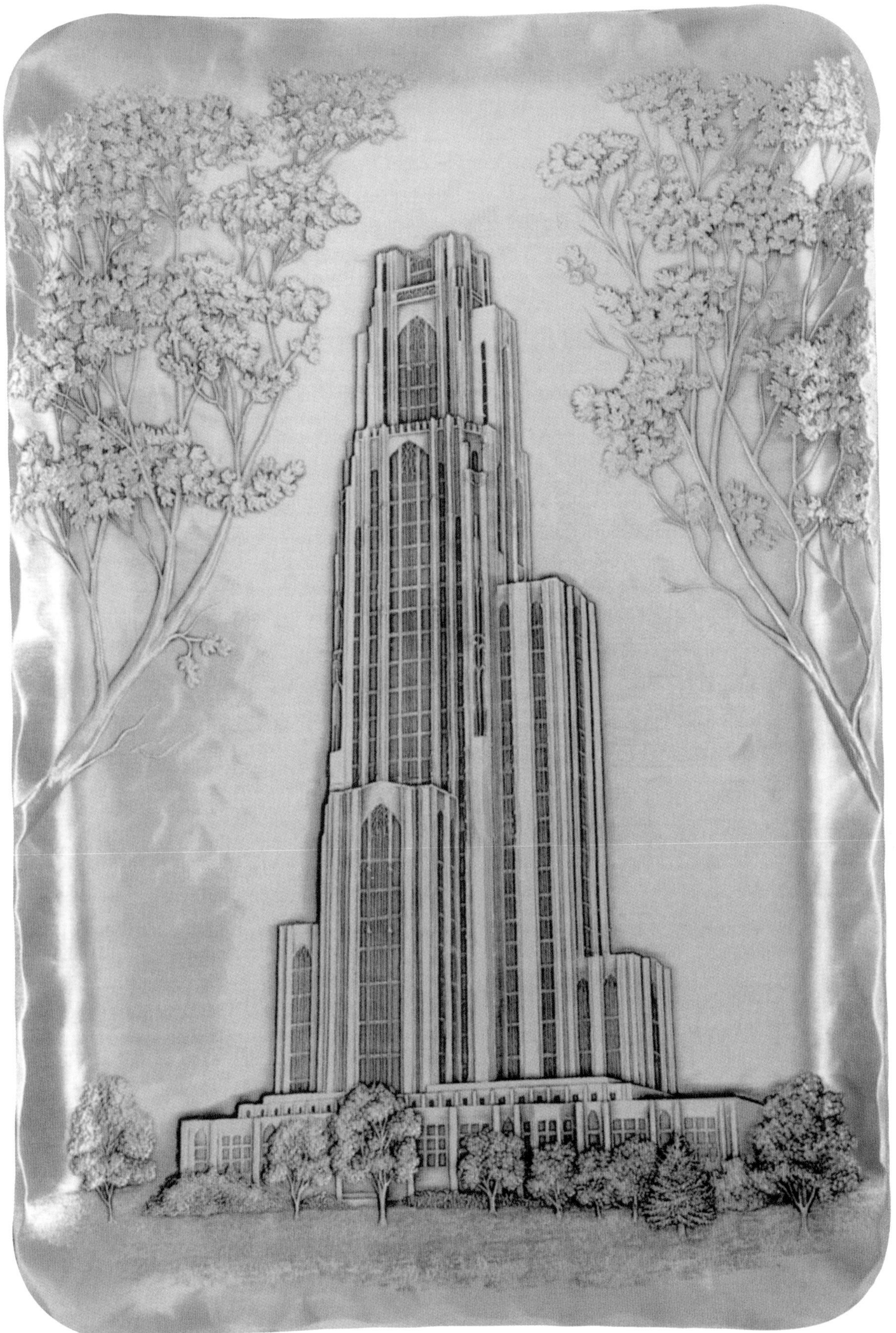

Bronze University of Pittsburgh Cathedral of Learning *from custom-engraved die* (note 6)

revenues from the sales of small hand made Wendell August Forge items, such as pins, pendants, magnets, mini-plates, and ornaments. In an exceptionally creative way to "give back something" to the local community, the Forge's Grove City sales manager, David Shively, experimented with a unique version of the fundraising program in 1997.[8] Children enrolled in the Grove City Area School District in kindergarten through the sixth grade were invited to participate in a contest to design Easter Egg ornaments.[9] Dies for the two winning designs were engraved at the Forge, and egg-shaped ornaments then produced for sale at both the Grove City and the Holmes County facilities, with a portion of the proceeds being returned to the school district.[10] Because children's art is so frequently treated as being transitory and disposable, one of the more intriguing aspects of this approach was that the relative permanence of the Wendell August Forge wares provided a means of preserving children's art.

The Forge's fourth thrust, direct marketing, has been expanding significantly for more than a decade, to the point that the annual direct-mail circulation in 1997 was in excess of 120,000. Starting in 1995, the direct marketing effort was enhanced by the issue of multiple catalogs and mailers during the year, emphasizing the many different types of needs that Wendell August handcrafted metalwares can meet. Despite the fact that the Wendell August Forge continues to create its giftwares much as it has since the early 1930s, the firm availed itself of more contemporary technology by establishing an international presence with an electronic direct marketing site on the World Wide Web.

With Will Knecht assuming the responsibilities of being the President of Wendell August Forge, it was now possible for Bill Knecht to give some attention to the creation of an additional venture, one in keeping with a long-held dream. An avid collector of "everything from clocks, pocket watches and art glass, to lamps, coins and hammered aluminum,"[11] Bill Knecht is personally familiar with countless people throughout the world who collect Wendell August Forge metalwares, and he was frequently approached about the possibility of developing some sort of "collector club." Although interested in the concept, he wanted to ensure that if he did something like that, he could do it the right way. Thus, for Bill Knecht, the formation of the Wendell August Collectors Guild in August of 1995 represented the achievement of a part of his vision for the future of the company.

Noting customer expressions of "their desire to collect special Wendell August pieces because they love the unique beauty, quality and exclusivity of our line,"[12] he kicked off his new endeavor with grand opening festivities at the Wendell August Forge in Grove City and in Berlin. The initial Guild lines included *Touched by an Angel* ornaments, *Kissin' Bridge* wooden curio boxes with bronze lid inserts, *Archive* reissues of vintage Wendell August Forge pieces in traditional aluminum, *Barns of Yesteryear* framed bronze Metalart plaques, and *Norman Rockwell's Finest Moments* bronze plates, using Norman Rockwell illustrations that had been featured on covers of *The Saturday Evening Post*.[13] [14]

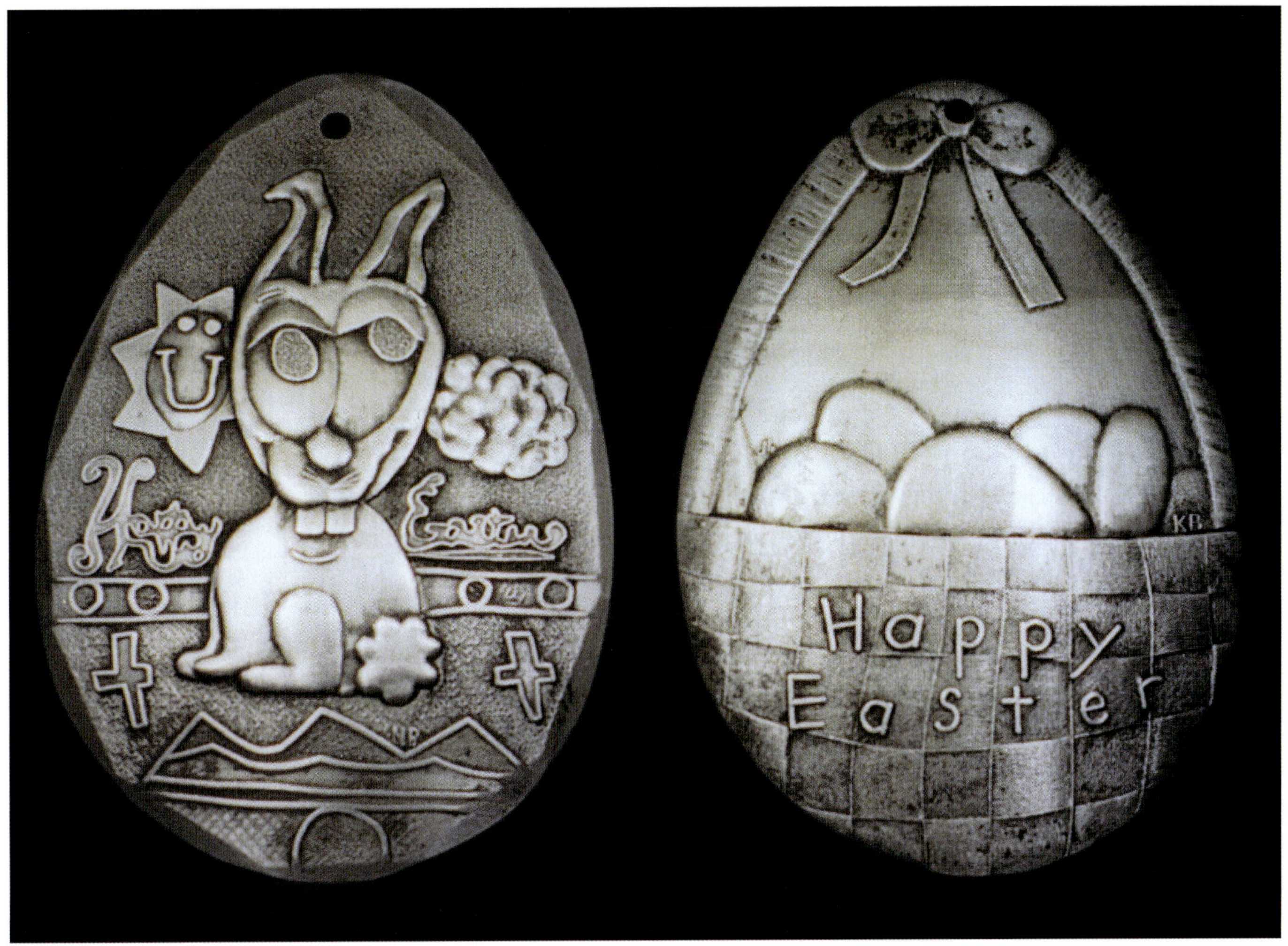

Easter egg ornaments designed by Grove City School District elementary school students (note 9)

The Curtis Publishing Co., owner of the rights to the Norman Rockwell covers, formally announced the selection of the Forge as "its exclusive licensee."[15] In March of 1996, Collectors Guild members were privileged to attend presentations in Grove City, by Marshall Stoltz, curator of the Norman Rockwell Museum in Philadelphia, and in Berlin, Ohio by Carol Brown, the *Saturday Evening Post* Archivist for Curtis Publishing.[16]

The increasingly important position of the Wendell August Forge in the antiques and collectibles world was also underscored in 1995 on television and in museums. In June, although "the show rarely visits businesses," the Forge was featured on the cable television show, *Personal fX: The Collectibles Show*.[17] Viewers were able to see how the Forge's metalwares are made, and were also provided with the opportunity to learn about some of the vintage pieces in the Wendell August Forge Collection. In addition to the attention brought by the television show, the Forge also received recognition when vintage Wendell August aluminum artifacts were featured in gallery and museum exhibitions. For several years, two Wendell August Forge pieces have been on display at the National Museum of American History of the Smithsonian Institution.[18] In 1995, a wide variety of vintage

First Editions

B. "First Love" Collectors Plate
First Edition
Norman Rockwell *Post* Collection

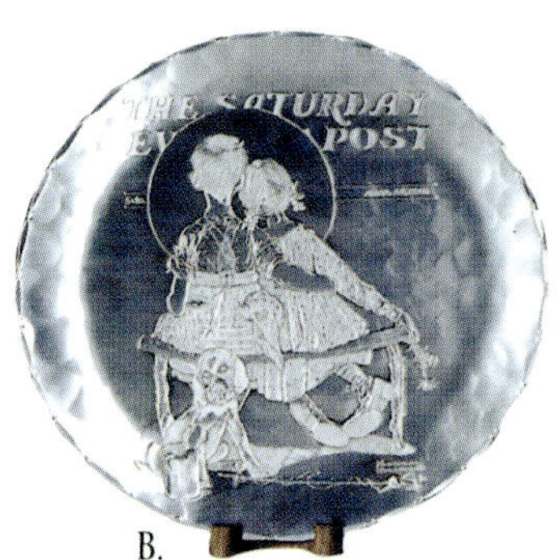

B.

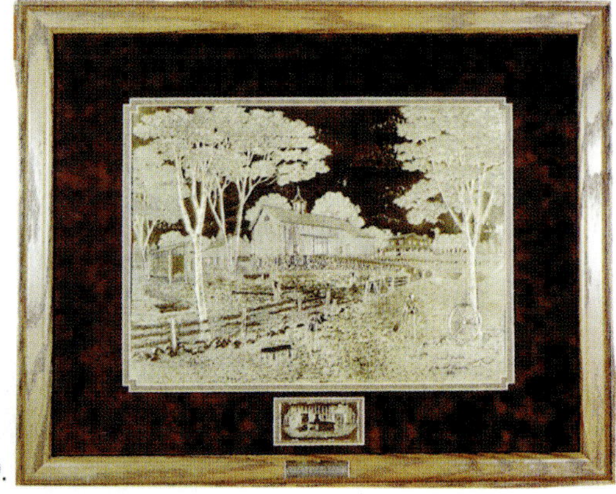

D.

C. "Knecht's Bridge" Curio Box
First Edition
Kissin' Bridge Collection

D. "Grandpa's Farm" Framed Metal Art
First Edition • Barns of Yesteryear Collection

C.

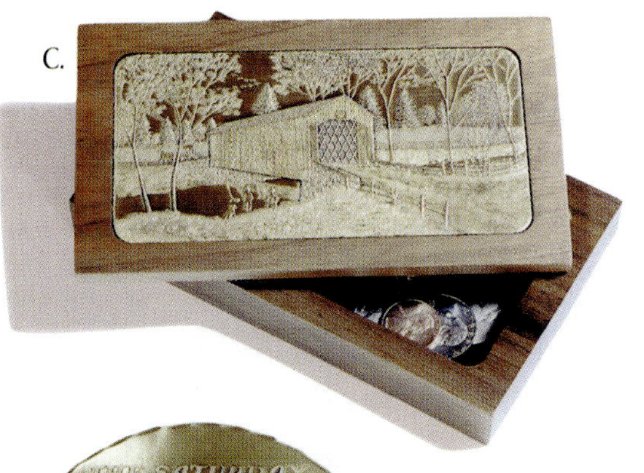

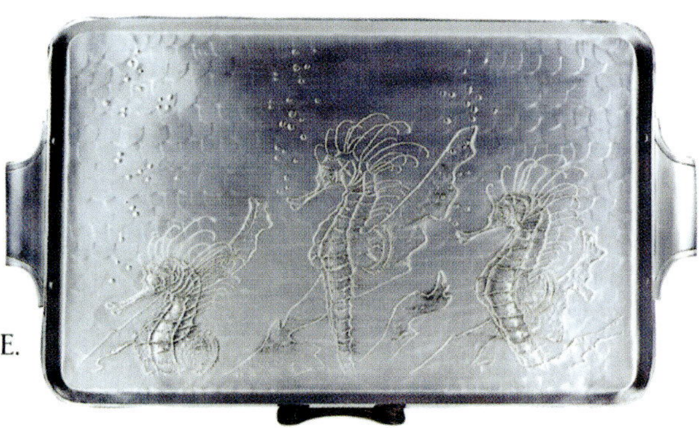

E.

E. 1932 "Children of the Sea" Serving Tray
First Edition • Archive Collection

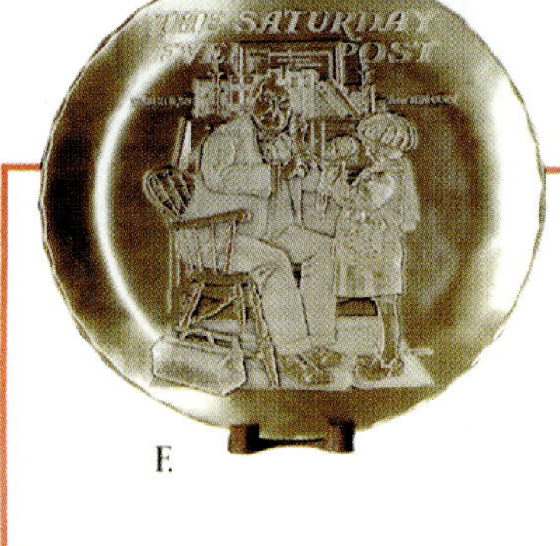

F.

Second Editions

F. "Doctor and Doll" Collectors Plate
Second Edition Norman Rockwell *Post* Collection

G. "Appleblossom" Scroll Footed Bowl
Second Edition • Archive Collection

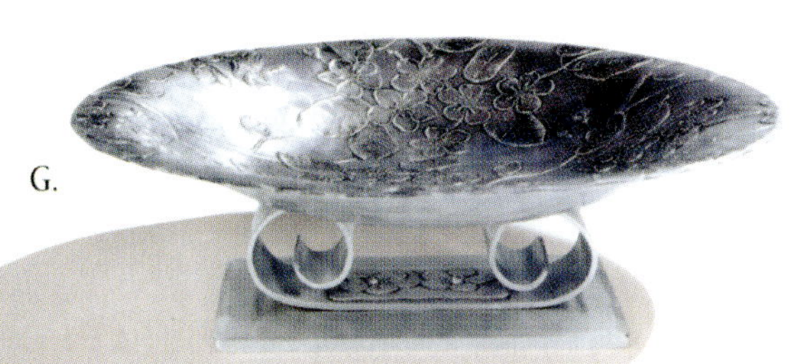

G.

Portions of Wendell August Forge catalog page illustrating selected Collectors Guild items (note 13)

Forge artifacts were brought to the public's attention on both the East and the West coasts. Aluminum items that had been produced by the Forge from the 1930s through the 1950s, ranging from furniture to jewelry, were featured in a California exhibition, *Depression Silver: Machine Age Craft and Design in Aluminum*, from April through July.[19] In addition, a Wendell August Forge desk set, selected from the holdings of the Wolfsonian Foundation, was included in the American Craft Museum's New York City exhibition, *Craft in the Machine Age*, from October of 1995 through February of 1996.[20]

Bill Knecht's ultimate dream is the creation of a museum in Grove City to perpetuate the hand craftsmanship and preserve the heritage of the hand-hammered aluminum industry that was founded by the Wendell August Forge. This is a lofty goal, because there is nothing about any museum that is either simple or inexpensive. Over the past 20 years, he and Catherine Youngo,[21] now the Assistant to the CEO, have worked together to accumulate selected pieces of the earlier handcrafted metal work made by the Forge and some other

Catherine Youngo

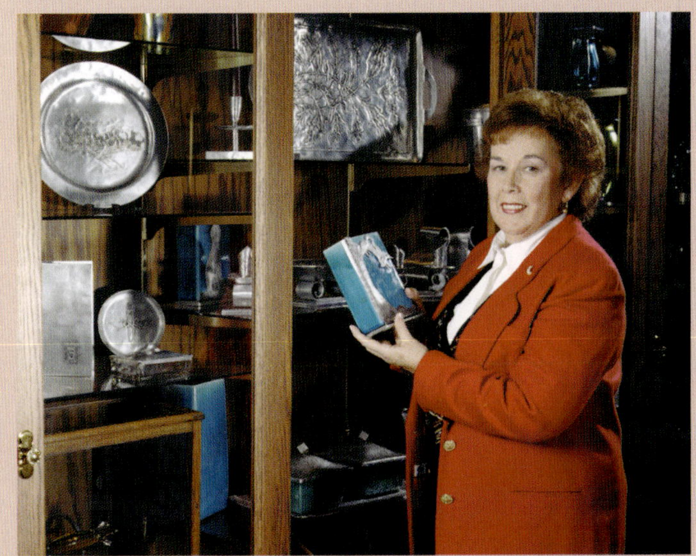

Catherine "Cathy" (Ross) Youngo was born and raised in Pittsburgh, Pennsylvania, and worked for the Pittsburgh Press after finishing school. In 1975, she started in a part-time position with the Wendell August Forge, doing front line sales, phone sales, and clerical work. Her part-time status lasted for only two weeks.

During her 23 years with the Wendell August Forge, Cathy worked her way up to retail manager and then office manager, and now serves as the Assistant to the Chief Executive Officer. She is deeply involved in the development and critique of the art work and the engraved dies for the Forge's showroom and custom metal wares. Since 1992, she has also been involved in the acquisition of vintage hammered aluminum artifacts for the Wendell August Forge Collection, as well as the accumulation of materials relating to the history of the company. She serves as the curator and historian for the Wendell August Forge museum collection at the Berlin, Holmes County, Ohio facility.

Cathy is also involved in community affairs, serving as an active member of the Mercer County Convention and Visitor's Bureau and the Board of the Grove City Chamber of Commerce. She has been a resident of Grove City since 1967. She has three sons, Leonard, Joseph and Christopher, all of whom are married, and six grandchildren. Two of her sons, Joe and Len, are also employed at the Wendell August Forge.[21]

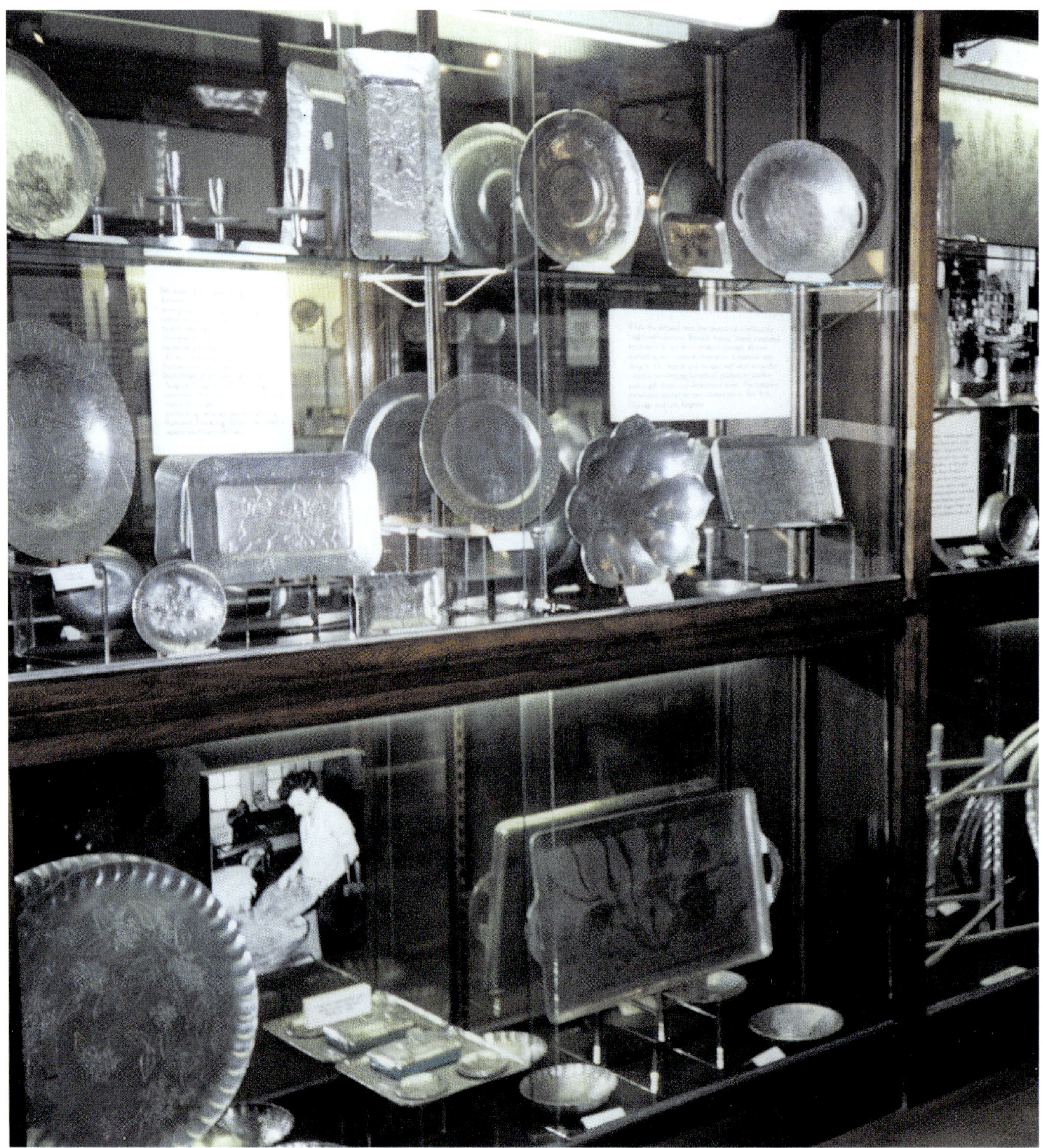

Portion of the Wendell August Forge museum area at Berlin, Holmes County, Ohio (note 22)

American producers. Their first museum effort was the development of a display of some of the Forge's older creations in a small area of the Grove City facility. The construction of the Holmes County facility provided an opportunity for a more expansive display area, for which Youngo has the curatorial responsibilities.[22] Even with the increased display capacity, visitors are able to view only a small fraction of the classic pieces in the

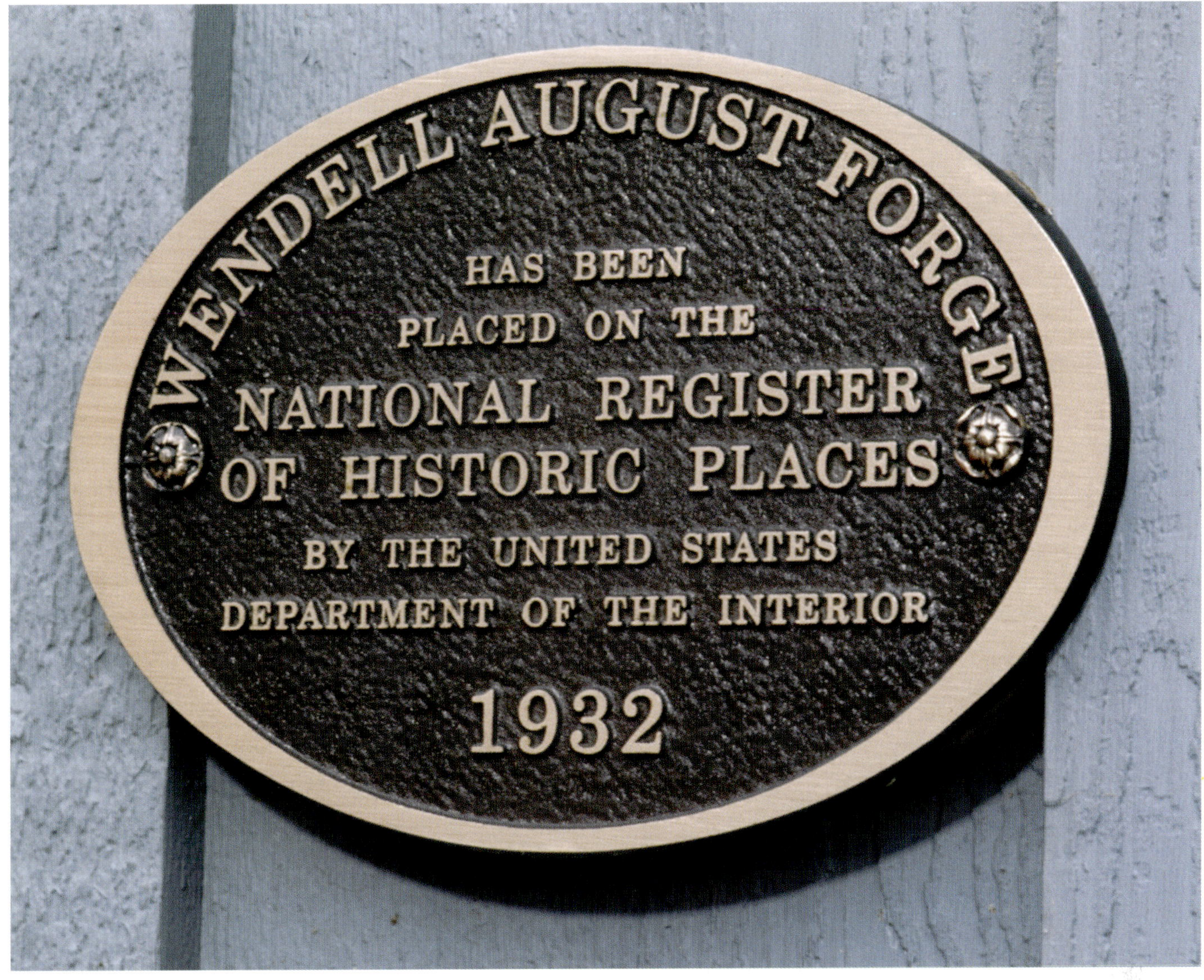

National Register of Historic Places plaque at Grove City facility (note 23)

continually expanding Wendell August Forge Collection. In conjunction with the thrust to preserve the Forge's history, the Grove City facility itself was recently added to the National Register of Historic Places.[23] The Forge's selection for inclusion on the roster was based on its "historic architecture and its continuation of a tradition of craftsmanship," recognizing its "commitment to a manufacturing process established by founder Wendell August in 1923 in Brockway, Pa."[24] What better location for a museum than near an historic site!

Will Knecht's commitment to the continued longevity of the Wendell August Forge was evident when he observed that "We look at everyone here as part of our family. Our number one goal is to keep employees and customers for life, not just for one transaction." Will envisions a national expansion of the Wendell August Forge, with facilities in several historical locations in the United States, so that more people can enjoy watching the hand crafting of the firm's metal wares. In so doing, he intends to maintain the focus on the "people and product" that guides the management of the firm. "Above all, we will not sacrifice what has gotten us here: Hand craftsmanship, American-made; that's what we're all about."[25]

Simply perusing a contemporary catalog allows you to create your own personal Wendell August Forge history. "Little round treasure boxes," you think, "they started making those not too many years ago . . . and a pewter *Apple Blossom* pendant . . . did they make any of those in aluminum before World War II?"[26] Later on a crimped bowl catches your eye, and you think, "This reminds me of the eight-sided pieces that they were making in the early 1940s and, of course, the *Pine* motifs have been popular since the late 1930s."[27] You find yourself chuckling at a mouse pad,[28] conjuring up images of trying to explain such a creature to blacksmith "Tony" Pisoni back in the 1930s, when electronic computers were still dreams and dead computer mice weren't even nightmares. And then you see the candle snuffer, which reminds you that Elaine Pompa had some like that made by Arthur Armour for banquet favors, and then after Armour retired more than 20 years ago, he eventually passed the design on to the Forge.[29] "Hmm, look at that casserole set with the *Barn Raising* motif," leads you to contemplate why such scenes weren't produced by the Forge in the 1940s.[30] "Oh, and look at the page of *Dogwood* pieces," you think, "the Forge has been making pieces with a *Dogwood* motif on them for a long time . . . like at least 60 years!"[31] The catalog items may be new ones, yet they seem as old as the Forge itself.

The Wendell August Forge is well poised to continue to share its rich American heritage with its hundreds of thousands of visitors, who arrive on foot, or by motor vehicle, or on one of more than 400 buses each year, or who visit by mail or electronically by cyberspace. The meticulous metalwares, still made from artisan-engraved dies, still made by hand, still made one at a time, remain as enticing as they were when they first launched a new industry.

Chapter Notes

1 "Knecht is WAF President," *Allied News*, 21 June 1995.

2 F.W. "Will" Knecht IV, personal interview, 28 July 1997.

3 Knecht IV, 28 July 1997.

4 Exterior, Wendell August Forge, Grove City, PA; 1991 photograph courtesy of Dennis Wildnauer.

5 Gift shop interior, Wendell August Forge, Berlin, Holmes County, OH; 1998 photograph courtesy of Lisa Grossenbacher, Wendell August Forge.

6 Bronze 6"x9" tray made from custom-engraved die; *University of Pittsburgh Cathedral of Learning*; featured in late 1990s Wendell August Forge literature; courtesy of the Wendell August Forge.

7 Quotes excerpted from customer letters, http://www.wendell.com/Corporate.html, 19 Jan. 1998.

8 David Shively, personal correspondence, 20 Jan. 1998.

9 Aluminum Easter egg ornaments, 1-3/4"x 3-1/2" each; designed by Nate Brubaker (fifth grade) and Kevin Baker (first grade) of the Grove City School District. Collection of William Freeman; photograph courtesy of William Freeman.

10 Chauntel Gallagher, "Easter forged through kids' eyes," *Allied News*, 26 Feb. 1997.

11 "Forging Friendships," newsletter, Issue 1, Fourth Quarter 1995, Wendell August Collectors Guild, Wendell August Forge.

12 "WAF has new collectibles club," *Allied News*, 20 Sept. 1995.

13 Portions of page 35 of the Holiday 1996 Wendell August Forge catalog; catalog loaned by William Freeman.

14 "Forging Friendships," Issue 1, Fourth Quarter 1995.

15 " 'First Love' Comes to Forge," *Allied News*, 20 Mar. 1996.

16 "Forging Friendships," newsletters, Issue 3, First Quarter 1996, and Issue 4, Second Quarter 1996, Wendell August Collectors Guild, Wendell August Forge.

17 Apryl Flynn, "Forging into show biz," *Allied News*, 8 June 1995.

18 Two small ashtrays made in the early 1930s are included in the "Material World" display at the Smithsonian Museum. Detailed documentation pertaining to these pieces is provided in Chapter Two.

19 Bonita Campbell, *Depression Silver: Machine Age Craft and Design in Aluminum*, exhibition catalog, California State University, Northridge, 1995.

20 Janet Kardon, ed., *Craft in the Machine Age: 1920-1945*, exhibition catalog (New York: Harry N. Abrams, Inc., 1995).

21 Catherine "Cathy" Youngo, with a small selection of the artifacts from the Collection of the Wendell August Forge. Information sources include Catherine Youngo, personal correspondence, 27 July 1997, 19 Feb. 1998, 4 March 1998. Photography by Richard Smaltz, Smaltz Studio, West Middlesex, PA; photograph courtesy of the Wendell August Forge.

22 Portion of the museum area at Berlin, OH, featuring a selection of *circa* 1930-1960 Wendell August Forge metalwares, including several pieces produced by Pine Grove Craftsman in 1940-1941. Collection of the Wendell August Forge; photograph courtesy of the Wendell August Forge.

23 Plaque designating placement of the Wendell August Forge in Grove City, PA on the National Register of Historic Places. Photography by Richard Smaltz; photograph courtesy of the Wendell August Forge.

24 Nick Hildebrand, "Forge named to list of historic places," *Allied News*, 23 April 1997.

25 Erin Behan, "Forge celebrating its 75th anniversary," *Allied News*, 18 Feb. 1998.

26 Holiday 1997 Wendell August Forge Catalog, p 5.

27 Holiday 1997 Catalog, p 15.

28 Holiday 1997 Catalog, p 18.

29 Elaine Pompa, personal interview, 25 July 1997; Holiday 1997 Catalog, p 23.

30 Holiday 1997 Catalog, p 25.

31 Holiday 1997 Catalog, p 26.

Appendix A

The People of the Wendell August Forge

This roster of the people who have been associated with the Wendell August Forge in Brockway, Grove City, and Berlin was compiled from a wide variety of sources, including directory searches, recollections of current and former Wendell August Forge employees, and contemporary personnel rosters provided by the Wendell August Forge. The roster was compiled prior to the opening of the Charleston, South Carolina facility. Unless otherwise noted, the location is Grove City. Unfortunately, it is inevitable that there may be some unintentional inaccuracies and omissions.

Adams, Catherine, 1994-present, shipping/receiving (Berlin)

Adams, Linda, 1991-95, order entry

Adams, Steven G., 1975-77, 1980-82, craftsman, die engraver

Aiken, Michael, 1978-79

Alcorn, Lisa, after 1978

Alessio, Linda, 1985-88

Alessio, Steven, 1977-78

Alexander, Julie, 1994-95, order entry

Alfreno, Mark, 1973, craftsman

Alger, Matthew W., 1982-88, sprayer

Allen, Steve, after 1978

Archer, Amy, after 1978

Armour, Arthur, 1932-33, designer

Arn, Rudy, 1977-78

Arnold, Max, 1980

August, Jessie (Palmer), 1933-42, corporate officer

August, Mary (Boquine), 1963-77, corporate officer

August, Robert E., 1945-78, corporate officer

August, Wendell M., 1923-63, corporate officer

August, Wendell M., Jr., 1946-78, corporate officer

Bable, Wendy, 1994-96, sales

Bachelder, Debbie, after 1978

Baer, Heather M., 1996-present, sales

Baily, Corey, 1996, craftsman (Berlin)

Baker, James A., 1985-95, craftsman

Barclay, Robert, 1994-95, finishing

Barger, John, 1995-96, custodian

Barker, Chris, after 1978

Barlow, Hugh, after 1978

Barnes, Leroy, 1932-35

Baron, Harry, 1938, sander, cleaner

Bartholomew, Larry, 1982

Bashman, Thad, 1994-95, craftsman (Berlin)

Batdorf, Brian, after 1978

Batdorf, Mike, after 1978

Baum, Tom, craftsman (Berlin)

Bell, Brian, 1982-83, spray specialist

Bell, Ken, after 1978

Bengs, Mary, 1987-present, sales

Bengs, Paul K., 1985-present, finisher

Bennett, Bill, after 1978

Bennett, Jeffrey, after 1978

Bennett, Lanette M., 1997-present, sales (Berlin)

Bennett, Susan E., 1988-present, sales

Berner, Diana I., 1992-present, sales

Best, Eric, 1980-82, material cutting

Bestwick, James, circa 1932-35, repoussé, finishing

Boleen, Gust, 1926 (Brockway), blacksmith's assistant

Bonanni, Albert M. Jr., 1973, sander

Bonanni, Bruce, 1976-81, sander

Bonanni, Mark, 1978-79, material cutting

Bordonaro, Bill, after 1978

Boughner, Ruthe K., 1986-88, sales

Bowen, Cheryl, 1994-95, sales

Boyd, John, circa 1936-39, coasters, ashtrays

Boyd, Willard W., circa 1939-41, 1946-54, coasters, ashtrays, jewelry

Boyer, Charles R., 1996-97, shipping/receiving

Boyles, Peggy, after 1978

Brant, Elaine, after 1978

Breese, Jim, circa 1939-41, flattening

Brinker, Kimberly, 1994-95, mail order

Bromley, Lloyd, after 1978

Brown, Gordon, circa 1934-41, 1947-49, repoussé, trays

Brown, Jeffrey S., 1974-present, master craftsman

Brown, Joyce, 1978-80

Brown, Richard, 1991-94, production manager

Brown, Robert, after 1978

Bruck, David A., 1976-77, 1980-present, finisher, die engraver

Bruner, George, after 1978

Buchanan, Barbara, 1985-86

Buchanan, Cynthia A., 1997-present, purchasing

Bulfone, Christopher E., 1986-present, engraver

Burns, Orace R. II, 1985-89, material cutter

Caccamo, Joseph Jr., 1946, metalworker

Calderone, Edward, 1946, metalworker

Calderwood, Tim, after 1978

Caldwell, Donald L., 1990-present, finishing

Camancho, Teresa, 1995, sales (Berlin)

Cameron, Donald, circa 1946-49, janitor

Cameron, Howard, circa 1932-34, finisher

Cameron, Thomas B., 1997-present, manufacturing director

Campbell, Raymond, 1975-82, welder

Campbell, Scott C., 1994-present, craftsman

Cannon, Kent D., 1997-present, finishing

Carothers, Matti, 1996-present, sales

Carter, Fred, 1987-88

Caulk, David H., 1994-present, engraving

Ceremuga, Elizabeth, 1994-present, shipping/receiving

Chapin, Howard J., 1923-31 (Brockway), 1932-34, sales, corporate officer

Chess, Jennifer A., 1996-present, sales

Chestnut, Carol A., 1995-present, sales

Childress, Tom, 1988-89

Chutz, Tana, after 1978

Cline, Danielle, 1995-present, sales

Coast, Darren L., 1986-88, sander

Cochran, Howard E., 1946, metalworker

Cokain, Charlene, 1991-94, sales

Coleman, Robert, 1995, craftsman (Berlin)

Collar, Steven, 1994-95, finishing

Collers, Jennifer C., 1994-present, sales

Collins, Stephen, 1994-95, finishing

Colosimo, Tom, 1983-85

Conner, Kevin, 1983-89, production director

Conto, Lisa, after 1978

Cooke, Bill, after 1978

Cookson, Lyle, circa 1934-38, sander

Corbett, Jeffrey, 1994-95, craftsman

Coulter, Dennis C., 1976-present, finisher

Coursen, Darlene, 1983-84

Courtney, Amy, 1996, sales (Berlin)

Coutsolioutsos, Terry, 1985-86

Covert, Linda, after 1978

Cowan, Estelle "Teddie", 1980-83

Creveling, Mike, 1978-81, repoussé

Croft, Leslie E., 1997-98, craftsman (Berlin)

Cross, Cristy, 1988

Crouser, Amy, 1993-95, shipping/receiving

Crum, Judy, after 1978

Cummings, Debbie, after 1978

Daniels, Alex, 1932-34, blacksmith

Daniels, Andrew, 1932-34

Davis, Jamie, 1996, sales (Berlin)

Davis, Joe, 1994-97, craftsman, cell leader (Berlin)

DeMatteis, Robert E., 1995-present, finishing

Deniker, Edward, circa 1932-34, repoussé, finishing

DePonceau, Arthur, 1933-36, finishing

DePonceau, James, 1929-32 (Brockway), 1932-34, carbon coloring

DePrano, Benjamin M., 1995-present, finishing

Dick, Marian J., 1966-77, office manager

Dickey, Linda L., 1984-87, sales

Digman, Alyson, 1997, sales (Berlin)

Dikeman, Samuel, circa 1932-39, janitor

Dikeman, Samuel Jr., circa 1933-39, repoussé

DiMichele, Lee, 1977, metalworker

Dlugozima, Diana L., 1994-97, shipping/receiving

Dolan, Dennis, 1972-77, sander

Donaldson, James E., 1977-78, sander

Donaldson, Patrick, 1994-present, finishing

Donato, Everl, circa 1937-39, sander

Donato, Louis W., 1932-41, die engraver

Dorr, Tom, 1988-89

Duda, Mike, after 1978

Dupree, Nancy, after 1978

Dydek, Frank, 1990-present, craftsman

Eakin, Jim, 1994-95, craftsman

Eakman, Robert, after 1978

Elder, Heather A., 1995-present, sales

Ellenberger, Sharon, 1995, sales

Ely, Sally M., 1986, clerk

Englehart, Danelle, after 1978

Erlandson, Sara, 1995, sales

Eshbaugh, Brett, 1986

Ewig, Galen C., 1986, sander

Faull, Marion M., 1979-present, accounting clerk

Fearon, Scott, 1994-95, craftsman (Berlin)

Ferguson, Chuck, circa 1933-38, welder

Fetter, Deborah (Knecht), 1989-95, marketing

Firster, Helen, 1992-94, shipping/receiving

Fisher, Dick, 1974-77, 79-84, projects manager

Fithian, Edwin J., 1932-33, corporate officer

Flickinger, Gary D., 1992-present, finishing

Flynn, Gary, 1983-86, material cutting

Flynn, Michael E., 1993-present, craftsman

Flynn, Patrick J., 1997-present, finishing

Ford, Shawn L., 1986, sander

Ford, Tyler, 1993-94, sales

Forese, Teresa, 1947-51, office support

Formani, Benjamin, circa 1928-32 (Brockway), 1932-41, 1946-79, blacksmith, craftsman

Forrer, Brian, 1997, sales (Berlin)

Foster, Bill, circa 1939-41, repoussé

Frampton, Judy, 1994-95, shipping/receiving

Frautschy, Derek W., 1997-present, craftsman (Berlin)

Fullerman, Anne L. (Berlin)

Fustine, Bill, circa 1932-34, metalworker

Gallagher, Dolores, 1984-89, sales

Galloway, Grant, 1995-present, die engraver (Berlin)

Gatewood, Glenda, 1992-94, sales

Geog, Mandy, 1995-96, sales (Berlin)

Gillmore, Bill, circa 1938-41, coasters

Gladden, William, 1985-87

Glessner, Dianne, 1986-87

Goelz, Marguerite L., 1996-present, database marketing manager

Grabb, Russ, 1982-84

Graham, Roberta, 1994-96, sales

Graham, William C., circa 1936-41, 1946-54, metalworker

Gregory, Rose M., 1949, clerk

Greenwood, Debra, 1989-93, sales

Gribben, Ken, 1994-96, craftsman (Berlin)

Griffith, Stacy, 1995, sales (Berlin)

Grossenbacher, Lisa M., 1996-present, admin. asst. (Berlin)

Grossman, Tina, 1994-95, clerical

Grove, Doyle, 1995, craftsman (Berlin)

Groves, E. Stillman, circa 1947-52, sales

Guarnieri, Mike, 1977-79, craftsman

Hague, Gary, 1983-84

Haizlett, Bruce, 1982-83

Hale, Richard, 1975-83

Hale, Ronald E., 1982-89, sander

Hamilton, Odis, 1985-89, craftsman

Hancock, Carla, 1992-94, sales

Harvey, Candace, 1991-95, receptionist

Hasenplug, Charlene A., 1996-present, sales

Hawkins, Carol, 1996, admin. assistant (Berlin)

Hawkins, Jennifer, 1995-96, group leader (Berlin)

Hershberger, Jenesia R., 1997-present, sales (Berlin)

Highland, Irene, 1988

Hile, Joe, 1986-88

Hines, John, 1995-96, craftsman (Berlin)

Hines, Theresa M., 1995-96, sales (Berlin)

Hobart, Dan, 1981-84, craftsman

Hochstetler, Jennifer, 1995, sales (Berlin)

Hodge, Edward P., 1985-present, craftsman

Hoffman, Alice, 1932-41, secretary, office manager

Hoffman, Frank, circa 1950s, finisher

Hoffman, Joseph, 1994-95, finishing

Hogg, Chuck, circa 1939, repoussé

Howes, Sara, 1996, sales (Berlin)

Hughes, Michael C., 1979-82, glass cutter

Hutson, William L., Jr., 1997-present, craftsman (Berlin)

Iliff, Richard, 1994-95, finishing

Isacco, Anthony P., circa 1946-49, metalworker

Isacco, Patsy, 1946, clerk

Jervis, Marcy J., 1993-present, sales

Johnson, Keith A., 1979-89, 91-present, systems manager

Johnson, Mindy, 1995-97, sales

Johnston, William B., 1995-present, proprietor (Berlin)

Kearney, Michael, 1994-95, finishing

Keith, Donna, 1996-present, sales

Kelly, Don, circa 1964-88, craftsman

Kilgore, Geraldine, circa 1946-52, office

Kilpatrick, Mary Ann, 1994-present, mail order

Kimes, Patty, 1988-95, clerical

King, Brian E., 1976-79, craftsman

King, Craig, 1978-79, repoussé

King, Scott, 1994-95, finishing

Klingler, Jon, 1994-97, lead craftsman (Berlin)

Knauff, Clare, circa 1939-41, 1946-48, cigarette boxes

Knauff, Dennis, 1965-82, craftsman

Knecht, F. W. III "Bill," 1978-present, CEO

Knecht, F. W. IV "Will," 1989-present, President

Knight, Michael, 1986-87

Kramer, Matthew W., 1995-present, fundraising sales

Kunselman, William A., 1995-present, production manager

Labor, Guyla, 1990-96, order entry

Lanier, Danielle M., 1995-97, collectibles sales

Larimore, Kelly, 1995, craftsman (Berlin)

Laycock, Charles, circa 1926-30 (Brockway), sales

Lee, Kathy A., 1995-present, shipping/receiving

Lenkner, Robert F., 1982-86, engraver

Leone, James, 1946, metalworker

Leone, Joseph F., 1933-41, 1969-70, finisher, installer

Lester, Larry, 1986-87

Linn, Robin, 1979-81

Locke, Dave, 1982-85

Lunn, Salley A., 1997-present, shipping/receiving

Mace, Dorrin K., 1996-present, buyer

Macri, Bob, 1982-83

Maddalena, Armand, 1946, metalworker

Magee, Daniel J., 1980-present, craftsman

Magee, Karen S., 1995-97, shipping/receiving

Manaia, Felix, 1980-present, craftsman

Marshall, Thomas J., Jr., 1997-present, craftsman (Berlin)

Martin, Christine, 1991-93, sales

Martin, Marcia J., 1994-present, engraver (Berlin)

Martin, Pat, 1980-84

Mast, Eli H., 1996-present, craftsman (Berlin)

Mast, Henry, 1994-present, craftsman (Berlin)

Mast, Nettie, 1994-95, sales (Berlin)

Masters, Dave, craftsman (Berlin)

Maurer, Trish, 1996-97, sales (Berlin)

Mayne, Justin, after 1978

Mays, Lee, circa 1933-35, blacksmith

McBride, Dave

McBride, Edward D., 1982-present, finisher

McBride, Jim, 1946, metalworker

McCabe, Kevin R., 1976-96, inventory manager

McCall, Philip, after 1978

McCallister, Charles, 1978-79

McCallister, Susan, 1992-93, sales

McCarl, Edward, 1946, apprentice

McCarthy, Michael, 1994-95, finishing

McCarthy, William J. Jr., 1946-49, apprentice

McCausland, James, 1928-32 (Brockway), 1932-41, 1946-58, designer, superintendent

McClimans, Robert, 1994-95, systems

McCloud, Jack, circa 1926 (Brockway), blacksmith

McConnell, Mary L., 1964-79, office manager

McGill, R. Kent, 1969-88, craftsman

McGivern, Robert E. Jr., 1990-94, sales manager

McIntire, Donald, 1978-79, repoussé

McKinley, Casey, 1996-present, sales

McKinney, David L., 1997-present, finishing

McKnight, Robin, 1990-94, mail order

McLaughlin, Milford, 1936-50, repoussé

McLaughlin, R. Leslie, c.1929-32 (Brockway), 1932-41, blacksmith

McLaughlin, Warren, c.1929-32 (Brockway), 1932-41, blacksmith

McQueeny, Patricia, 1990-94, sales

McTaggart, John, 1976-79

Mellor, Tom, 1997, craftsman (Berlin)

Meyerl, Richard, 1991-97, craftsman

Mickley, Linda L., 1995-present, housekeeping

Miller, Cary M., 1995-present, showroom supv. (Berlin)

Miller, David, 1985-86, die engraver

Miller, Deborah L., 1986, clerk

Miller, Joseph E., 1997-present, craftsman (Berlin)

Miller, Karen A., 1991-present, shipping/receiving

Miller, Shannon, 1994-97, sales (Berlin)

Miller, Steven, 1995, craftsman (Berlin)

Miller, Sue, 1996, sales (Berlin)

Miller, Tara M., 1997, sales (Berlin)

Miller, William, 1923-32 (Brockway), 1932-41, 1946-49, shipping

Minor, Ardeth, 1992-95, shipping/receiving

Miscimarra, Jonathan R., 1994-present, staff accountant

Monroe, Ted, circa 1933-41, inspection

Moore, William, 1993-96, sales manager

Morrison, Nancy, 1996, sales (Berlin)

Mullet, Rachel, 1995-97, sales (Berlin)

Mullet, Rita, 1995, admin. assistant (Berlin)

Northcott, Lee, 1994-95, finishing, craftsman

Nulph, Randall L., 1986-87, sander

O'Connell, April, 1994-95, sales

Oliphant, Amy E., 1995-present, sales

Oliphant, Jessica, 1995, sales

O'Mahony, Michael J., 1978-present, master craftsman

Ondo, Kathy, 1988-89

Opitz, Debra, 1986-87

Orsillo, Mark R., 1986, sander

Orsillo, Rick, 1986-89, production coordinator

Osborne, Marc, 1976-77

Otema, Malinda, 1996, sales (Berlin)

Palmer, Arthur, 1932-35, sales manager

Panada, Bob, circa 1936-40, metalworker

Panada, George, circa 1936-40, metalworker

Parenti, Louie, 1946, apprentice

Parquette, Delbert E., 1971-77, finisher

Parquette, Fred L., 1948-89, carbon coloring

Patrick, John, 1932-33, coasters

Pauley, Larry, 1994-97, maintenance (Berlin)

Peale, George, 1982-84

Pedlar, Benny, circa 1926 (Brockway), blacksmith

Perlin, Heather, 1995-present, sales

Perry, Jim, circa 1940-41, finisher

Peterson, Mindy K., 1994-present, mail order

Pisano, Erin L., 1995-present, sales

Pisoni, Otto, 1933, sander

Pisoni, Ottone I. "Tony", 1923-32 (Brockway), 1932-34, blacksmith

Pittock, James, 1978-83, craftsman

Polding, Donald W., 1990-present, finishing

Pompa, Anthony C., 1932-41, 1946-52, craftsman

Pompa, Nicholas, 1933-38, 1949, craftsman

Porterfield, George, 1994-95, finishing

Pryor, Mark E., 1977-present, finisher, asst. foreman

Puntureri, Vincent, 1946, metalworker

Raber, Lisa, 1995-96, sales (Berlin)

Ramsey, David, 1980-83, press operator

Ramsey, Tricia, 1994-95, sales

Rausch, Sara, 1992-94, purchasing agent

Reckhart, James R., 1994-present, engraving

Reddick, Georgetta, 1990-93, mail order

Redmond, Jeannette, 1994-present, master scheduler

Reinhart, Jason A., 1996-present, systems administrator

Ribaudo, Susan, 1984-86, sales

Ricci, Erik, 1996, craftsman (Berlin)

Richardson, Peg, 1979-85

Riddle, Clarence, circa 1950s, finisher

Riddle, Doug, 1984-86, repoussé

Robertson, Rosa, 1994-95, admin. assistant (Berlin)

Rockburn, Kimberly L., 1994-97, sales

Roddy, Sharon, 1994-95, shipping/receiving

Rodgers, Ruth A., 1997-present, shipping/receiving

Rodgers, William C., 1986-88, maintenance

Rogers, Elizabeth, 1994-95, engraver (Berlin)

Romine, Craig, 1995, craftsman (Berlin)

Rossi, Christopher, 1974-79, craftsman, die engraver

Rossi, Frank, 1937-41, 1946-47, sanding, jewelry

Rossi, Lewis M. "Doc", 1930-32 (Brockway), 1932-41, 1946-80, craftsman

Rossi, Natale L., 1927-32 (Brockway), 1932-41, 1946-80, blacksmith, die engraver, sales

Rossi, Pietro, 1950-60, janitor

Rowe, Bobbie, 1995, sales (Berlin)

Royal, (unknown), 1938-41, packing

Royer, Jeffrey K., 1975-present, master craftsman

Royer, Marilyn, 1994-95, sales

Rummes, Chris, 1994-95, craftsman (Berlin)

Runion, Laura K., 1997-present, mail order

Russell, John, 1982-83

Russell, Rick L., 1978-94, craftsman, shop supervisor

Russell, Rob, 1979-80, craftsman

Russo, Rocco, R., 1991-present, sales manager

Sabousky, Randall J., 1996-97, custodian

Sansom, Douglas, 1992-95, shipping

Sansom, Jamie, 1992-94, finishing

Sansotta, Joseph, 1949, craftsman

Sansotta, Rudolph Jr., 1973, sander

Saylor, Jerry A., 1980-88, craftsman

Scala, Gene, 1984-93, glass engraver

Schell, Megan, 1981-83

Schilling, Brad, 1995-96, craftsman (Berlin)

Schultz, David G., 1991-present, craftsman

Schur, John, 1988

Schutter, Bruce, 1988

Scott, Cheryl M., 1994-present, sales (Berlin)

Scott, Lori, 1996-97, sales (Berlin)

Seabright, Cheryl A., 1994-present,
inventory control (Berlin)

Setti, Thomas, circa 1939, craftsman

Sherman, Lyle M., 1991-present, finishing

Shetler, Eli I., 1996-present, craftsman (Berlin)

Shively, David S., 1995-present, sales manager

Shoenfelt, Heath, 1995-present,
craftsman, cell leader (Berlin)

Shoenfelt, Heather, 1997-present, sales (Berlin)

Shook, Eric, 1995-96, craftsman (Berlin)

Simon, William, 1976-77

Smith, Cynthia, 1994-95, sales

Smith, Damione L., 1997-present, custodian

Smith, Deborah J., 1991-present, receptionist

Smith, Laura, 1995, sales (Berlin)

Smith, Mary, 1994-95, shipping/receiving

Snow, Jamie, 1996, craftsman (Berlin)

Snyder, Bill, 1983-84

Snyder, Carol A., 1982-present, sales

Snyder, William S., 1949, apprentice

Sommers, Mary, 1996-present, sales (Berlin)

Sommers, Scott, 1994-95, craftsman (Berlin)

Sopher, Stacy L., 1994-96, shipping/receiving

Sowers, Corey, 1983-84

Spatara, Anthony P., circa 1954-63, sander

Sponseller, Robert, 1996-97, craftsman (Berlin)

Staff, Paul F., 1982-present, finisher

Stamets, Bob, craftsman (Berlin)

Stevenson, Adrian, 1977-78

Stewart, Carolyn S., 1986, sales

Stewart, Shirley, 1995-97, sales (Berlin)

Stivason, Tabatha, 1996-97, sales

Stone, Bradley, 1976-79, craftsman

Stull, Steven, 1994-95, finishing

Summerville, Greg P., 1990-present, craftsman

Surrena, Amy, 1989

Sutherland, Lester, circa 1934-41,
blacksmith, craftsman

Swartzentruber, Amber, 1994-96, sales (Berlin)

Swartzentruber, Cindy, 1994-96,
showroom supv. (Berlin)

Swartzentruber, Esther, 1994-95, sales (Berlin)

Swyers, Henry, 1931 (Brockway), finishing

Tavares, Karen, 1977-80

Taylor, Paul E., 1979-88, craftsman

Taylor, Rick, 1978-79, craftsman

Terwilliger, Gloria, 1996-present, sales

Thomas, Andrew A., circa 1939-41, 1946-75, craftsman

Thomas, Louis, circa 1939-41, 1946-65, wastebaskets

Thompson, Daniel J., 1994-present, repoussé

Thompson, Robert Jr., 1992-present, controller

Thompson, Ty L., 1985-present, craftsman

Thompson, Wilbur, 1949, apprentice

Todarello, Fred P., 1941, 1946-78,
welding, shop supervisor

Todarello, Joseph, circa 1948-49, finishing

Todarello, Todd, 1965, welding

Town, Eugene, 1933-39, repoussé

Towne, Jack, circa 1928-30 (Brockway), welding

Townsend, Frank, circa 1936-38, sales

Travis, Steven L., 1997-present, finishing

Trepasso, John A., 1946, apprentice

Trepasso, Ralph, 1940s, craftsman

Trepasso, Thomas M., circa 1939-41, 1946-49, craftsman

Trepasso, William, 1949, craftsman

Trowbridge, Richard W., 1984-88, craftsman

Troyer, Abe, 1994-present, maintenance (Berlin)

Tudor, Michael J., 1994-present, craftsman

Turner, Jay W., 1982-85, craftsman

Tuscano, Anthony, 1993-94, die engraver

Urban, Eric, 1998-present, craftsman (Berlin)

Urey, David A., 1983-present, craftsman

Vallely, Stacy, 1986-88, sales

Varner, Joy, 1996-97, sales (Berlin)

Vinton, James M. "Jimmy", circa 1932-41, 1946-54, craftsman

Vinton, William, circa 1936-39, sander

Vodenichar, Amy, 1992-94, sales

Voorhees, Mark B., 1990-present, finishing

Vroom, Lisa, 1982-85

Wallwork, John, circa 1933-41, craftsman

Wallwork, W. Robert, circa 1933-41, 1946-49, carbon coloring

Walmsley, Howard, 1982, shipping

Wardell, Nicki, 1994-98, sales (Berlin)

Watters, Todd W., 1997-present, finishing

Weaver, Dennis, 1986

Weaver, Greg, 1986-88, showroom manager

Weaver, Jolene F., 1996-98, sales (Berlin)

Weaver, Randall D., 1994-present, repoussé

Wengard, Crystal, 1995, sales (Berlin)

Wengard, Stephanie, 1995, sales (Berlin)

Werstler, Neal, 1995, sales (Berlin)

White, Christopher, 1975-77

White, Joseph, 1946, apprentice

Wigton, Cynthia J., 1997-present, shipping/receiving

Williams, Leroy, circa 1939-41, tool room, repoussé

Williams, Michael, 1981-83, sanding

Wilson, Mark, 1994-96, master scheduler

Winder, Ronald E., 1972-present, master craftsman

Winger, Wes, circa 1932-35, repoussé

Winkler, Glenn, 1979-81, sanding

Wirick, Linda, 1985-86, glass cleaner

Wise, Janet, 1993-94, sales

Wise, Patrick J., 1973, sanding

Wiser, Vernon, circa 1936-39, metalworker

Wolfe, Clara E., circa 1946-1962, office manager

Woodworth, Linda, 1991-94, sales

Yoder, Carol, 1997-present, cleaning (Berlin)

Yoder, Erma, 1997-present, cleaning (Berlin)

Yoder, Tim, craftsman (Berlin)

York, Joe, 1978-79, repoussé

Young, Curtis L., circa 1986-91, craftsman

Young, John F., 1995-present, sales

Young, Sara, 1995-present, sales

Youngo, Catherine E., 1975-present, assistant to CEO

Youngo, Joseph, 1994-present, retail director (Berlin)

Youngo, Leonard J., 1981-present, finisher, die engraver

Ziccardi, Anthony, 1946, apprentice

Ziccardi, Dennis, circa 1960s, finishing

Ziccardi, Joseph P., 1940-41, 1946-77, finisher, craftsman

Appendix B

United States Patent

For many years, there has been confusion about the meaning of the "patent applied for" notations found on many Wendell August Forge items. The few relevant records available at the Forge were catalog materials that made vague references to a patent (or patents). Some of these statements, such as "the finish is obtained by our own process for which patent rights have been filed," suggested the existence of a patent for the carbon coloring process used by the Forge.

The volumes and indexes of the *Official Gazette* of the United States Patent Office were thoroughly searched for any trademark registrations, design patents, or utility patents potentially related to the Wendell August Forge or any of its personnel. Only one patent, a utility patent for the repoussé process used by the Forge, was located.

As noted in the patent document, application was made on July 25, 1932. The patent was published more than two years later, on August 28, 1934. However, one of the basic requirements for a utility patent is that the invention qualify as something that is new. Objections were raised regarding many of the patent claims, since a variety of generic repoussé processes have been in use since antiquity, and thus are neither new nor patentable. On February 5, 1935, the *Official Gazette* published disclaimers from the Wendell August Forge for portions of the patent claims. Although the patent still remained, it now had little value.

It is likely that the impetus for undertaking the patent process came from Edwin J. Fithian, who fully recognized the value of claiming intellectual property rights for new processes. Much of his wealth had been derived from patents held by the Bessemer Gas Engine Company, of which he had been a principal and the President.

The patent as it appears in this Appendix was reproduced from the microfilm of the original.

Patented Aug. 28, 1934 **1,971,700**

UNITED STATES PATENT OFFICE

1,971,700

METHOD OF FORMING ORNAMENTAL RELIEF FIGURES

Wendell August, Grove City, James McCausland, Falls Creek, and Howard J. Chapin, Brockway, Pa., assignors to Wendell August Forge Incorporated, Grove City, Pa., a corporation of Pennsylvania

Application July 25, 1932, Serial No. 624,640

7 Claims. (Cl. 41—24)

The present invention is designed to form ornamental relief figures on metal plates which plates may be used for various purposes. The method gives to the plates a very artistic effect and can be readily and rapidly practiced so that the plates may be formed in production rapidly enough to make such plates readily salable. The plates may be used for various purposes, as for example, they may be shaped into ordinary trays. Features and details of the invention will appear more fully from the following specification and claims.

A preferred manner of carrying out the method and the tools for practicing it are illustrated in the accompanying drawing as follows:—

Fig. 1 shows a plan view of a pattern block.

Fig. 2 a section on the line 2—2 in Fig. 1.

Fig. 3 the ornamented plate as it comes from the block.

Fig. 4 a section on the line 4—4 showing a retouching, or forming tool more definitely outlining the design.

Fig. 5 a plan view of the finished plate.

Fig. 6 a rear view of the finished plate.

1 marks the plate. This may be of ductile metal. We have found that ordinary aluminum is well adapted for the purpose. A pattern block 2 has a pattern formed in its face. For very many designs the pattern may be formed by cutting the design through the block and in the block 2 portions of the pattern 3 are formed by simply sawing out, or removing the material entirely through the block in the ordinary manner of scroll sawing. Other parts of the pattern may be worked into the surface of the block giving more definitely a surface design. Such portions of the pattern are indicated at 4. The parts 4 are depressed and with the parts 3 give the completed pattern. With this pattern block the plate to be ornamented is laid upon the block and hammered by a hammer, such as the hammer 5. The operator merely hammers the back of the plate. The yielding of the metal indicates to the operator those portions of the plate which are over the parts of the design having the higher relief and having some idea of the pattern the operator can manipulate his hammer using the peening end, if desired, forcing the metal to follow the design. For very many parts of the design no particular attention is required by the operator. Merely following over the surface with the hammering operation will form a sufficient relief to carry the pattern. The amount of relief and nature of the design controls to some extent the hammer blows, but generally speaking it is a very simple operation which can be readily followed.

In the hammering of the metal the metal partly flows and partly bends to form the design. The parts of the design having the least relief are formed largely through the flowing of the metal while the deeper portions are formed by flowing and bending. The design as it comes from the block is indicated in Fig. 3, having the relief portions 1a, and the outlines of the design may be sharpened to give it a slightly more pleasing effect, if desired, by a forming tool 6 which is operated along the lines to bring out more definitely certain of the lines desired by depressing portions 1b along the line of the relief. This is just an ordinary forming tool operated upon by a hammer and worked along the lines desired. The face of the material may be given an added mottled appearance, as indicated in Fig. 5, by hammering, but it gets some of this effect from the hammering on the back, so that this step is unnecessary.

The rear of the plate has the hammered surface with the hammer indentations 7 and the higher parts of the relief are indicated to some extent, but with no definite degree of similarity by depressions 8 which are formed by the hammer, or peening end of the hammer, the peening portion of the hammer being small enough to follow these depressions.

What we claim as new is:—

1. The method of forming ornamental plates which consists in forming a depressed portion in a pattern block; placing the plate thereupon and hammering the back of the plate forcing the material of the plate into the depressions formed in the block.

2. The method of forming ornamental plates which consists in forming a depressed portion in a pattern block; placing the plate thereupon and hammering the back of the plate; forcing the material of the plate into the depressions formed in the block forming a hammered surface on the plate.

3. The method of forming ornamental plates which consists in forming a depressed portion in a pattern block; placing the plate thereupon and hammering the back of the plate forcing the material through such hammering of the plate into the depressions formed in the block by metal flow and bending.

4. The method of forming ornamental plates which consists in forming a depressed portion in a pattern block; placing the plate thereupon and hammering the back of the plate forcing the material through such hammering of the plate into

the depressions formed in the block; reversing the plate; and sharpening the outline of the design by a forming operation along the lines of the design.

5. The method of forming ornamental plates which consists in forming a depressed portion in a pattern block by extending openings entirely through the block; placing the plate thereupon and hammering the back of the plate; and forcing the material through such hammering of the plate into the depressions formed in the block.

6. The method of forming ornamental plates which consists in forming a depressed portion in a pattern block by extending a part of said depressed portion entirely through the block and other parts of the depressed portion partially through the block; placing the plate thereupon and hammering the back of the plate; and forcing the material through such hammering of the plate into the depressions formed in the block.

7. The method of forming ornamental plates which consists in forming a depressed portion in a pattern block; placing the plate thereupon and hammering the back of the plate forming a hammered surface on the plate and forcing the material through such hammering of the plate into the depressions formed in the block; reversing the plate; and sharpening the outline of the design by a forming operation along the lines of the design.

WENDELL AUGUST.
JAMES McCAUSLAND.
HOWARD J. CHAPIN.

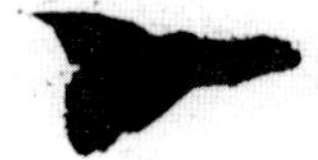

DISCLAIMER

1,971,700.—*Wendell August*, Grove City, *James McCausland*, Falls Creek, and *Howard J. Chapin*, Brockway, Pa. METHOD OF FORMING ORNAMENTAL RELIEF FIGURES. Patent dated August 28, 1934. Disclaimer filed January 15, 1935, by the assignee, *Wendell August Forge Incorporated.*

Therefore, disclaims from each of claims 1, 2, and 3 any method of forming ornamental plates except a method in which the hammer blows upon the back of the plate are selectively governed with relation to the depressed portion of the pattern block.

[*Official Gazette February 5, 1935.*]

Aug. 28, 1934. W. AUGUST ET AL 1,971,700

METHOD OF FORMING ORNAMENTAL RELIEF FIGURES

Filed July 25, 1932

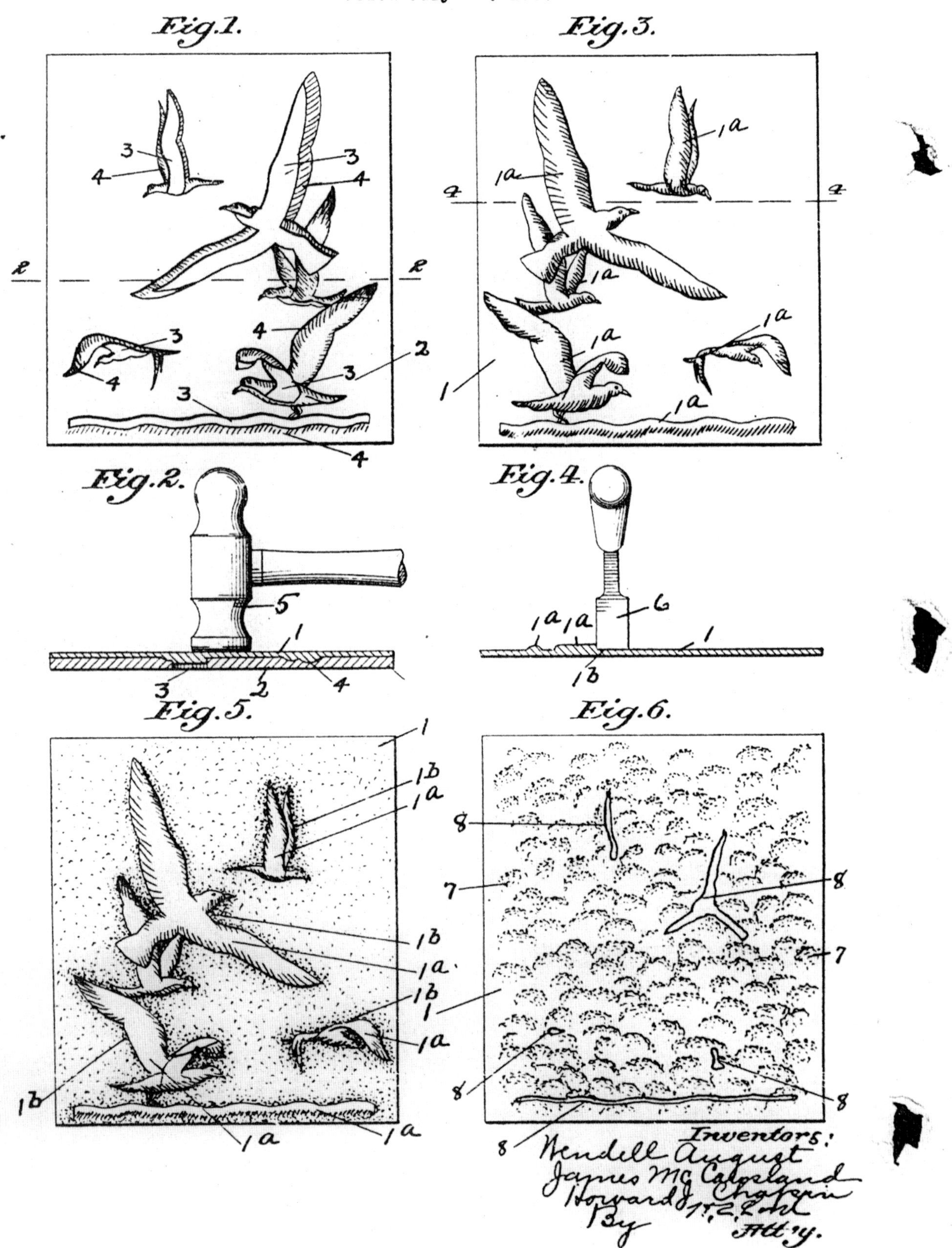

Appendix C

Wendell August Forge Trademark Handstamps

At least 25 forms of trademark handstamps that appear on Wendell August Forge products are useful for dating and other identification purposes. The representations illustrated here were hand-drawn by the author from Wendell August Forge items and represent generic types of trademark handstamps used. It is likely that there exist some additional trademark handstamps (e.g., for pewter) for which documentation has not yet been acquired.

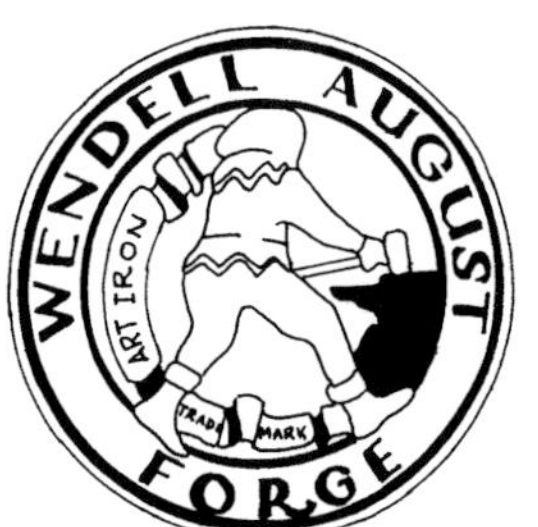

Representation of the Wendell August Forge trademark as it appears on the 1932 brochure. Searches of U.S. Patent Office records have thus far revealed no formal registration of this trademark.

The selection of many of these marks, together with the probable time frames of their use, is based on an analysis of the author's data base of Wendell August Forge metal wares, primarily aluminum. This data base, which contains approximately 2,300 records, was constructed over a period of more than two years from printed documentation and from pieces in private collections. Regardless of the size of the data base, it is important to realize that the Wendell August Forge produced a wide variety of product, much of which remains undocumented. In addition, production was low during some time periods, and available records (e.g., advertising materials) are sparse.

It is helpful in assessing some of the circular trademark handstamp variations that were used during the 1932-1978 time frame if one has a basic understanding of the manner in which the Wendell August Forge acquired and disposed of its handstamps. A certain number of handstamps were made at a given time. Because they were made by hand, *each of these handstamps had slight variations.* When the workforce expanded, as it did frequently in the years preceding World War II, additional sets of handstamps were made to be used by the additional craftsmen. However, the older handstamps were not discarded until such time as they were too worn to continue in use. Also, those that were used by craftsmen who made the larger-volume and more heavy-duty items (e.g., trays), wore out much more rapidly than those used either less frequently or for more delicate items (e.g., jewelry). Further, each time an additional set of handstamps was made, the rendition of the trademark that resulted was usually a slight corruption of the rendition that was used as a model. As a result, there were several varieties of trademark handstamps in use at any given time.

It should also be noted that some Wendell August Forge items will be found that have been dated by either engraving the date on the piece or adding a custom handstamp mark having a date to the back of the piece. The custom handstamps frequently include the name of an organization. In both cases, the engraving and the custom handstamp mark may have been applied some considerable time subsequent to the actual production of the item.

An additional point of interest is that, even though many custom-designed dies have been engraved that include dates, it is typical that the die was engraved and the items actually produced prior to the date included in the die engraving. Furthermore, there are a few well-documented Wendell August Forge items that were made, from older dies having dates engraved in them, a considerable number of years after the dies were actually engraved.

The trademark handstamp facsimiles presented in this Appendix are divided into four groups as follows:

1. **Text Marks** - These marks consist primarily of text and do not contain Wendell August Forge trademark representations. Some of the marks in this group (e.g. REA Metals) are not those of the Wendell August Forge *per se*.
2. **Circular Wendell August Forge, Inc. Marks** - These are Wendell August Forge trademark representations, approximately 5/8" in diameter, that contain the *Inc.* designation in the mark. The time frame of their use ranges from *circa* 1932-1950.
3. **Circular Wendell August Forge Marks** - These are Wendell August Forge trademark representations that do *not* include the *Inc.* designation in the mark and that were discontinued when the Forge was acquired by F.W. "Bill" Knecht III. The time frame of their use ranges from *circa* 1936-1978.
4. **Knecht Marks** - These are Wendell August Forge trademark handstamps that were introduced from 1978 through 1998.

Additional information pertinent to the dating of the metal wares associated with the Wendell August Forge can be found in Appendix D (Die Engraver and Craftsman Identification), Appendix E (Limited Edition and Collector Series), and Appendix F (Wendell August Forge Dating Clues).

Text Marks

The trademark handstamps illustrated here consist predominantly of text and as such are readily identified. Three of the marks (i.e., "Wendell," "Pinegrove," and "REA Metals") are technically not marks associated with items produced by the Wendell August Forge. However, they are included here because of the close relationship between the Wendell August Forge, Pine Grove Craftsman, and REA Metal Products, Inc. The relationship of the Wendell August Forge to the other two firms and their products is discussed in Chapter Three.

WENDELL AUGUST FORGE

T-1

T-1 – is the earliest Wendell August Forge trademark handstamp, and was used primarily in Brockway. Its use appears to have been discontinued *circa* 1932, when the Forge moved to Grove City and applied for the patent on the repoussé process. Aluminum items bearing this mark could not have been produced before 1930.

WENDELL AUGUST FORGE

PAT. APL'D FOR

T-2

T-2 – is T-1 with the addition of a separate handstamp mark indicating that application had been made for the patent. Its use was confined primarily to the *circa* 1932-1935 time frame.

WENDELL

T-3

T-3 – is the trademark handstamp that appears on the Pine Grove Craftsman product that was hand-hammered in mid-1940.

PINEGROVE

T-4

T-4 – is the trademark handstamp that appears on the Pine Grove Craftsman product made in 1940-1941 for which the motifs were rolled.

H. POMERANTZ INC. N.Y.

PLATANELLE

T-5

T-5 – is the trademark handstamp that appears on jewelry that was made by the Wendell August Forge, *circa* 1946-1947, and marketed by Herman Pomerantz of New York City. One item has been reported on which the word "Platanelle" does not appear. Currently-available information pertaining to Pomerantz is addressed in Chapter Three.

HAWAIIAN
PATTERNS

T-6

T-6 – is the trademark handstamp that appears on some jewelry and other items, such as trays and coasters, that were made by the Wendell August Forge, *circa* 1947-1949. Although most of the items bear only the "Hawaiian Patterns" handstamp, a few also bear the F-4 Wendell August Forge trademark handstamp. A selection of "Hawaiian Patterns" items is depicted in Chapter Three.

REA METALS
GROVE CITY, PA.

T-7

T-7 – is the trademark handstamp that appears on items made by REA Metal Products, Inc. Although the handstamp was used from 1946 on into the 1970s, the estimated production time frame for the florist vases and planters was *circa* 1947-1949.

Circular Wendell August Forge, Inc. Marks

The trademark handstamp representations grouped together here all state "Wendell August Forge, Inc." (rather than simply "Wendell August Forge"), in the outside ring of the mark. These trademark handstamps are variations, over time, of the October, 1932 rendition of the Wendell August Forge trademark shown at the beginning of this Appendix. For many years, collectors speculated as to the nature of the objects that appear between the feet of the blacksmith figure that is in the center of the mark, one of the prevailing speculations being that they were representations of chunks of the coal used in the forge. As can be seen in the October, 1932 rendition, the objects are corrupted versions of loops of a banner bearing the words "trade" and "mark" between the figure's feet.

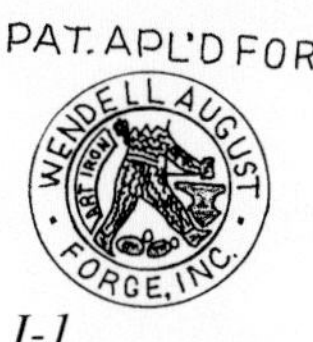

I-1

I-1 – consists of two separate handstamps, one for the trademark and one to indicate the patent application. Because the two handstamps were separate, the marks may appear in configurations other than the one illustrated here. It is sufficient for identification and dating purposes to note that the outside ring of the circular mark includes "Wendell August Forge, Inc." and that the "Pat.Apl'd For" consists of a separate mark. The use of this configuration was confined primarily to circa 1932-1935.

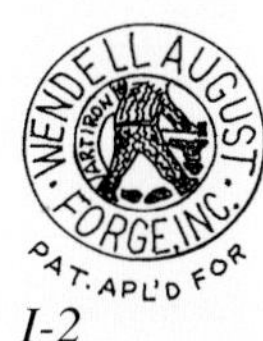
I-2

I-2 – consists of a single trademark handstamp with "Wendell August Forge, Inc." in the outside ring and that incorporates the "Pat.Apl'd For" notation in small letters around the lower edge of the circular trademark. It is sufficient for identification and dating purposes to note that the trademark handstamp incorporates the "Pat.Apl'd For" notation in small letters around the lower edge of the mark. This trademark handstamp was in use *circa* 1932-1935.

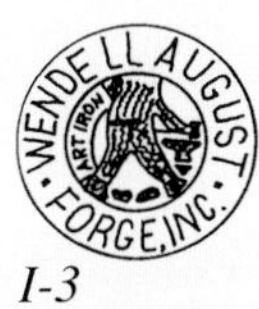
I-3

I-3 – is the same as I-2, but with the "Pat.Apl'd For" notation removed. The width of the ring which includes "Wendell August Forge, Inc." is 1/8" wide, and the two objects between the feet of the blacksmith are not merely outlines but have some texture and depth. This trademark handstamp was in use *circa* 1935-1939.

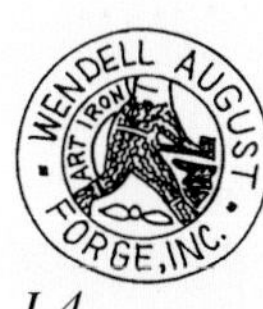
I-4

I-4 – has three outlined objects between the feet of the blacksmith figure as well as the "Wendell August Forge, Inc." in the outside ring of the mark. Although there are variations in this handstamp, the outline of three objects between the feet is consistently sufficient for identification and dating purposes. The use of this trademark handstamp was *circa* 1935-1941.

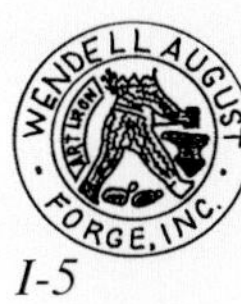
I-5

I-5 – is I-1 without the additional application of the "Pat.Apl'd For" handstamp. Mark I-5 can be readily distinguished by the fact that the outside ring containing "Wendell August Forge, Inc." is only 3/32" wide. The use of this trademark handstamp spanned *circa* 1935-1950 (excluding the *circa* 1942-1945 World War II period).

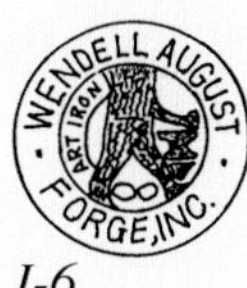
I-6

I-6 – can be distinguished by the fact that the two objects between the feet of the blacksmith are merely outlined, albeit somewhat irregularly at times. The inclusion of "Wendell August Forge, Inc." in the outside ring of the mark, together with the relatively circular outlines of two objects between the feet of the blacksmith, yield consistent referents for identification and dating purposes. This trademark handstamp was in use *circa* 1938-1948 (excluding the *circa* 1942-1945 World War II period).

There are variations of each of the generic handstamp representations shown above.

Circular Wendell August Forge Marks

(prior to the introduction of the Knecht Marks beginning in 1978)

The trademark handstamp forms considered here are closely related to those just considered; however, the marks in this group all state "Wendell August Forge" (rather than "Wendell August Forge, Inc.") in the outside ring of the mark. In addition, these trademark handstamp forms predate the Knecht Marks discussed in the next group.

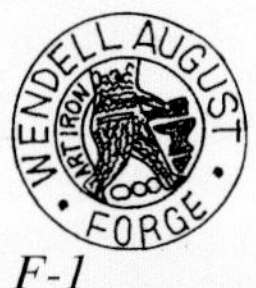

F-1

F-1 – can be distinguished by the three outlined "circles" between the feet of the blacksmith figure and the densely textured (or "striped") suit on the blacksmith figure. The texturing or "striping" in the upper portion of the figure may run either horizontally or vertically. This trademark handstamp was in use *circa* 1936-1941.

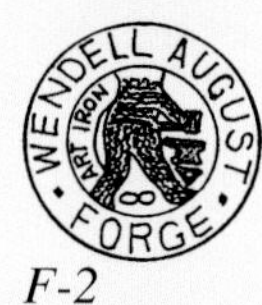

F-2

F-2 – can be distinguished by the two outlined "circles" between the feet of the blacksmith figure and the densely textured (or "striped") suit on the blacksmith figure. The texturing or "striping" in the upper portion of the figure may run either horizontally or vertically. This trademark handstamp was in use *circa* 1938-1948 (excluding the *circa* 1942-1945 World War II time period).

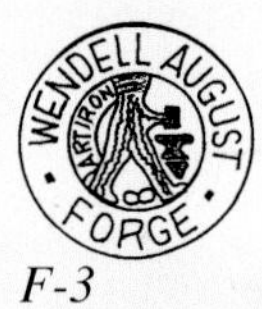

F-3

F-3 – can be distinguished by the two outlined "circles" between the feet of the blacksmith figure, and the highly abstracted blacksmith figure having no "waist" and relatively widely-spaced vertical "stripes" in its suit. This trademark handstamp was in use *circa* 1940-1950 (excluding the *circa* 1942-1945 World War II time period).

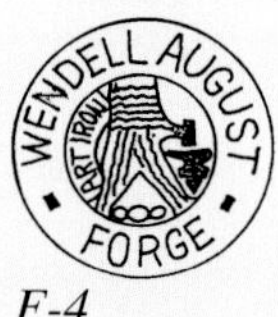

F-4

F-4 – can be distinguished by the three outlined "circles" between the feet of the blacksmith figure, and the highly abstracted blacksmith figure with relatively widely-spaced "stripes," consistently horizontal above the "waist" and consistently vertical below the "waist." This trademark handstamp was in use predominantly *circa* 1945-1961; however, some items from the 1960s and 1970s have been documented that bear this generic form of the trademark handstamp.

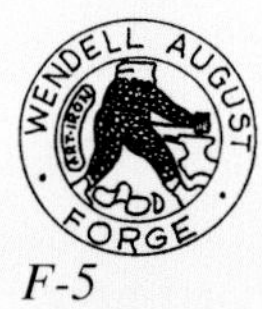

F-5

F-5 – is distinguishable by the relatively realistic blacksmith figure and the texturing of the blacksmith's suit with dots. This trademark handstamp was introduced *circa* 1961, and continued in use into 1978.

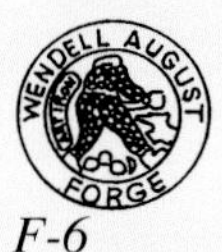

F-6

F-6 – is similar to F-5, in that the blacksmith figure is relatively realistic, and the suit of the blacksmith is textured with dots. However, F-6 is only 1/2" in diameter, whereas F-5 is 5/8" in diameter. This trademark handstamp was definitely in use throughout the 1970s; however, the data pertaining to the use of this mark during the 1960s are as yet inconclusive.

There are variations of each of the generic handstamp representations shown above.

Knecht Marks

The trademark handstamps implemented since the time that the Wendell August Forge was acquired by F.W. "Bill" Knecht III are easily distinguishable from one another and yield relatively straightforward dating information regarding the Forge's product since the latter part of 1978. There are, however, some approximations required at the very beginning and end of the periods of use of each different trademark handstamp. For example, K-1 did not come into use until the latter part of 1978, and K-2 actually came into use near the end of 1979.

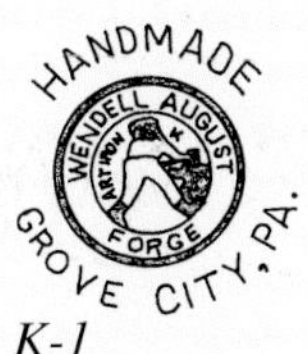

K-1

K-1 – was implemented in the latter part of 1978. This was the first trademark handstamp for the Wendell August Forge that incorporated "Handmade" and "Grove City, PA." into a single mark.

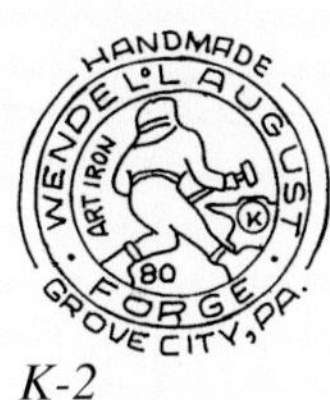

K-2

K-2 – came into use near the end of 1979 and was used throughout the 1980s.

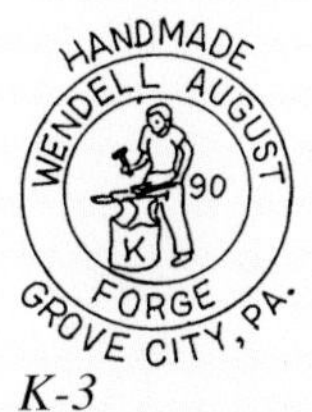

K-3

K-3 – was in use *circa* 1990-1994.

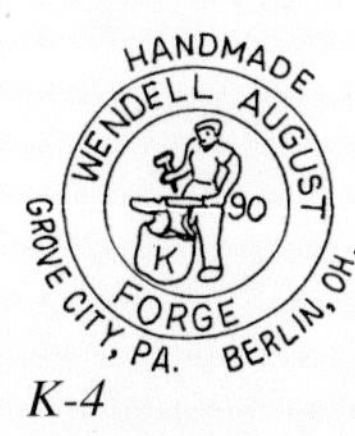

K-4

K-4 – is essentially K-3 with the addition of "Berlin, OH." and was introduced in 1994 when the gift shop and forge was opened in Berlin, Holmes County, Ohio. This trademark handstamp remained in use into 1998, when K-6 was introduced.

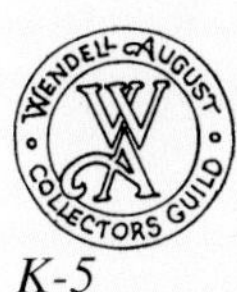

K-5

K-5 – is the trademark handstamp that is used on Wendell August Forge Collectors Guild items. It made its appearance in late 1995 when the Collectors Guild was formed and it continues in use.

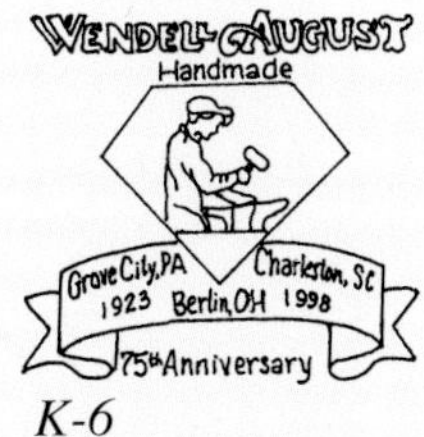

K-6

K-6 – is the trademark handstamp introduced in 1998 to commemorate the 75th Anniversary of the Wendell August Forge and, at the same time, the opening of the Charleston, South Carolina gift shop and forge.

Appendix D

Die Engraver and Craftsman Identification

Die engravers: The practice of including die engraver identification in each engraving started in about 1969 when the Wendell August Forge became involved in the creation of Limited Edition items. Dies engraved prior to this time bear no engraver identification. Oral history suggests that dies created between 1930 and mid-1932 were engraved by several artisans, including Natale Rossi. With only a few possible exceptions, Louis Donato engraved all of the dies from mid-1932 through 1941, and Natale Rossi engraved all of the dies from 1945 to about 1969.

Because a die can be used for several years, the usefulness of die engraver identification for nothing more than dating purposes is typically limited to determining the earliest year in which a particular item might have been made. A more important use of die engraver identification is that it allows one to discern the differences in technique and execution used by the different artisans. In some cases, die engraver identification also facilitates the identification of the artist or designer responsible for creating the original art work for the engraving.

Die engraver identification usually consists of stylized versions of engraver initials embedded in the engraving. Die engraver identifications, together with the years in which they were used, are provided below. It should be noted that these are not facsimiles.

The Wendell August Forge also employs machine engravers, who engrave custom lettering on completed metal wares but do not engrave dies.

Name	Dates	Identification
Steven Adams	1980-1982	sa
David Bruck	1980-present	db
Grant Galloway	1995-present	gg
David Miller	1985-1986	dm
Christopher Rossi	1974-1979	cr
Natale Rossi	1969-1979	nr, By Natale, N. Rossi
Anthony Tuscano	1993-1994	tt
Leonard Youngo	1982-present	ly

Craftsmen: The use of craftsman touchmarks was introduced in January of 1977, while Robert August was President of Wendell August Forge. The touchmarks are applied at the time that an item is made, although there has been some inconsistency in this practice. Items produced prior to January of 1977 bear no touchmarks and thus cannot be attributed to individual craftsmen.

The information provided here has been constructed from numerous sources of touchmark facsimiles and data, and thus there may be some error in representation or attribution. The dates provided refer to the years in which the touchmark has been reported as being used by a particular person. In some cases, there may be variations observed between the facsimiles and the actual touchmarks as they appear on an item. It is possible that touchmarks exist for which no information was available when this reference was compiled.

Facsimiles of Craftsman Touchmarks

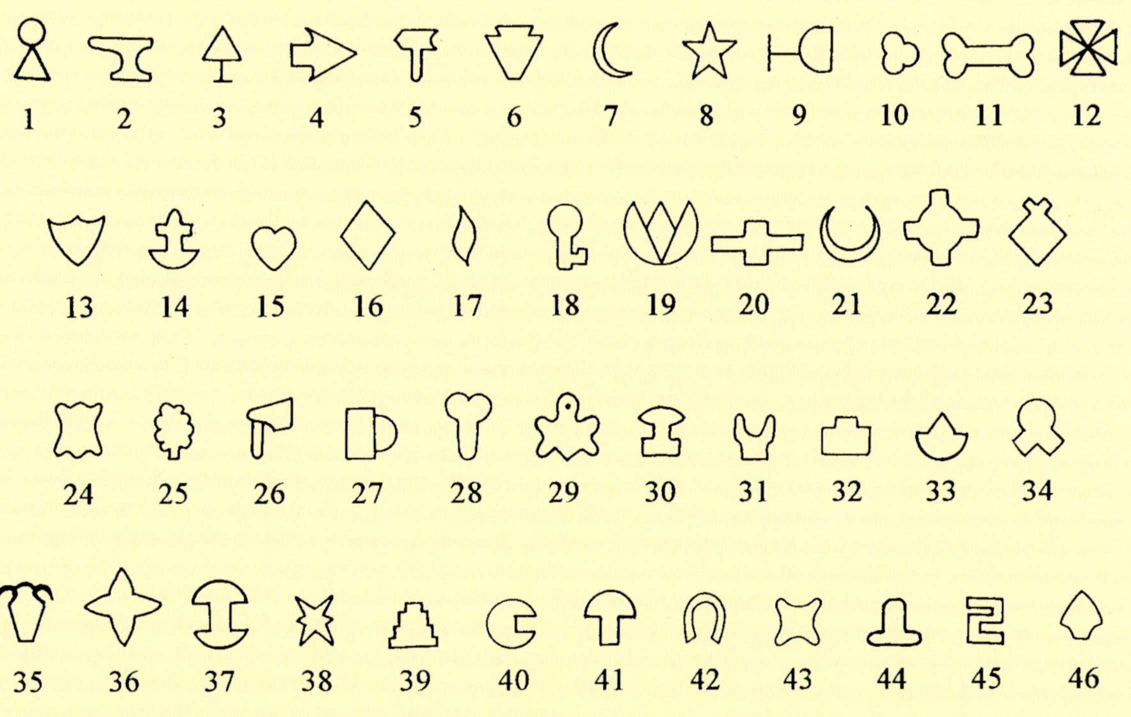

Wendell August Forge Craftsmen

1. *Grove City: Jeff Brown (1977-present)*
2. *Grove City: Jeff Royer (1977-present)*
3. *Grove City: Ron Winder (1977-present)*
4. *Grove City: Ben Formani (1977-79)*
5. *Grove City: Mike Guarnieri (1977-79)*
6. *Grove City: Don Kelly (1977-85)*
7. *Grove City: Dennis Knauff (1977-82)*
8. *Grove City: Kent McGill (1977-88)*
9. *Grove City: Lewis "Doc" Rossi (1977-80)*
10. *Grove City: Chris Rossi (1977-79)*
11. *Grove City: Brad Stone (1977-79)*
12. *Grove City: Brian King (1977-1979), Steve Collar (1994-95)*
13. *Grove City: Mike O'Mahony (1978-present), Rick Taylor (1978-79)*
14. *Grove City: Rick Russell (1978-94)*
15. *Grove City: Rob Russell (1979-80), Scott Campbell (1994-present)*
16. *Grove City: Paul Taylor (1979-88)*
17. *Grove City: Dan Magee (1980-present)*
18. *Grove City: Jerry Saylor (1980-88)*
19. *Grove City: Dan Hobart (1981-84), Ty Thompson (1985-present)*
20. *Grove City: Jay Turner (1982-85)*
21. *Grove City: Dave Urey (1983-89)*
22. *Grove City: Odis Hamilton (1985-89)*
23. *Grove City: Ed Hodge (1985-present)*
24. *Grove City: Curt Young (1986-91)*
25. *Grove City: Greg Summerville (1990-present)*
26. *Grove City: Frank Dydek (1990-present)*
27. *Grove City: Dave Schultz (1991-present)*
28. *Grove City: Richard Meyerl (1991-97)*
29. *Grove City: Mike Flynn (1993-present)*
30. *Grove City: Lee Northcott (1994-95), Randy Weaver (1994-present)*
31. *Grove City: Mike Tudor (1994-present)*
32. *Grove City: Dan Thompson (1994-present)*
33. *Grove City: Jim Eakin (1994-95)*
34. *Grove City: Jeff Corbett (1994-95)*
35. *Grove City: Doug Flickinger (1990s)*
36. *Berlin: Henry Mast (1994-present)*
37. *Berlin: Kenny Gribben (1994-96), Tom Baum (1990s)*
38. *Berlin: Heath Shoenfelt (1995-present), Jon Klingler (1994-97)*
39. *Berlin: Joe Davis (1994-97)*
40. *Berlin: Eli Mast (1996-present), Corey Baily (1996), Chris Rummes (1994-95), Dave Masters (1990s)*
41. *Berlin: Eric Urban (1998-present), Leslie Croft (1997-98), Brad Schilling (1995-96)*
42. *Berlin: Joe Miller (1997-present), Scott Sommers (1994-1995)*
43. *Berlin: Eli Shetler (1996-present)*
44. *Berlin: Robert Sponseller (1996-97), Thad Bashman (1994-95), Tim Yoder (1990s)*
45. *Berlin: Tom Marshall (1997-present), Scott Fearon (1994-95)*
46. *Berlin: Derek Frautschy (1997-present), Tom Mellor (1997), Jon Hines (1995-96), Craig Romine (1995), Bob Stamets (1990s)*

Appendix E

Limited Edition and Collector Series

(in order of initial issue year of item or series)

The information provided here was compiled from numerous sources, and there may be some unintentional inaccuracies or omissions. The item sizes that are provided are nominal sizes; some actual item sizes may differ slightly. To the extent that it has been possible to do so, the artist and the die engraver have been identified for each of the items.

Great Moments in History

Limited Edition Plates; Issued in Aluminum, Bronze, Pewter and Sterling; Art and Engraving by Natale Rossi

1971 9" diameter, Columbus - First Edition (die destroyed by 1982)

1972 9" diameter, Landing of Pilgrims (die destroyed by 1982)

1973 9" diameter, Thanksgiving (die destroyed by 1982)

1974 9" diameter, Patrick Henry (die destroyed by 1982)

1975 9" diameter, Paul Revere (die destroyed by 1982)

1976 9" diameter, Signers of the Declaration (die destroyed between 1982 and 1985)

1976 16" diameter, Signers of the Declaration - Last Edition (die destroyed between 1982 and 1985)

Presidents

Issued in Aluminum, Bronze, Pewter and Sterling; Retired between 1985 and 1987; Engraved by Natale Rossi

1971 John Fitzgerald Kennedy Plate, 9" diameter

1972 Abraham Lincoln Plate, 9" diameter

Peace on Earth

1973 Peace on Earth, also referred to as Doves of Peace; 9" Sterling Silver Plate framed in walnut; Retired between 1985 and 1987; Art by Stanley W. Galli; Engraved by Natale Rossi

Model T Ford

1974 Model T Ford; 9" diameter Plate; issued in Aluminum, Bronze, Pewter and Sterling; Retired between 1985 and 1987; Engraved by Natale Rossi

Tribute to American Transportation

Limited Edition; Retired between 1989 and 1990

1974 9" diameter Best Friend Plate and 16 oz Best Friend Tankard; Engraved by Natale Rossi

1974? 9" diameter Conestoga Wagon Plate and 16 oz Conestoga Wagon Tankard; Engraved by Natale Rossi

1980 Homeward Bound; Revision of Conestoga Wagon; Engraving started by Natale Rossi and completed by David Bruck

Christmas

Limited Edition 8" diameter Plates; Issued in Bronze, Pewter and Sterling

1974 The Carolers - First Edition (unavailable by 1982), Engraved by Natale Rossi

1975 Christmas in the Country (die destroyed by 1982), Engraved by Natale Rossi

1976 The Lamplighter (die destroyed by 1982), Engraved by Natale Rossi

1977 The Covered Bridge (die destroyed by 1982), Engraved by Natale Rossi

1978 Night Before Christmas (die destroyed by 1984), Engraved by Natale Rossi

1979 The Grist Mill (die destroyed in 1984), Engraved by Natale Rossi

1980 Christmas in Old Town (die destroyed in 1985), Art and Engraving by David Bruck

1981 The Skating Pond (die destroyed in 1986), Art and Engraving by Steven Adams

1982 The Sleigh Ride (die destroyed in 1987), Art and Engraving by David Bruck

1983 The General Store (die destroyed in 1988), Art and Engraving by Leonard Youngo

1984 Journey to Bethlehem (die destroyed in 1989), Art and Engraving by David Bruck

1985 Santa's Express (die destroyed in 1990), Art and Engraving by Leonard Youngo

1986 Sounds of Christmas (die destroyed in 1991), Art and Engraving by David Bruck

1987 The Homecoming (die destroyed in 1992), Art and Engraving by Leonard Youngo

1988 Santa (die destroyed in 1993), Art and Engraving by David Bruck

1989 Building a Snowman (die destroyed in 1994), Art and Engraving by Leonard Youngo

1990 Trimming the Tree (die destroyed in 1995), Art and Engraving by David Bruck

1991 Grandma's Christmas Kitchen (die destroyed in 1996), Art by David Bruck, Engraved by Leonard Youngo

1992 A Savior is Born (die destroyed in 1997), Art and Engraving by David Bruck

1993 A Victorian Christmas, Art by Linda Hoover, Engraved by Leonard Youngo

1994 The Christmas Carol, Art by Linda Hoover, Engraved by David Bruck

1995 Holiday Express, Art by Linda Hoover, Engraved by Leonard Youngo

1996 Sleigh Bells Ring, Art by Linda Hoover, Engraved by David Bruck

1997 A Country Christmas, Art by Linda Hoover, Engraved by Leonard Youngo

Wildlife

Limited Edition 9" diameter Plates; Issued in Aluminum, Bronze, Pewter and Sterling

1977 On Guard (Deer) - First Edition (die destroyed in 1982), Engraved by Natale Rossi

1978 Thunder Bird (die destroyed in 1983), Engraved by Natale Rossi

1979 Honey Bear (die destroyed in 1984), Engraved by Natale Rossi

1980 Prairie Lord - Last Edition (die destroyed in 1985), Engraving started by Natale Rossi and completed by Steven Adams

Neuschwanstein Castle

Plates Issued in Aluminum, Bronze, Pewter and Sterling; Tankards Issued in Aluminum and Bronze; Retired between 1985 and 1987; Engraved by Natale Rossi

1979	7"	Neuschwanstein Castle Plate
1979	8"	Neuschwanstein Castle Plate
1979	16 oz	Neuschwanstein Castle Tankard
1979	20 oz	Neuschwanstein Castle Tankard

Sterling Eagle on Bronze

1980 Sterling Eagle on Bronze; 6" Plate, Signed and Numbered; "A Wendell August First"; Special Limited Edition, 15 known to be issued; Retired by 1985; Art and Engraving by David Bruck

Christmas Heirloom Ornaments

Dated, 2-1/8" Diameter, Issued in Aluminum, Bronze and Sterling

1980 Holly - First Edition (retired in 1996), Engraved by David Bruck

1981 Partridge in the Pear Tree (retired in 1996), Engraved by Steven Adams

1982 Christmas Bells (retired in 1996), Engraved by David Bruck

1983 Christmas Ornaments (retired in 1996), Engraved by Leonard Youngo

1984 Season's Greetings (retired in 1996), Art and Engraving by David Bruck

1985 Rocking Horse (retired in 1997), Engraved by Leonard Youngo

1986 Christmas Teddy Bear, Art and Engraving by David Bruck

1987 The Sleigh, Engraved by Leonard Youngo

1988 Peace Dove, Art and Engraving by David Bruck

1989 The Church, Art by David Bruck, Engraved by Leonard Youngo

1990 The Angel, Art and Engraving by David Bruck

1991 Santa Claus, Engraved by Leonard Youngo

1992 Candlelight, Art and Engraving by David Bruck

1993 Snowman, Engraved by Leonard Youngo

1994 Gingerbread House, Art by Linda Hoover, Engraved by Anthony Tuscano

1995 Old-Fashioned Sleigh Ride, Art by Linda Hoover, Engraved by David Bruck

1996 The Littlest Angel, Art by Linda Hoover, Engraved by Leonard Youngo

1997 The Drummer Boy, Art by Linda Hoover, Engraved by David Bruck

In 1996 only, Limited Edition Set of the 1980-1984 ornaments was issued; In 1997 only, Limited Edition Set of the 1985 and 1997 ornaments was issued in Pewter

The Animal Kingdom

Limited Edition 9" diameter Plates; Issued in Aluminum, Bronze, Pewter and Sterling

1981 Raccoons - First Edition (die destroyed in 1986), Art by Earl C. Martz, Engraved by David Bruck

1982 Canadian Geese (die destroyed in 1987), Art by Earl C. Martz, Engraved by Steven Adams

1983 The Lions (die destroyed in 1988), Art by Earl C. Martz, Engraved by David Bruck

1984 Big Horn Sheep (die destroyed in 1989), Art by Earl C. Martz, Engraved by Leonard Youngo

1985 Saw-Whet Owl (die destroyed in 1990), Art by Earl C. Martz, Engraved by David Bruck

1986 Koala Bear (die destroyed in 1991), Art and Engraving by Leonard Youngo

1987 The Elephants (die destroyed in 1992), Art and Engraving by David Bruck

1988 The Fox (die destroyed in 1993), Art by David Bruck, Engraved by Leonard Youngo

1989 Scarlet Macaws (die destroyed in 1994), Art and Engraving by David Bruck

1990 The Leopard (die destroyed in 1995), Art by David Bruck, Engraved by Leonard Youngo

1991 The Seals (die destroyed in 1996), Art and Engraving by David Bruck

1992 The Kangaroos (die destroyed in 1997), Art by David Bruck, Engraved by Leonard Youngo

1993 Timber Wolves - Last Edition, Art and Engraving by David Bruck

Olde World Santa

Limited Edition Christmas Ornaments, issued only in year specified; Oval; Issued in Bronze, Pewter and Sterling

1993 Father Christmas - First Edition, Art by Linda Hoover, Engraved by David Bruck

1994 Kris Kringle, Art by Linda Hoover, Engraved by Leonard Youngo

1995 Nicholas, Art by Linda Hoover, Engraved by David Bruck

1996 The Bell Ringer, Art by Linda Hoover, Engraved by Grant Galloway

1997 The Lantern Bearer - Last Edition, Art by Linda Hoover, Engraved by David Bruck

The Spirit of the Holiday

Limited Edition Christmas Ornaments, issued only in year specified; Tear drop shape; Issued in Bronze, Pewter and Sterling

1994 Immanuel - First Edition, Art by Linda Hoover, Engraved by Leonard Youngo

1995 Star of Bethlehem, Art by Linda Hoover, Engraved by David Bruck

1996 Journey of the Magi, Art by Linda Hoover, Engraved by Grant Galloway

1997 Escape into Egypt, Art by Linda Hoover, Engraved by David Bruck

1998 Joseph's Dream, Art by Linda Hoover, Engraved by David Bruck

Touched by an Angel

Limited Edition Collectors Guild Ornament Collection; 1-7/8"x2-3/4" oval; Issued in Bronze, Pewter and Sterling

1996 The Heavenly Harpist - First Edition, Art by Linda Hoover, Engraved by Leonard Youngo

1997 Angelique, Art by Linda Hoover, Engraved by Leonard Youngo

1998 Celeste, Art by Linda Hoover, Engraved by Leonard Youngo

Archives

Collectors Guild; Marked re-issues of earlier Aluminum pieces

1996 Children of the Sea Serving Tray - First Edition (retired in 1996) (9"x15", Seahorse motif), Art probably by Louis Donato, Engraved by Louis Donato

1997 Appleblossom Scroll Footed Bowl (10-1/2" dia, 3-1/2" high), Art by James McCausland, Engraved by Louis Donato

1998 Bittersweet Bon Bon Dish (5"x7-1/2", 1" deep), Art by James McCausland, Engraved by Louis Donato

Kissin' Bridge

Collectors Guild Limited Editions

1996 Knecht's Bridge Curio Box - First Edition (retired in 1996) (3-1/2"x6" velvet-lined walnut box with bronze insert in lid), Art by Linda Hoover, Engraved by David Bruck

1997 Robert's Bridge Cut Out (2-3/4"x7", 4-1/2" tall, bronze mounted on walnut), Art by Linda Hoover, Engraved by Leonard Youngo

1998 Philippi Covered Bridge Curio Box (7"x9", 2-1/3" high, wood box with bronze insert in lid), Art by Linda Hoover, Engraved by Leonard Youngo

Norman Rockwell's Finest Moments Saturday Evening Post

Collectors Guild; 9" Diameter Plate Series; Issued in Bronze, Pewter and Sterling; Numbered pieces

1996 First Love - First Edition (retired March, 1997), Art work by Linda Hoover, Engraved by Leonard Youngo

1997 Doctor and Doll, Art work by Linda Hoover, Engraved by David Bruck

1998 Before the Shot, Art work by Linda Hoover, Engraved by Leonard Youngo

In 1997, Doctor and Doll also issued as 14-1/2"x15-1/2" framed metalart

Barns of Yesteryear

Collectors Guild; Metalart in bronze framed in wood; Offered in oak or fruitwood framing, with and without artist's remarque, double matting in burgundy/brown marble, shamrock green/wine, or smoky blue/wine

1996 Grandpa's Farm - First Edition (11"x15" bronze image in 18"x22" frame), Art by Linda Hoover, Engraved by David Bruck

1998 Creekside Memories - Second Edition (9"x11" bronze image in 12"x14" frame), Art by Linda Hoover, Engraved by Leonard Youngo

Collectors Guild Members Only Piece

1998 Hobnail Bowl; Copper; One-year availability; Art by James McCausland, Engraved by David Bruck (Original engraving by Louis Donato)

Appendix F

Wendell August Forge Dating Clues

Many people want to determine when a particular Wendell August Forge item was made. Of additional interest is determining the time frame during which the form of the piece and/or its motif were first designed and developed.

There is no definitive listing or reference for all Wendell August Forge items produced, and there has been a great deal of variety in the items produced over an extended period of time. The author's data base, which focuses predominantly on the *circa* 1930-1967 time frame, currently identifies more than 2,300 different items, yet accounts for a relatively small percentage of the different items produced.

Nevertheless, certain attributes of a given piece of Wendell August Forge metalware can generally be utilized, *in combination with one another*, to place the item within a reasonable time frame of both design and production. These attributes range from the trademark handstamp that was used to the type of metal of which the piece is made.

Certain characteristics of the Wendell August Forge trademark handstamps that appear on most pieces have been found useful in identifying approximate time frames during which an item was made. In several cases, production dating based on the trademark handstamp in and of itself is both highly reliable and rather narrowly defined. Trademark handstamps are documented in Appendix C.

Die engraver identifications (in use since about 1969) and craftsman touchmarks (in use since January of 1977) can frequently be used in combination with the trademark handstamp to narrow the time frame of production of an item. These identifiers are documented in Appendix D.

Pieces in the Limited Edition and Collector series generally carry the year of original issue, thereby making it relatively straightforward to determine the time frame of design. However, they may have been actually produced at a later time. In addition, some of the items do not carry the year of original issue. Thus a combination of information may also be necessary in order to date such items. Limited Edition and Collector series issues are documented separately in Appendix E.

What remains to be used as a guide in dating pieces of Wendell August Forge metalwares are such things as the design and production characteristics of the item, together with additional marks and labels that may be found on the item. Changes in these attributes were not typically discrete; that is, most changes tended to occur gradually over a period of time. The total number of die engravings is quite large, and documentation is incomplete. In addition, a motif or a form might be retired from the product line for some period of time, and then emerge again later on, occasionally with a different name.

What follows is a selection of pertinent information in the form of a time line, indicating the likely time frames appropriate to the attribute(s) identified. It must be emphasized that most of the *circa* dates provided are, by definition, approximations, and that they are based on the correlation of actual pieces with those dated materials that have emerged from research. Nevertheless, the information provided in this time line of dating clues, taken together with the information provided in the other Appendices and the illustrations provided in the text, should facilitate the dating of the vast majority of Wendell August Forge items within reasonable time frames.

Circa Dates	Clues
1923-1930	Wrought iron items, some incorporating brass, predominantly architectural
1930-1940	Aluminum architectural items, most frequently in the style of wrought iron work; some with repoussé motifs. First period of production of bank furnishings (e.g., tables), with bank lamp production beginning *circa* 1933.
1930-1931	Aluminum giftware items generally lack technical refinement; few were produced; motifs primarily of "silhouette-style" design
1932	Giftwares motif names as stated in literature include *Acorn, Anchor, Archery, Astronomy, Bee, Beetle, Cactus, Camels, Cherry, Coat of Arms, Crab, Craw Fish, Cricket, Dachsund, Dragon Fly, Duck, Fish, Floral, Fox, Frog, Geese, Golf, Gun, Horse, Hunting, Maple, Moon Fish, Oak, Police, Polo, Residence, Rose, Sail, Scarab, Scroll, Sea Horse, Selyham, Setter, Star Fish, Stony Brook, Swords, Terrier, Thistle, Turkey, Turtle, Vine, Violet, Wasp, Wreath*
1933	Increasingly rapid production of aluminum giftware items begins, continuing until World War II
1933-1937	Period of aluminum chair production (excluding 1980s youth chairs)
1933-1940	Production of aluminum table lamps and floor lamps
1934-1938	Production of ecclesiastical architectural items, such as grills, gates and light fixtures
1934-1938	Giftware increasingly refined; some items (or parts of items, such as handles) have 1/16"-wide decorative grooves; many rivets have "X" marks on the heads

1935	Appearance of aluminum jewelry
1935	Giftwares motif names as stated in literature include *Airplane, Barley, Butterfly, Ducks, Fantail Fish, Gun and Ducks, Initials, Polo, Sail Boat, Sail Fish, Sea Horse, Thistle, Tropical Fish, Yachts*
1935-1938	Cambridge glass used with some items
1936	Giftwares motif names as stated in literature include *American, Angel Fish, Barley, Bass, Birds, Butterfly, Cactus, Camels, Cape Cod, Cherry, Chinese Lily, Cock Fight, Country Estate, Crab, Crossed Swords, Ducks, Elephants, Fantail Fish, Flight of Geese, Gloucester Fisherman, Golf, Goose, Grape, Grecian, Gull and Sailboats, Gun and Ducks, Hobnail, Horse, Hunter, Initials, Italian Floral, Jumping Trout, Larkspur, Lobster, Maple Leaf, Mauritius Victoria, Medallion, Oak Leaf, Persian Leaf, Pheasant, Polo, Rod & Creel, Rose, Sail Boat, Sail Fish, Sea Horse, Sea Subjects, Shasta Daisy, Shell Design, Solomon Seal, Star Fish, Sword Fish, Thistle, Tree, Tropical Fish, Trout, Turkey, Turtle, Water Lilies, Wheat, Wheat & Grapes, Wistaria, Wreath, Yacht Race, Zinnia*
1937	Giftwares motif names as stated in literature include *Angel Fish, Barley, Bass, Cactus, Camels, Cape Cod, Cherry, Chickadee, Chinese Lily, Cock Fight, Crab, Dogwood, Ducks, Flight of Ducks, Goose, Grapes, Grecian, Gull and Sail Boats, Heraldry, Hobnail, Horn Crop and Stirrup, Horse Head, Hummingbird, Italian Floral, Jumping Trout, Larkspur, Lily of the Valley, Lobster, Medallion, Modern Bird, Oak Leaf, Palm Trees, Persian Leaf, Pine, Robin, Rose, Sail Boats, Sail Fish, Sea Horse, Shasta Daisy, Shell Pattern, Solomon Seal, Starfish, Swallow, Sword Fish, Thistle, Trailing Arbutus, Tree, Tropical Fish, Trout, Turkey, Turtle, Water Lilies, Wheat, Wheat and Grapes, White Pine, Wistaria, Woodpecker, Wren, Yacht, Yacht Race, Yacht Wheel, Zinnia*
1938	Giftwares motifs appearing in literature include *Ducks, Elephants, Horse Head, Pine, Sail Fish, Sea Horse, Sword Fish, Wheat and Grapes, Yacht Wheel*
1938	Significant product line redesign; trays with self-handles having a slightly convex surface introduced
1939	Printed documentation of aluminum jewelry appears
1939	Pieces with sharply-fluted corners appear; giftwares having scallop-shell forms make initial appearance; allover motifs introduced (e.g., *Apple Blossom)*

1939	Approximate time that a separate handstamp for MADE IN USA appears to have been in use
1939-1941	Production of aluminum giftware items incorporating crackle-glazed La Mirada pottery
1940	Giftwares motif names as stated in literature include *Acorn and Grape, Apple Blossoms, Arbutus, Bird Dog, Bittersweet, Cactus, Cattails, Chinese Lily, Dogwood, Ducks, Educational Design, Geese, Grape, Harvest, Iris, Peacock, Pine, Poppy, Rhododendron, Sailboats, Sailfish, Shasta Daisy, Solomon Seal, Turkey, Wheat, Yacht, Zinnia*
1940	Notching (referenced in print as "tool marked") of edges of some pieces begins, particularly #830 bowls
1941-1942	Some pewter giftwares produced
1942-1945	Production of aluminum items ceases near the end of 1941 due to World War II and does not resume until about mid-1945
1945-1948	High level of production of aluminum giftwares, predominantly tablewares
1945-1952	Edges of most of the aluminum giftware items are notched
1946-1950	Additional production of aluminum giftware items incorporating La Mirada pottery
1946-1950	Significant amounts of aluminum jewelry produced
1946-1970	Aluminum architectural work, primarily in banks, decreasing over time, with decrease in use of repoussé work; some items gold-anodized. Second period of bank lamp production confined primarily to *circa* 1946-1956
1948	Giftwares motif names as stated in literature include *Acorn and Grape, Apple Blossoms, Bird Dog, Bittersweet, Cactus, Dogwood, Duck, Hobnail, Iris, Pine, Sailboat, Shasta Daisy, Vegetables*
1948-1956	Gummed paper labels affixed to most pieces (blue and white oval labels, approximately 7/8"x1-1/4")
1949	Introduction of separate GROVE CITY, PA handstamp
1949-1958	Decline in production of giftware items
1950	Giftwares motif names as stated in literature include *American Elm, Apple Blossoms, Bird Dog, Bittersweet, Cactus, Dogwood, Duck, Educational Design, Hobnail, Oriental Tree, Pine, Poppy, Sailboat, Shasta Daisy, Strawberry, Turkey, Vegetables, Water Lily, Wheat*

1950-1951	Blenko green glass used with some items
1952	Product line significantly redesigned; beveled edges used on most giftware items; trays with convex-surface self-handles phasing out
1953-1964	1/8"-wide groove used as part of the decoration of many giftware pieces
1953-1959	Period of use of "Coffee Bean" motif (name given by collectors)
1956	Introduction of separate HANDMADE handstamp
1958-1970	Very low level of production of giftware items
1960	Beginning of significant increase in production of commercial pieces, both stock items with special handstamps and items produced from custom-engraved dies
1960	Use of increasingly simplified forms begins; motifs having a primitive artistic style introduced, continuing through the late 1970s; motif subjects primarily wildlife; retirement of dies from the *circa* 1939-1950s period begins and is completed by the late 1960s.
1960-1961	Bronze and stainless steel giftware items introduced
1962	Decline in use of style numbers on backs of items begins; practice appears to have been completely discontinued by about 1967
1967	Introduction of electric stylus for some die engraving
1969	Die engraver identification is started
1970	Limited Edition plates introduced into the product line
1970	Increase in production begins
1970s	Metals listed in literature include aluminum, bronze, stainless steel, stainless clad aluminum. Giftwares motifs illustrated in literature include Bear, Colonial Eagle, Deer (jumping fence), Dogwood, Ducks, Golfer (Male), Pheasants, Pine, Schooner, Train, Turkey, Wild Turkeys ("official" motif names have not been located in print)
1977	Craftsman touchmarks introduced in January of 1977
1980	Removal of primitive-style motifs from the product line begins
1980	Increase in intricacy of die engravings and decline in the use of unembellished space in the motifs

early 1980s	Increase in use of juvenile and Holiday motifs begins. Motif names documented include *American Eagle, Amish, Angel, Angus Bull, Antique Bicycle, Apple Blossom, Arabian Horseheads, Baby Tray Design, Bass, Bear, Birdfeeder, Bittersweet, Bunny and Fawn, Butterfly Daisy, Caduceus, Cardinal, Chessie, Clown Face, Colonial Eagle, Cornucopia, Deer, Dogwood, Duck, Ducks in Flight, Eagle, Frogs, German Shepherd, Golfer, Grouse, Harvest, Holly, Indian, Irish Setter, Jumping Deer, Liberty Bell, Lily of the Valley, Lobster, Longhorn, Mallard Duck, Maple Leaf, Mushrooms, New England Clipper, Nittany Lion, Oriental Tree, Owls, Phoenix Bird, Pine, Pineapple, Pittsburgh, Poodle, Puppy, Raspberries, Ringneck Pheasant, Rose, Runner, Santa Mouse, Schooner, Stag, Strawberries, Sweet Pea, Teddy Bear, Thistle, Thunderbird, Train, Tropical Fish, Turkey, Unicorn, Victorian, Water Lily, Wedding Bells, Wheat, Wild Turkeys, World Map*
mid 1980s	Metals listed in literature include aluminum, bronze, copper, pewter, stainless clad aluminum, stainless steel
late 1980s	Metals listed in literature include aluminum, bronze, copper, pewter, stainless clad aluminum, stainless steel, sterling silver. Giftwares motif names listed in literature include *American Eagle, Amish, Angel, Angus Bull, Antique Bicycle, Antique Doll, Applebasket, Apple Blossom, Aquarius, Aries, Aztec Calendar, Basketball Player, Bear, Beehive, Birdfeeder, Bittersweet, Boxer, Butterfly Daisy, Cardinal, Carousel, Cherries, Cornucopia, Daisy, Daffodils, Deer, Dogwood, Double Wedding Bells, Eagle, Football Player, Geese and Teddy Bear, Gemini, German Shepherd, Golfer, Grapes, Harvest, Horse Heads, Hot Air Balloons, Hummingbird, Independence Hall, Indian Head, Iris, Irish Setter, Kitten, Lady Justice, Leo, Leprechaun, Liberty Bell, Libra, Lighthouse, Lion, Love Heart, McConnell's Mill, Mercury Winged Foot, Monticello, Motocross, Mount Rushmore, Mountain Laurel, Music Note, New England Clipper, Nittany Lion, Oilwell, Old School House, Owls, Penguin, Pine, Pineapple, Pisces, Pittsburgh, Poodle, Rooster, Rose, Runner, Saggitarius, Sailboats, Santa Mouse, Schooner, Scottie Dog, Skating Partners, Snowman, Soaring Eagle, Statue of Liberty, Strawberries, Swimmer, Teddy Bear, Tennis Player, Thistle, Train, Tropical Fish, Tulips, Turkey, Virgo, Water Lily, Wedding Bell, Wheat, Windy Day, World Map, Wreath*
1988	Use of power hammering begins
1989	Remainder of primitive-style motifs removed from the product line
1990	Metals listed in literature include aluminum, bronze, pewter, sterling silver

early 1990s	Metals listed in literature include aluminum, bronze, pewter. Giftwares motif names listed in literature include *American Flags, American Quarter Horse, Amish, Angel, Antique Car, Antique Doll, Applebasket, Apple Blossom, Ballet Shoes, Barber Shop Quartet, Barn, Basketball Player, Bear, Birdfeeder, Black Labrador, Blacksmith, Bunnies, Butterfly Daisy, Carousel, Chickadee, Cleveland, Clown, Cornucopia, Daffodils, Deer, Dogwood, Double Wedding Bells, Duck Family, Eagle, Erie, Farm Scene, Football Player, Frogs, Gathering, Golfer, Graduation Diploma, Grouse, Harvest, Horse Farm, Hot Air Balloons, Hummingbird, Indian on Horseback, Iris, Kitten in a Basket, Lighthouse, Lily of the Valley, Lovebirds, Love Heart, Manger Scene, McConnell's Mill, Mushrooms, Music Note, Nittany Lion, Old School House, Owls, Pine, Pineapple, Pittsburgh, Poinsettia, Poodle, Praying Hands, Puppy, Raspberries, Ringneck Rose, Sailboats, Satellite Drive-In, Scottie Dog, Soaring Eagle, Star of David, Strawberries, Teddy Bear, Thistle, Train, Tulips, Turkey, Unicorn, Waiting for Santa, Wedding Bell, Wheat, World Map, Wreath, Youngstown Skyline*
1992	Use of panagraph-type milling machine for some die engraving begins
1995	Chemical coloring of some items begins
late 1990s	Metals listed in literature include aluminum, bronze, pewter, sterling silver. Giftwares motif names listed in literature include *Angel, Applebasket, Apple Blossom, Apples, Baby Plate Design, Barn, Barn Raising, Basketball Collage, Bears, Birdfeeder, Blessing, Blueberries, Bow, Bunnies, Butterfly, Cardinal, Carousel, Cherries, Cleveland, Cleveland Jacobs Field, Coming Home, Cornucopia, Daffodils, Deer, Dogwood, Double Wedding Bells, Duck, Eagle, Fall Harvest, Fishing Boat, Flowering Evergreen, Fly Fisherman, Football Collage, Footprints in the Sand, Frogs, Gardening, Gathering, Golf Collage, Graduation Diploma, Grapes, Grouse, Harvest, Horse Farm, Hummingbird, Indian on Horseback, Kitten, Lady Justice, Lemons, Leprechaun, Lily of the Valley, Lovebirds, Love Heart, McConnell's Mill, Mother's Bouquet, Mountain Laurel, Music Collage, Noah's Ark, Nursery Plate, Nursing Collage, One Room School House, Peaches, Pears, Pine, Pineapple, Pittsburgh, Poinsettia, Praying Hands, Puppy, Raspberries, Ringneck Pheasant, Santa, School Days, Shamrock, Skating Angels, Soaring Eagle, Soccer Collage, Special Delivery, Star of David, Strawberries, Sunflower, Sweet Pea, Thistle, Toyland, Train, Tulips, Turkey, Victorian Rose, Water Lily, Wedding Bell, Wheat, Wild Turkeys, Wreath*

Index

Selected Chapter notes and Appendix data have been indexed
Page numbers in ***italics*** refer to illustrations

I

J

K

L

M

N

O

P

R

S

T

U

V

W

Y

Z

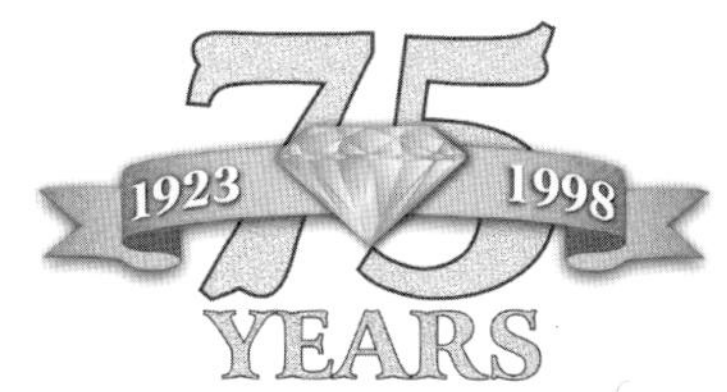

WENDELL AUGUST FORGE

Wendell August Gift Shoppe & Forge
620 Madison Avenue
Grove City, PA 16127
1 (724) 458-8360

Wendell August Gift Shoppe & Forge
7007 Dutch Country Lane & Rt 62
Berlin, OH 44610
1 (330) 893-3713

Wendell August of Charleston
309 King Street
Charleston, SC 29401
1 (843) 722-1911

For Retail Catalog Orders Call:
1 (800) WAF-GIFT
1 (800) 923-4438

Retail Fax
1 (724) 458-0906

Business to Business
1 (800) 923-1390

Visit Our Website at http://www.wendell.com
E-mail: gifts@wendell.com

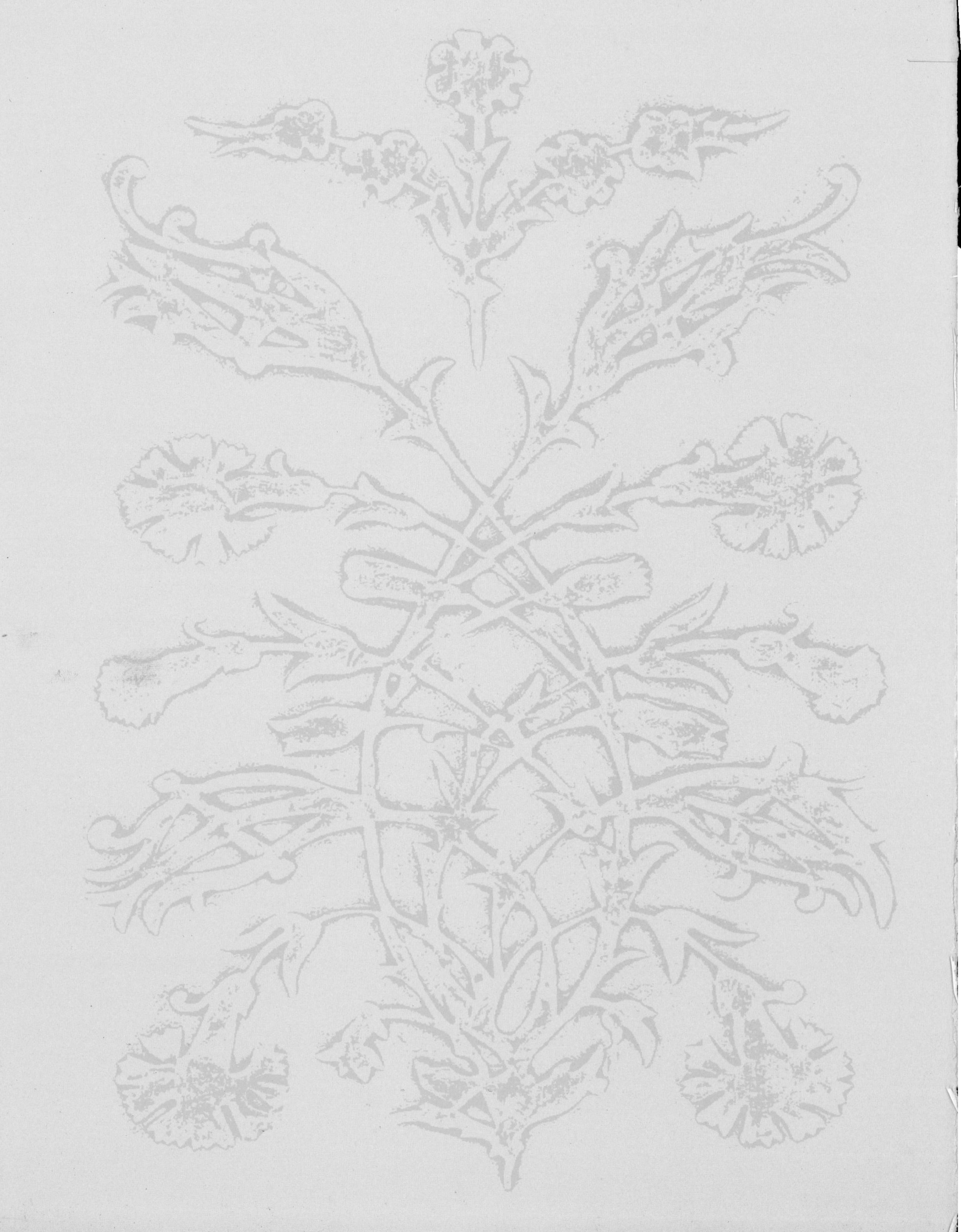